Historiography and Mythography in the Aristotelian *Mirabilia*

This is the first full-length volume in English that focuses on the historiographical section of the *Mirabilia* or *De mirabilibus auscultationibus* (*On Marvelous Things Heard*), attributed to Aristotle but not in fact by him.

The central section of the *Mirabilia*, namely §§ 78–151, for the most part deals with historiographical material, with many of its entries having some relationship to ancient Greek historians of the 4th and 3rd centuries BC. The chapters in this volume discuss various aspects of this portion of the text, including textual issues involving toponyms; possible structural principles behind the organization of this section; the passages on Theopompus and Timaeus; mythography; the philosopher Heracleides of Pontos; Homeric exegesis; and the interrelationship between pseudo-Plutarch's *On Rivers*, a section of the historian Stobaeus' *Geography*, and the *Mirabilia*.

Historiography and Mythography in the Aristotelian Mirabilia is an invaluable resource for scholars and students of this text, and of Greek philosophy, historiography, and literature more broadly.

Stefan Schorn is Professor of Ancient History at KU Leuven. He has published extensively on fragmentary Greek historiography and he is Editor-in-Chief of *Die Fragmente der Griechischen Historiker Continued. Part IV: Biography and Antiquarian Literature*.

Robert Mayhew is Professor of Philosophy at Seton Hall University. He has published extensively on ancient philosophy, and especially on Aristotle and other Peripatetics. His most recent book is *Aristotle's Lost Homeric Problems: Textual Studies*, and among his current projects is an edition of the fragments of Aristotle's lost *Zoïka*.

Rutgers University Studies
in Classical Humanities

Historiography and Mythography in the Aristotelian *Mirabilia*

Edited by Stefan Schorn and Robert Mayhew

Routledge
Taylor & Francis Group

LONDON AND NEW YORK

First published 2024

by Routledge
4 Park Square, Milton Park, Abingdon, Oxon OX14 4RN

and by Routledge
605 Third Avenue, New York, NY 10158

Routledge is an imprint of the Taylor & Francis Group, an informa business

British Library Cataloguing-in-Publication Data
A catalogue record for this book is available from the British Library

Library of Congress Cataloging-in-Publication Data
Names: Aristotelian Mirabilia (2022 : Nice, France) | Schorn, Stefan, editor. | Mayhew, Robert, editor.
Title: Historiography and mythography in the Aristotelian Mirabilia / Stefan Schorn, and Robert Mayhew.
Description: Abingdon, Oxon ; New York, NY : Routledge, 2024. | Series: Rutgers University studies in classical humanities | Includes bibliographical references and index.
Identifiers: LCCN 2023022237 (print) | LCCN 2023022238 (ebook) | ISBN 9781032569505 (hardback) | ISBN 9781032569512 (paperback) | ISBN 9781003437819 (ebook)
Subjects: LCSH: De mirabilibus auscultationibus Historiography—Congresses. | Curiosities and wonders—Early works to 1800—Congresses. | Science—Early works to 1800—Congresses.
Classification: LCC Q171 .A665 2024 (print) | LCC Q171 (ebook) | DDC 509—dc23/eng/20230803
LC record available at https://lccn.loc.gov/2023022237
LC ebook record available at https://lccn.loc.gov/2023022238

ISBN: 978-1-032-56950-5 (hbk)
ISBN: 978-1-032-56951-2 (pbk)
ISBN: 978-1-003-43781-9 (ebk)

DOI: 10.4324/9781003437819

Typeset in Times New Roman
by Apex CoVantage, LLC

Contents

Preface

In April 2022, a number of scholars met in Nice, at the Université Côte d'Azur, for a *Project Theophrastus* conference called *The Aristotelian Mirabilia*. There for three days we discussed drafts of our papers, on various aspects of the *Mirabilia* or *De mirabilibus auscultationibus*, a work attributed to Aristotle but certainly not by him.

The timing of this conference was perfect: many of the principals of Project Theophrastus were looking for new directions to go in, beyond volumes focused on collections of fragments of early Peripatetics, and the pseudo-Aristotle opuscula seemed like one excellent option. Ciro Giacomelli had recently finished his superb critical edition of the *Mirabilia* and was enthusiastic about participating in the conference and sharing his edition with the group prior to its publication. One of us (Mayhew) was working on an English translation of the *Mirabilia* (based on Ciro's text) for a new Loeb edition of the pseudo-Aristotle opuscula (in preparation) and was happy to share it with the conference participants (for which he received excellent feedback), while the other (Schorn) had just put the finishing touches on the edition (since published) of *Die Fragmente der Griechischen Historiker Continued IV E 2: Paradoxographers of the Imperial Period and Undated Authors*, which included contributions from five of the authors also appearing in the present volume. All of this is evidence of a renewed interest in both ancient paradoxography and its connection to ancient Greek historiography, and a reappraisal of the *Mirabilia* in that context has proven to be an excellent idea.

Although the two of us, along with Arnaud Zucker and Oliver Hellmann, were the conference organizers, the organizers and hosts on the ground, so to speak, were Arnaud Zucker and graduate student Anaelle Broseta. It was regrettable that the founder of *Project Theophrastus*, William W. Fortenbaugh (Rutgers University, Classics, emeritus) was unable to join us. But we were pleased that Tiziano Dorandi, long time participant in and supporter of *Project Theophrastus*, was able to attend and make introductory remarks to open the conference.

The number and length of the essays proved too much for one book, and so the essays that originated in that conference (as well as some others) appear in two separate volumes – in addition to this one, *The Aristotelian* Mirabilia *and Early Peripatetic Natural Science* (edited by Arnaud Zucker, Oliver Hellman, and Robert Mayhew).

SS
RM

Contributors

Charles Delattre is Professor of Ancient Greek Language and Literature at the University of Lille. His research focuses on the mythographic corpus and on ancient narrative practices involving the modern notion of "mythology," from a perspective of literary theory and cultural anthropology.

Ciro Giacomelli is a researcher in Classical Philology at Padua University. He received his PhD in 2018 with a dissertation on the pseudo-Aristotelian *Mirabilia*, part of which was published in 2021 (Berlin-Boston, De Gruyter). His critical edition of the Greek text saw the light in 2023 (Rome, Accademia Nazionale dei Lincei).

Robin J. Greene is Associate Professor of History and Classics at Providence College. Her critical edition of the treatise *Paradoxographus Florentinus* was recently published by Brill in the *Die Fragmente der Griechischen Historiker Continued* series.

Robert Mayhew is Professor of Philosophy at Seton Hall University. He has published extensively on ancient philosophy, and especially on Aristotle and other Peripatetics. His most recent book is *Aristotle's Lost Homeric Problems: Textual Studies*, and among his current projects is an edition of the fragments of Aristotle's lost *Zoïka*.

Irene Pajón Leyra is Assistant Professor of Greek at the University of Seville. She has had research positions at the Spanish High Council of Research and the University of Nice and has been a visiting researcher at Oxford University and F.U. Berlin. She has studied paradoxography since her PhD (2008), and has published a monograph on it (2011).

Stefan Schorn is Professor of Ancient History at KU Leuven. He has published extensively on fragmentary Greek historiography and he is Editor-in-Chief of *Die Fragmente der Griechischen Historiker Continued. Part IV: Biography and Antiquarian Literature*.

Kelly Shannon-Henderson is Associate Professor in the Department of Classics at the University of Cincinnati. She is the author of *Religion and Memory in Tacitus' Annals* (OUP), a commentary on Phlegon of Tralles' *On Marvels*

(Brill), and various articles on aspects of Greek and Roman historiography, religion, and paradoxography.

Søren Lund Sørensen studied Classics at the University of Copenhagen, Jewish Studies at the University of Oxford, and obtained his PhD in ancient history from the University of Southern Denmark. He is currently a research assistant at Freie Universität Berlin and publishes, *inter alia*, on paradoxography, the *Physiologus*, and the relations between Ancient Yemen and the Greco-Roman world.

Pietro Zaccaria, PhD (2018), is Postdoctoral Fellow of the Research Foundation – Flanders (FWO) at KU Leuven, Ancient History Research Unit. He has published on ancient biography, historiography, and the history of ancient philosophy.

Introduction

Pseudo-Aristotle's *De mirabilibus auscultationibus* and historiography

Stefan Schorn and Robert Mayhew

The pseudo-Aristotelian work Περὶ θαυμασίων ἀκουσμάτων, *On Marvelous Things Heard*,[1] is a collection of 178 for the most part unrelated chapters containing reports on strange phenomena regarding nature, animals, and human culture. The work belongs to a group of texts generally labeled as paradoxography, a type of literature that has not enjoyed a high reputation in modern scholarship. Wilhelm von Christ, Wilhelm Schmid, and Otto Stählin call the genre "ein Parasitengewächs am Baum der historischen und naturwissenschaftlichen Literatur,"[2] as these works are usually not based on autopsy of the author of the collection; rather, all the information has been excerpted from earlier literature. Another feature of these works, as noted by scholars, is a lack of explanation of marvelous facts, even if these are present in their sources, in order to increase the *thauma*, which has contributed to the reputation of such works as unscientific. Even among more recent literature, the entry on paradoxographical works in *Brill's New Pauly* states: "on the one hand they might have been collections of material (perhaps only for personal use), on the other hand entertainment literature,"[3] which does not much recommend these works for study. So, is it worthwhile spending one's time studying this literary genre (if we may speak of literature at all)?

One may argue that *Mir.* constitutes a valuable document of Peripatetic thought, a collection of material compiled for further studies which we know were often made by Peripatetics. The manuscripts attribute the work to Aristotle, but it is missing in the catalogues of his works in Diogenes Laertius (5.22–27) and Ptolemy (43 Rashed), and is included only in the list in the *Vita Hesychii* (10, line 200 Dorandi).[4] It is first attested as a work of Aristotle in Athenaeus around 200 AD.[5] As some of the information it contains clearly dates after the death of Aristotle, there can be no doubt that it is spurious; but one may suspect that it originates, as do other pseudo-Aristotelian works, in the early or later Peripatos. However, it

---

1 On the variant forms of the title, see Giacomelli (2021) 42–44. It is best known by one of its standard Latin titles, *De mirabilibus auscultationibus* or *Mirabilia* (hereafter abbreviated *Mir.*).
2 Christ, Schmid and Stählin (1920) 237.
3 Wenskus (2006).
4 See, for example, Giacomelli (2021) 11 with n. 39 (literature).
5 Athenaeus 12.541a: Ἀλκισθένην δὲ τὸν Συβαρίτην φησὶν Ἀριστοτέλης ἐν τοῖς Θαυμασίοις κτλ. which refers to *Mir.* 96.

DOI: 10.4324/9781003437819-1

is a matter of debate whether this is the case or not here. While Hellmut Flashar hypothesizes that the earliest version of the collection may have been a 3rd century BC Peripatetic competitor to Callimachus' collection of *mirabilia*,[6] the most recent editor, Ciro Giacomelli, does not link it to this philosophical school at all and, following a suggestion by Valentin Rose, tentatively explains its attribution to Aristotle by the fact that its first chapters consist of excerpts from *Historia animalium* 8(9), which might have led to the wrong ascription.[7] Even among the contributors to this volume, there is no consensus in this regard. So again the question imposes itself: what are we supposed to make of such a work?

Inspiration may come from studies of other paradoxographical collections which have attracted some interest in recent years. They have shown that these works are not, or not always, careless collections of material, but may structure their information by linking the entries by way of association or applying a "hodological" principle, that is, presenting the *mirabilia* in the form of a journey.[8] Their authors were thus no mere compilers but may have had a literary agenda and applied literary strategies. In a seminal article, Guido Schepens and Kris Delcroix (1996) addressed the question of the *Sitz im Leben* of such works and their potential readership, and Irene Pajón Leyra (2011) produced the first history of this kind of literature, which includes many interesting interpretations. The new editions with commentary of all the paradoxographical works, included in *Die Fragmente der Griechischen Historiker Continued. Part IV: Biography and Antiquarian Literature*, take the authors of such collections seriously and try to reconstruct their literary aims and strategies, the structural elements of the collections, and the ways in which they have selected and dealt with the material lifted from their sources.[9] All these studies show that paradoxographical works can be interesting objects of study and that many topics still need to be addressed or scrutinized in more detail.

There is one aspect of paradoxography that has always been appreciated by scholars: these collections have preserved many fragments of philosophical and historical works that have otherwise been lost. The recent studies on them have also shown that for a correct evaluation of these decontextualized pieces of information and their use as historical and philosophical sources, an in-depth knowledge of the collection that transmits them is necessary. We have to know which direct and indirect sources the authors of paradoxographical collections used, and how the original pieces of information may have been modified in the course of transmission. Still too much is claimed on the basis of such "fragments" that is in reality based on old-fashioned 19th- and early 20th-century *Quellenforschung*, whose basic principles may or may not be valid any longer.

6 Flashar (1972) 52; cf. Vanotti (2007) 31–32, 40, 52.
7 Giacomelli (2021) 14–16.
8 The term "hodological principle" was coined by Geus (2016); cf. Sørensen (in this volume) 261 for examples. For the associative principle, see Spittler's commentary on Apollonius, *FGrHist* 1672 and Greene's commentary on *Paradoxographus Florentinus*, *FGrHist* 1680 in Schorn (2022).
9 The authors of Imperial Roman times and the undated authors have been published in Schorn (2022); many Hellenistic authors are already available in the parallel digital edition in Jacoby Online: https://scholarlyeditions.brill.com/bnjo.

Mir. has so far not yet benefitted from this new interest in paradoxography. The text edition most widely used today is still that by Alessandro Giannini (1965), which is notoriously problematic. There exists a commentary by Hellmut Flashar (1972) which carefully collects parallel traditions to those in *Mir.*, explains *realia*, and discusses the sources on which the single chapters may be based. It is often very helpful, as is the more recent brief commentary by Gabriella Vanotti (2007). But an in-depth study of this work is still lacking, and many of its chapters have been investigated thoroughly only in far-scattered studies that take only single chapters into consideration. With Ciro Giacomelli's monograph on the textual transmission of *Mir.* (2021) and his 2023 edition with philological commentary, to which the contributors to this volume had access, the basis for an ambitious study and a reassessment of *Mir.* has been laid and it is thus time to have a fresh look at *Mir.* and to ask new questions – or to ask old questions in a new way. This is best done through an interdisciplinary approach, because *Mir.* contains chapters on a wide variety of topics that exceed the competencies of any single scholar.

The goal of the chapters in this volume is thus twofold: (1) to contribute to a better definition of the specific characteristics of the collection, and (2) to determine how *Mir.* may be used as a source of works that are lost in their entirety or how the information it contains may otherwise be assessed. For reasons of space, the chapters here deal only with those aspects that are related to history and historiography in a broad sense, thus also including mythography and ethnography.[10]

In the following, we address some general aspects of *Mir.* that the reader should be aware of, and especially zoom in on those sections of the work in which historiographical sources have been used or which contain historical information. We highlight the importance of *Mir.* as an historical source and sketch what new insights the studies in this volume provide in this regard and what still remains to be done.

Mir. divides into two major parts, an older main part (1–151) and a later "Appendix" (152–178), which was compiled between 6th and 12th centuries and in which mostly material from late sources has been assembled.[11] The exact date of the first part is uncertain, but it may be the work of one compiler to which some chapters may or may not have been added in the course of transmission. As Giacomelli argues, the historical information it contains and the first mention by Athenaeus allow us to date this part to the period from the end of the 4th century BC to ca. 200 AD, he himself tending toward a date in early Imperial times.[12] However, the historical situation mirrored in *Mir.* is mainly that of the 3rd century BC,[13] so one wonders if a compiler active in Imperial times would have taken over information from his sources without adapting it, presenting Carthage as flourishing and Italy as not yet completely under Roman rule. Thus, an earlier date is still an option, but this question requires further research.

---

10 The sections of *Mir.* dealing with natural science are discussed in another volume in the RUSCH series.
11 See Giacomelli (2021) 22–42.
12 See Giacomelli (2021) 17–21.
13 See, for example, Flashar (1972) 45.

The main part of the work (1–151) divides into three clear-cut parts if we take its *sources* as the basis of division. Its first section (1–77) contains excerpts from Aristotle's *Historia animalium* 8(9) and various works of Theophrastus,[14] and the second section (78–138) consists of excerpts from historiographical works. While the exact identity of the sources in the first part of this section (78–121) is a matter of discussion (see below), it is certain that the second part (122–138) mainly consists of excerpts from Theopompus' *Philippica*. In the third part (139–151), Theophrastus is again *Mir.*'s source. It needs to be stressed that in all these sections, the sources indicated above are only the main source and that occasional additions from other sources may have been inserted into a string of excerpts from a certain text (see below). Such a combination of Peripatetic and historical sources is less surprising than it may seem, and scholars have often referred to pseudo-Antigonus' similar use of series of sources.[15] If we take the *primary content* of the chapters as the basis of division, we get a slightly different disposition because only the first part of the historiographical section mainly deals with ethnographic and man-made *mirabilia*, while the second part, Theopompean, contains primarily *mirabilia* on nature, or at least such *mirabilia* in which the focus is clearly on natural phenomena and not on how humans react to them. Seen from this perspective, it is more similar to the material from Theophrastus that follows. So one may speak of a ring composition both regarding the sources and the content but with slightly different borders: Peripatetic sources (1–77) – historiographical sources (78–138) – Peripatetic sources (139–151) *versus mirabilia* on nature (1–77) – ethnographical *mirabilia* (78–121) – *mirabilia* on nature (122–151). Since historical elements are present in the parts on natural phenomena and natural *mirabilia* appear in the ethnographic part, the different parts do not appear as monolithic and completely separate blocks.

The amount of historical material in a broad sense, including mythology and ethnography, in *Mir.* is much higher than in any other paradoxographical collection. We find it especially in 78–121, where it often becomes so dominant that it moves the *paradoxon* to the background.[16] But also those parts of *Mir.* that are based on philosophical sources contain such material to some extent. The historical information in *Mir.* is often unique and the same is true for its mythological accounts, which usually treat less prominent heroes or report less-known stories about well-known heroes, with a strong focus on the Greek West, a feature already detected by Vanotti.[17] *Mir.* is thus a highly relevant source of Western Greek mythology and of

---

14 On this section, see Zucker, Mayhew and Hellmann (forthcoming).

15 For the similarities, see Flashar (1972) 54–55. *Pace* Flashar, the similarities need not point to a similar date for the two collections, and even if it did, as there is now agreement among most scholars that Antigonus of Carystus is not the compiler of the collection transmitted under the name of Antigonus (without further specification), we do not have a certain date for this collection. Further, Apollonius combines Peripatetic with historiographical sources: see Spittler on Apollonius, *FGrHist* 1672 in Schorn (2022) 412 (list of sources). Theophrastus' *On Waters* is an important source of *Paradoxographus Florentinus* together with historiographical works, as well as those dealing with the Greek West: see Greene on *FGrHist* 1680 in Schorn (2022) 652 (list of sources).

16 Cf. Vanotti (2007) 38 making this claim for 78–136, but after 122, with the Theopompean section, this aspect becomes less relevant.

17 Vanotti (2007) 26, 38–39.

Greek and non-Greek history from the Archaic to the Hellenistic times, especially of local history. No other paradoxographical collection contains so many mythological stories as *Mir.* It seems paradoxographers usually avoided them, so their prominence in *Mir.* is clearly due to the personal interest of our author. But at the same time, their distribution within *Mir.* is remarkably uneven: myth dominates the historiographical passages 78–121,[18] but it is very rare in the Peripatetic parts (only in 51, 58) and totally absent from the chapters based on Theopompus, though we know that this historian included many myths in his *Philippica*.[19] Robin Greene, who has studied all these aspects, has also compared the myths in *Mir.* with those in other paradoxographical collections and has clearly identified the differences in the use and in the character of the myths found in *Mir.* and elsewhere. Another aspect she correctly highlights is our author's interest in crafted items, man-made marvels, which reveals a concept of the marvelous that is unique in paradoxography, and she shows how the myths create the impression of the West as a Greek country.

One of the main problems *Mir.* causes for scholars is that it generally does not name the sources of individual chapters. In the few cases in which a source is mentioned, it is rightly assumed by scholars that these were the sources already referred to in the works excerpted by our author.[20] As is the case with some other paradoxographical collections,[21] *Mir.*'s old part consists of series of excerpts from different works, but into these series single chapters from other sources may occasionally be inserted. These additions are sometimes clearly identifiable if they interrupt a geographically or thematically coherent group of chapters (e.g., 83; 99); but the example of a chapter (126) from Aristotle's *Historia animalium* in a Theopompean series, discussed by Pietro Zaccaria,[22] should remind us to keep in mind that such additions may also be completely coherent with the surrounding material and not identifiable as such unless we happen to know their provenance, which is usually not the case. Furthermore, a principal problem in identifying the source of a chapter, often not sufficiently taken into account in studies on sources, has become obvious in the same contribution and in that of Stefan Schorn (2022)[23]: scholars often come to a conclusion about the source of a chapter if part of its information has a parallel in the named fragment of a fragmentary author. There is, however, one case in which it is clear that Theopompus and Lycus must have had very similar accounts[24] and in another Timaeus and Lycus.[25] In the latter case, it is even very

---

18 More than 35% of the chapters contains myth: Greene (in this volume) 170.
19 A list of all mythological accounts in paradoxography can be found in Greene (in this volume) 206–208.
20 See, for example, Flashar (1972) 46 (on Polycritus); Vanotti (2007) 39–41.
21 See, for example, Antigonus with Ziegler (1949) 1145–1149; *Paradoxographus Vaticanus, FGrHist* 1679 with Sørensen in Schorn (2022) 599; *Paradoxographus Florentinus, FGrHist* 1680 with Greene in Schorn (2022) 652.
22 Zaccaria (in this volume) 101–105.
23 Zaccaria (in this volume) 107–108; Schorn (in this volume) 64.
24 Theopompus, FGrHist 115 F 274b and Lycus, FGrHist 570 F 4 quoted consecutively in Ael. *NA* 17.16.
25 *Schol. Lycophr. Alex.* 615a Leone with the "collective reference" to Timaeus, *FGrHist* 566 F 53 and Lycus, *FGrHist* 570 F 3.

probable that one author was used by the other. This shows that it can be dangerous to identify the source of a report based on similar information in the fragment of an author unless there are, in addition, significant verbal similarities. And even then, it cannot always be asserted with certainty whether an author was used directly or indirectly by the compiler of *Mir.*

As for the Theopompean section (122–138), Zaccaria has shown much clearer than earlier research which chapters can be claimed for this author with certainty and which only with probability. It has also become clear from the comparisons of *Mir.* with named fragments of Theopompus that our compiler did not manipulate the information on a large scale or combine information from different sources into new accounts, although he did heavily (and often misleadingly) condense what he read in his source and that he adapted it to some extent.[26] In this regard, it remains a problem to assess to what extent our compiler invented place names to "singularize" certain *mirabilia* and to lend credibility to them, as was argued by Flashar,[27] because Giacomelli shows in his contribution that there may sometimes be other explanations for place names which are unique or contradict the parallel tradition.[28] Zaccaria was also the first to recognize a geographical arrangement of the material in this section and to show that *Mir.* contains more material from Theopompus than reflected in the collections of his fragments.[29]

The identification of the sources is most problematic in the ethnographical passage 78–121. It was claimed for Timaeus (Valentin Rose), Timaeus and Lycus (Karl Müllenhoff), or Timaeus, Lycus and Posidonius (and possibly other authors) (Hellmut Flashar).[30] Here Irene Pajón Leyra's chapter constitutes a big step forward, since she has shown that it is extremely likely that this section contains excerpts from a work *On Islands*, into which two series of thematically coherent excerpts from other sources were inserted. The first is a geographically structured passage that follows the itinerary of Heracles' tenth Labor in reverse order (85/86–98),[31] and the second is a group of chapters connected by the traces left by Homeric heroes in the West (106–110). In addition, the Heracles passage also seems to contain insertions from the work *On Islands*, added in the geographically correct places (here again we see that a series of excerpts may contain interpolations from elsewhere). This analysis has been confirmed by Schorn's study of Timaean elements in *Mir.*, who argues that Timaeus was always used indirectly by *Mir.* through this work *On Islands*, while he was probably the direct source of those reports on the Western islands in Diodorus' book 5 that show thematic overlap with *Mir.* He concludes that the author of *On Islands* has seriously reworked the Timaean accounts and that

---

26 See already Flashar (1972) 44 and Wilson (forthcoming).
27 Flashar (1972) 43, 49–50. See also Zaccaria (in this volume) 101–105 on the name in *Mir.* 126 and Giacomelli's somewhat different perspective, on pp. 242–243.
28 Giacomelli (in this volume) *passim.*
29 Zaccaria (in this volume) 99–101, 133–134.
30 A history of scholarship of this question in Schorn (in this volume) 32–33 and in Pajón Leyra (in this volume) 11–15.
31 This thematic unity had already been recognized in earlier scholarship; see, for example, Flashar (1972) 46–47.

Mir. can be used only in a very limited way as evidence for Timaeus and any other author that entered the collection via this intermediary source. It is unfortunately impossible to establish its identity.

The hypothesis, still advocated as a possibility by Flashar,[32] that Posidonius (135–51 BC) may be among the sources of *Mir.* has not been confirmed by the studies in this collection. The reason for such an assumption was that the Heracles section has parallels in passages of Strabo that have been assigned to Posidonius by *Quellenforschung*. The parallel accounts in *Mir.* and Strabo studied by Schorn do not, however, show significant enough overlap to necessitate assuming the use of Posidonius in *Mir.*, and he shows that Posidonian authorship of Strabonian passages is not always as certain as was assumed by earlier research. Here Vanotti's skepticism has proven to be justified and her main argument against the use of Posidonius in *Mir.* is still valid: Posidonius would not have left unchanged the historical information found in his sources if it was out of date at the time he wrote his *Histories*. As the relevant section of *Mir.* still contains information that was anachronistic at the time of Posidonius, *Mir.* probably does not rely on him here.[33]

Another possible source for one chapter of the ethnographical passage (95) and some other chapters outside it (31, 32, 178) has been proposed by Kelly Shannon-Henderson: Heraclides Ponticus. It is indeed possible that he was among the direct or indirect sources of the book *On Islands* and of other works used by our compiler, so her thesis is compatible with the results of the above-mentioned studies of *Mir.*'s sources.

Against this background, we may again look at the historical information in *Mir.* and try to identify the specific interests of our author. We limit ourselves to select aspects. Heracles is surely a favorite topic of our author. Not only is the Heracles section (85–98) based on one of his Labors, the hero also appears in 51, 58, 84, 100, 118, 121, 133, and 136. Chapter 51 is clearly an insertion into a series from Theophrastus.[34] So one might suspect this chapter to go back to a historical source, perhaps the work on Heracles used elsewhere, and the same may be suspected for (at least part of) 58. These two chapters are the only ones in the whole collection outside section 78–121 in which myth appears. Within this more general topic of mythology, *Mir.*'s interest in inscriptions by mythical figures, which are regularly quoted, is especially remarkable. We find them in different parts of the collection: 58, 110, 116, 118. Phoenicians and Carthaginians are popular topics in all three parts of the ethnographical section with a clustering at its end: 84, 88, 96, 100, 113, 117, 119–121. The Persians (especially their king) are only found in chapters taken from Theophrastus (27, 35, 49) and briefly reappear in the ethnographic section (96) without being at the center of interest. In general, we recognize that the focus of interest of our author is on the Mediterranean and adjacent areas and that the typical *thaumasia* of the lands in the East are rare (e.g., Indian marvels only in 49, 61, and 71). The author is interested in reports related to Homer especially in

32 Flashar (1972) 46–47; but see 52 n. 3.
33 Vanotti (2007) 43–44.
34 See Flashar (1972) 90.

the series 106–110, but verses of the poet are also quoted in the context of the journey of the Argonauts in 105; and a Homeric passage may be the basis of the long description of the Straits of Messina in 115, as is argued for at length by Charles Delattre. Growing metal takes center stage in the section based on Theophrastus' *On Metals* (42–44; 47), but is again thematized in the ethnographical section in a chapter on mining on Elba (93). Poison is the uniting feature of another series of excerpts from Theophrastus (139–151), but is not absent in the ethnographical part either (78, 86). All these features, and others, may tell us something about the literary personality of our author, or at least about his interests; but they can sometimes also be relevant for the identification of the source of a certain passage. Here more research may lead to a deeper understanding of our author and his text.

The Appendix (152–178) is much less relevant from a historical point of view, although it also contains, besides some chapters dealing with natural phenomena, some mythical, historical, and ethnographical information. Its sources are much later than those of part one,[35] and most of them are still preserved. As its compiler is different from that of the first part, we unfortunately cannot learn from the Appendix, by comparing *Mir.* with the original texts, how the author of the first part dealt with his sources. The most remarkable feature of the Appendix, studied by Søren Lund Sørensen, is the fact that it contains 11 excerpts from pseudo-Plutarch's *On Rivers*, around 40% of the chapters, and that the compiler very probably made direct use of that work. Here it is important to note that the compiler leaves aside the very peculiar mythical *aitia* always present in the original accounts.[36] This shows that his approach is very different from that of the compiler of the first part, whose selection is characterized by a strong interest in little-known mythological stories. It is also remarkable, as is stressed by Sørensen, that the compiler of the second part and Stobaeus, who also excerpted chapters from *On Rivers*, did not realize or care that the stories in pseudo-Plutarch were "pseudo-paradoxography," freely invented *mirabilia* combined with mythological *aitia* made up by using elements of existing myths, and probably meant as some kind of parody of the several paradoxographical works titled *On Rivers*.[37]

Works cited

Dorandi, T. 2006 [2009]. "La *Vita Hesychii* d'Aristote." *Studi classici e orientali* 52: 87–106.

Flashar, H. 1972. *Aristoteles. Mirabilia, übersetzt von H. F. De audibilibus, übersetzt von U. Klein*. Berlin: Akademie-Verlag; ²1981 with corrigenda to *Audib.*, but not to *Mir.*

Geus, K. 2016. "Paradoxography and Geography in Antiquity. Some Thoughts about the 'Paradoxographus Vaticanus'." In González Ponce, F.J. *et al.* eds. *La letra y la carta. Descripción. verbal y representación gráfica en los diseños terrestres grecolatinos. Estudios in honor de Pietro Janni*. Sevilla: Universidad de Sevilla. 243–257.

Giacomelli, C. 2021. *Ps.-Aristotele, De mirabilibus auscultationibus. Indagini sulla storia della tradizione e ricezione del testo*. Berlin; Boston: De Gruyter.

---

35 For a list of the sources, see Giacomelli (2021) 24.
36 On this aspect, see Sørensen (in this volume) 272.
37 See Sørensen, (in this volume) 260, 272.

Giacomelli, C. 2023. *Pseudo-Aristotele. De mirabilibus auscultationibus. Edizione critica, traduzione e commento filologico*. Rome: Accademia Nazionale dei Lincei.

Giannini, A. 1965. *Paradoxographorum Graecorum reliquiae recognovit, brevi adnotatione critica instruxit, latine reddidit*. Milan: Istituto Editoriale Italiano.

Pajón Leyra, I. 2011. *Entre ciencia y maravilla: el género literario de la paradoxografía griega*. Zaragoza: Prensas Universitarias de Zaragoza.

Rashed, M. 2021. *Ptolémée 'al-gharīb'. Épître à Gallus sur la vie, le testament et les écrits d'Aristote. Text établi et traduit*. Paris: Les Belles Lettres.

Schepens, G. and Delcroix, K. 1996. "Ancient Paradoxography: Origin, Evolution, Production and Reception." In Pecere, O. and Stramaglia, A. eds. *La letteratura di consumo nel mondo greco-latino. Atti del Convegno internazionale. Cassino 14–17 settembre 1994*. Cassino: Università degli Studi di Cassino. 373–460.

Schorn, S. ed. 2022. *Die Fragmente der Griechischen Historiker* Continued *IV E: Paradoxography and Antiquities. Fascicle 2. Paradoxographers of the Imperial Period and Undated Authors [Nos. 1667–1693]*. Leiden; Boston: Brill.

Vanotti, G. 2007. *Aristotele. Racconti meravigliosi. Testo greco a fronte. Introduzione, traduzione, note e apparti*. Milan: Bompiani.

von Christ, W., Stählin, O. and Schmid, W. ⁶1920. *Geschichte der griechischen Literatur. II. Die nachklassische Periode der griechischen Literatur*. München: C.H. Beck.

Wenskus, O. and Daston, L. 2006. "Paradoxographi." In Cancik, H. and Schneider, H. eds. *Brill's New Pauly, Antiquity volumes*. Leiden; Boston: Brill. Consulted online on 16 November 2022.

Wilson, M. (forthcoming). "The Lives of Metals." In Zucker, A., Hellmann, O. and Mayhew, R. eds. *The Aristotelian* Mirabilia *and Early Peripatetic Natural Science*. London: Routledge.

Ziegler, K. 1949. "Paradoxographoi." *Paulys Realencyclopaedie der classischen Altertumswissenschaft* XVIII(3): 1137–1166.

Zucker, A., Mayhew, R. and Hellmann, O. eds. (forthcoming). *The Aristotelian* Mirabilia *and Early Peripatetic Natural Science*. London: Routledge.

1 Islands and their marvels as structural principle in the so-called historiographical section of the *De mirabilibus auscultationibus**

Irene Pajón Leyra

Like most of the extant paradoxographical collections, the *De mirabilibus auscultationibus* (henceforth *Mir.*), which circulated in antiquity under the authority of Aristotle, is a work defined by the confluence of materials coming from sources of different kinds and origins, arranged according to a structure and a general purpose that is not always obvious at first sight. In this context, the main aim of this chapter is to call attention to an aspect that has received little or no attention from scholars, that is, the importance of islands as a possible element of coherence underlying the apparent lack of a clear structure in items 78–121 of this collection of curious anecdotes.[1]

We will focus our attention, therefore, on a series of 44 anecdotes defined by a thematic interest in ethnographical and geographical curiosities, which appear after a long series of *paradoxa* dedicated to natural science – the more "peripatetic" section of the collection – and before a section, whose main source seems to be Theopompus of Chios,[2] in which natural phenomena are once again an important topic. Scholars have repeatedly observed the special nature of this series of chapters,[3]

* This study was carried out in the context of the following research projects, directed by Francisco J. González Ponce: "El prisma romano: ideología, cultura y clasicismo en la tradición geo-historiográfica, II" (PID2020–117119GB-C22); "Incognitae terrae, incognitae gentes. El conocimiento geográfico e historiográfico antiguo: formación, evolución, transmisión y recepción" (P20_00573), and "Hacia las fronteras del mundo habitado. Conocimiento y transmisión de la literatura geográfica e historiográfica griega" (US-1380757). I am indebted to Stefan Schorn and Robert Mayhew for their countless remarks and suggestions, which have greatly improved the final product. Alwyn Harrison's English copyediting assistance has been precious. Every remaining mistake, however, is entirely my responsibility.

1 Like the rest of the contributors to this volume, I follow the sequence of items adopted by Ciro Giacomelli in his edition of the text: see Giacomelli (2023) 146–171 and 296–297; see also Giacomelli (2021) 51–53. Chapters 78–121 are affected by the mechanical transposition that at some point moved items 115–122 from their original place and turned them into chapters 130–137 in some manuscripts, an order that later became standard in the extant editions, and that Giacomelli corrects. Therefore, the series of items that are the object of this study are those corresponding to chapters 78–114 and 130–136 in the old editions.

2 For a concise evaluation of the material related to Theopompus in *Mir.*, see Pietro Zaccaria's contribution to this volume.

3 Schrader (1868) 225; Müllenhoff (1890) 427–429; Geffcken (1892) 90; Giannini (1964) 135; Flashar (1972) 45; Vanotti (2007) 33, 37–38.

DOI: 10.4324/9781003437819-2

distinct from the rest of the collection of *paradoxa* because of its aforementioned interest in historical and ethnographical/geographical anecdotes, as well as because of the frequent references to mythological stories[4] related to the places it deals with, which often explain their remarkable features. However, the question of its internal organization and structure has not yet received a definitive answer, though some hypotheses have been suggested. It is possible to distinguish two main trends in the scholarship, the most prevalent of which construes the general distribution of contents within this section of *Mir.* as a series of diverse sources that the paradoxographer used. In general, scholars have attributed most of the material to Timaeus of Tauromenium and Theopompus of Chios: the latter being responsible for the details of Eastern locations, while the former for the information about places in the West. As corollary to this interpretation, scholars have maintained that a geographical distribution lies behind the apparent disorder.

I shall argue, however, that a closer examination of the sequence of anecdotes reveals a better way of understanding the structure of this central part of the paradoxographical collection.

1. *Mir.* 78–121: in search of its sources

Contrary to the common practice among paradoxographers, who normally show particular care and attention in explicitly mentioning the sources from which they have obtained the content of their compilations, the author of *Mir.* only names the origins of his anecdotes on five occasions.[5] This is a feature that distinguishes his work, for instance, from those of his closest contemporaries in paradoxographical literature, in particular Antigonus, whose collection of rarities is perhaps the most outstanding example of a paradoxographical text in which the sources are always mentioned – the catalogue is even organized into different sections dedicated to particular sources.[6]

Scholars in the 19th century interested in the identification of the sources of *Mir.* offered us valuable exercises in *Quellenforschung*, from which an image of the work emerged that brings it close to the common procedures among paradoxographers. They seem to have assumed that the work as a whole, and in particular the section we are dealing with, consists of a sort of chain of clusters of anecdotes drawn from a certain work or author, sometimes combined with isolated items

---

4 For a thorough study of the role of myths both in *Mir.* and in paradoxographical literature as a whole, see Robin J. Greene's contribution to this volume.

5 *Mir.* 37 (the Periplus of Hanno), *Mir.* 38 (Xenophanes), *Mir.* 112 (Polycritus, author of a poem on the history of Sicily; probably also the author referred to by οὗτος in 115 as well), *Mir.* 119 (Φοινικικαὶ ἱστορίαι), and *Mir.* 173 (Eudoxus). Callisthenes is mentioned in *Mir.* 117, but not as the source of the anecdote, but to deny his trustworthiness.

6 According to the testimony of Antigonus Paradoxographus *Mir.* 129–173, this seems to have been a common practice among paradoxographers since the very beginning of the genre, with the work of Callimachus. There are further instances in the texts of Apollonius the paradoxographer, Phlegon of Tralleis (*FGrHist* 1667), *Par. Flor.* (*FgrHist* 1680), *Par. Vat.* (*FgrHist* 1679), and *Par. Pal.* (*FgrHist* 1681). On these authors, see the contributions in Schorn (2022).

whose origins are unknown. This interpretation was suggested for the first time by Valentin Rose,[7] who attributed chapters 78–136 (of the old sequence) to two main sources: Timaeus of Tauromenium and Theopompus of Chios, the former responsible for the anecdotes corresponding to the western Mediterranean, items 78–114 and 115–121 (old 130–136), and the latter for those on eastern locations in items 122–138 (137, 115–129, 138 in the old sequence).

Later scholars continued on the path opened by Rose. Karl Müllenhoff[8] examined the sequence in detail and arrived at a more complex vision: in his opinion, every set of items with a certain coherence[9] should be attributed to a different source. As a result, the chapters Rose attributed to Timaeus are in Müllenhoff's analysis distributed between two main authors, Timaeus and Lycus of Rhegium, so he assigns items 82, 84–98, and 100–103 to Timaeus, and regards items 78–81 and 104–121 as coming from Lycus. Timaeus thus occupies a central place in the historiographical section, between two subsections consisting of material obtained from Lycus. Chapters 83 and 99, however, remain without any clear attribution, as they are not connected to the surrounding items.

Müllenhoff's interpretation was in general well received among contemporary scholars, even if in some cases it was also an object of criticism, mainly from Geffcken,[10] who returned to the idea of Timaeus as the principal source of the western anecdotes, this time establishing a method to identify Timaean material in *Mir.* through the coincidences between the items of our collection and the contents attributed to Timaeus in other authors, mainly Diodorus and Strabo.[11]

Even if Geffcken's ideas were strongly criticized in the first half of the 20th century,[12] they were accepted in general and upheld as the standard by some.[13] The last extant analysis, in the work of Hellmut Flashar,[14] agrees with Geffcken about the importance of Timaeus in this part of the paradoxographical collection, but Flashar also revives the old ideas of Müllenhoff and suggests a strong presence of Lycus as an authority, in particular in chapters 78–81, 111–114, and 130. Flashar is also skeptical about attributing items 104–105 to Timaeus. Moreover, in the series 85–87 and 89–94, he notes close parallels with texts in which Strabo relies on the authority of Posidonius, a fact already remarked on by Geffcken,[15]

7 Rose (1854) 55; cf. Rose (1863) 280. Criticism of Rose's opinion can be found in Schrader (1868) 230.
8 Müllenhoff (1890) 426–442.
9 The principle is not explicitly formulated by Müllenhoff himself, but it is described by Geffcken (1892) 84 as the attribution of "jede zusammenhängende Reihe" to a different source.
10 Geffcken (1892) 84 n. 4, also refers to von Gutschmid's recension of Müllenhoff (von Gutschmid [1871] 526–527), which criticized the suggestion of Lycus as an origin of some of the anecdotes, pointing out that perhaps Timaeus could have used him as his own source.
11 A full analysis of Geffcken's methodology and conclusions can be read in Schorn's contribution to this volume, pp. 33–34.
12 On the criticism to Geffcken in the review of Niese and in the works of Schwartz and Jacoby, see Schorn, in this volume, pp. 33, n. 7.
13 Geffcken's interpretation is fully accepted, for instance, by Pearson (1987) 54.
14 Flashar (1972) 45–48.
15 Geffcken (1892) 95. In his opinion, Posidonius had used Timaeus.

but that in the opinion of Flashar could suggest Posidonius as the source of these anecdotes.[16]

As we can see, the sources of the historiographical section of *Mir.* remain a matter of debate from which the only general consensus is the likely presence of material originating from Timaeus. However, the specific items that should be attributed to this author, the presence or not of other sources and their possible identification (mainly Lycus, but also Theopompus among others), whether they were consulted directly by the paradoxographer or reached the author of *Mir.* via Timaeus or other intermediaries, or even the possibility that Timaean material could have come to our paradoxographical collection through later authors, such as Posidonius, are aspects on which there is no agreement. Even Timaeus, as Stefan Schorn clearly shows in his contribution to this volume, can be identified with certainty as a source of *Mir.* in only a few chapters,[17] as most of the alleged Timaean materials are very uncertain.

2. The paradoxographer's plan: geographical explanation

It should be remarked that the main aim of the aforementioned scholars was not to explain whether the organization of this series of *paradoxa* corresponds to some plan on the part of the paradoxographer. Rather, their focus was the identification of the sources the compiler used. Nevertheless, as Schrader already noted,[18] the two problems, that of the sources and that of the compiler's plan in this section, are deeply connected. For Schrader, the lack of clarity regarding the "parts" of the series 78–138 affects our chances of identifying the sources of its individual items.

In comparison to the problem of the sources, that of the plan of the paradoxographer has received much less attention. However, some hypotheses have been formulated that deserve comment. First, Geffcken[19] glimpsed a structure that might reflect the organization of Timaeus' work. In his opinion, the chapters of *Mir.* offer traces of a plan, but they appear "in quite a motley order" ("in recht bunter Ordnung"). If they are read "in the right order," it is possible to appreciate significant parallels with the original sequence of Timaeus. Geffcken suggested reading 81–80–79, 102, 103; 104, 105; 106–110; 117–121 (132–136 of the old sequence). He also found that the series 84–94, if read backward, parallels the passage referencing Timaeus in Polybius 12,[20] which he interpreted as evidence of the sequence of content in the original Timaean text. Geffcken suggested that Timaeus probably dealt first with Sicily, his homeland, moving on to Illyria and northern Italy, Umbria, Daunia, Bruttium, Campania, Rome, Etruria, and Liguria; the land of the Celts, Iberia, and the Balearic Islands then followed, before the text went on to deal with places beyond the Pillars of Heracles, and later coming back to Libya,

---

16 Flashar (1972) 47 himself expresses his doubts about the Posidonius hypothesis. However, these doubts are less clear in his commentaries on *Mir.* 87 and 89–94; (1972) 110–115.

17 Indeed, he considers Timaean origins as certain only in the cases of *Mir.* 102 and (in part) *Mir* 88.

18 Schrader (1868) 225.

19 Geffcken (1892) 97–98.

20 Polyb. 12.28a.3: . . . καὶ πολυπραγμονῆσαι τὰ Λιγύων ἔθη καὶ Κελτῶν, ἅμα δὲ τούτοις Ἰβήρων . . ., see *FGrHist* 566 F 7.

Carthage, Sardinia, and Corsica. Roughly speaking, *Mir.* 78–121 does reflect this plan, even if Geffcken's theory requires drastic changes to the extant order of the anecdotes, without any attempt to explain the origins or causes of such changes.[21]

Flashar,[22] for his part, does not suggest such radical alterations of the extant order of the anecdotes, but he too sees a general periegetic intention, interrupted several times as a consequence of the frequent changes of source. He discerns a certain geographical structure in items 79–81 (eastern Italy described from south to north), followed by a set of items (82–84) which do not present geographical coherence. From item 85 on, however, a coherent periegetic sequence begins, roughly following the Heraclean Road from the Land of the Celts and Iberia to Liguria (86–92), to the island of Elba (93) and later passing through Italy from north to south (94 Etruria, 95 Cumae, 96 Sybaris and the Lacinian Promontory, 97–98 Iapygian Promontory). A change of source interrupts the sequence at this point and, after one chapter which does not fit the rest of the series (99, on Orchomenos in Boeotia), the compiler returns to Timaeus, even if the following items do not present any clear coherence with the end of the periegetic series in chapter 98. Chapters 100–103 deal with the Tyrrhenian islands (Sardinia, the Islands of Aeolus and Lipara, the Sirenusae); a new geographical jump follows, in chapters 104 and 105, which go back to eastern locations. In chapters 106–110, however, Timaeus returns to deal with southern Italy, again in a roughly periegetic structure. After this, a new change of source explains the section dedicated to Sicily in chapters 111–115 (old sequence 111–114, 130),[23] which Flashar attributes to Lycus. The last chapters of our series, 115–121 (130–137 in the old sequence) are assigned by Flashar to Timaeus or Lycus, but at this point he does not try to explain their geographical coherence with the rest of the sequence.

Flashar's construction does not entail the arbitrary changes of order that are assumed in Geffcken's view. Still, his interpretation frequently slips into circular reasoning: he explains the deviations from geographical coherence as the result of switches between the sources and simultaneously identifies where the sources change through the deviations from the alleged periegetic coherence.

Finally, a curious attempt to explain our section of *Mir.* according to a geographical order has also been made by Alessandro Giannini. In his famous 1964 paper on paradoxographical literature, he describes *Mir.* 78–138 of the old sequence (78–121 + 122–137 in Giacomelli's order) as divided into three subsections, respectively devoted to the western Mediterranean (78–114), the East, which also includes Greece (123–137, old 115–129), and to a mixture of both western and eastern locations (115–121, old 130–138).[24] It is surprising that Giannini includes chapters 115–129 of the old sequence in this geographical distribution, and he deals with them as if they were part of the paradoxographer's original plan, even if he is clearly aware of the fact that they are not in the right place.[25] Furthermore, the

---

21 Nevertheless, Geffcken never formulated in a clear or full way the precise order in which, in his opinion, *Mir.* 78–121 should be read.

22 Flashar (1972) 46–48.

23 Chapters 115–129 are excluded from the periegetic section by Flashar as belonging to Theopompus and corresponding to a different section of *Mir.*

24 Giannini (1964) 135.

25 Giannini (1964) 135 n. 219, attributing items 137–138 and 115–129 to Theopompus, in that order.

alleged mixture of eastern and western locations in items 115–121 is dubious: the reality is that western locations plainly predominate, the only clear exception being 116, on Eleusis. Chapter 118, which discusses an inscription found in Hypate, in the Aeniac country (that is, between Thessaly and Epirus), probably cannot be regarded as an example of a purely eastern location, given that the center of the anecdote refers to the explanation of the name of the island of Erytheia, which is connected to the far west of the world.[26]

As can be seen, the diverse attempts to explain our section of *Mir.* as having an internal geographical coherence agree in construing such coherence as a result of the sources used by the paradoxographer. Giannini's West–East division reflects the old idea of a distribution of items between Timaeus and Theopompus; Geffcken, for his part, made a great effort to adapt the paradoxographic series to the structure of Timaeus' work as it can be reconstructed, and Flashar systematically explained what he interpreted as geographical inconsistencies as due to interruptions to the use of the paradoxographer's main source, which is, again, identified as Timaeus. In order to establish a certain logic in the sequence of *paradoxa*, then, and to see a geographical/periegetic structure behind the apparent disorder, scholars have needed to draw on arbitrary changes in the order of the items or to assume as certain changes between sources that are far from clear, or even to neglect the history of the text, forgetting the transposition of items 115–122 and 123–137, of which they are clearly aware.

In my opinion, therefore, the periegetic/geographical hypothesis, at least in its current formulations, does not lead to solid conclusions about the paradoxographer's plan in compiling this part of his work. The study of the sources, as we have seen, has not led to hypotheses that have won widespread acceptance among scholars, and therefore this line of research does not provide us with a convincing basis to understand the internal structure of *Mir.* 78–121 either.[27]

Given the lack of any reliable basis on which to establish a neat distribution of the paradoxographer's sources and/or to understand his plan as a sort of periegesis, it is time to explore new means to interpret this central section of the pseudo-Aristotelian *Mirabilia*. I do not propose to dismiss *Quellenforschung* nor to abandon the search for a coherence that derives from geographical features. However, the insistence on searching for a "chain" of excerpted authors and a "general" periegetic organization, or a logical succession of geographical areas of interest – which, moreover, should derive from the sources – may have hindered other lines of research which could lead to better explanations.

Thus, perhaps a new examination of the set of anecdotes as a whole will reveal details that passed unnoticed in the extant scholarship. Specifically, I believe that information about islands is significant and can provide a clearer understanding of this part of the pseudo-Aristotelian *Mirabilia*.

---

26 Regarding Erytheia as an element coherent with the items that follow, which deal with Phoenician settlements in western Mediterranean, see Vanotti (2007) 202–203.

27 It is precisely this lack of agreement that led Vanotti (2007) 44 to describe *Quellenforschung*, at least as it was carried out among 19th-century scholars, as a fruitless effort producing only arguable results.

3. *Mir.* 78–121: a series of *paradoxa* on islands?

In my opinion, the series of items we are dealing with is defined by a deep interest in including *paradoxa* involving islands and their peculiarities. If the sequence *Mir.* 78–121 is examined in its entirety, it becomes self-evident, and this feature of the series can perhaps offer new clues to a better understanding of the section, its meaning within the paradoxographical work, and even lead to new hypotheses about the sources used by the compiler.

In order to provide the reader with a clearer view of the sequence of items and the presence of information regarding islands, the following table lists locations, content, and a quick survey of the sources scholars have suggested for chapters 78–121.[28] Items containing information relative to islands are marked in bold in the first column.

Item	Place	Content	Source
78	Italy, Circaean mountain	Strong poison used in attempted murder of Cleonymus of Sparta	Timaeus? Lycus? Theophrastus?
79	Island of Diomedes, in the Adriatic	Birds that attack barbarians and respect Greeks	Lycus? Timaeus?
80	Umbria	Extraordinary fertility of land, animals, and women	Timaeus? Lycus (perhaps using Theopompus)?[29]
81	Electrides Islands, Icarus Island, Adriatic	Myth of Daedalus and Icarus' flights from Sicily and Crete; amber from the Eridanus, a lake, fall of Phaethon	Timaeus? Lycus? Theopompus?[30]
82	Sicily	Peculiar cave with violets and their smell; myth of the rape of Persephone	Timaeus[31]

---

28 This table is aimed at offering a synthetic view of the debate regarding the sources of *Mir.* 78–121, without any attempt on my part to take sides. I rely mainly on the opinions of Müllenhoff (1890), Geffcken (1892), Giannini (1965), Flashar (1972), and Vanotti (2007). It does not reflect my opinion, but my intention is rather to stress the extent to which these attributions are subject to disagreement among scholars. The contributions of Stefan Schorn and Pietro Zaccaria to this volume offer in-depth revisions of the anecdotes which can be linked to Timaeus and Theopompus with certainty, as well as fresh perspectives on the discussion as a whole.

29 Nevertheless, Zaccaria, in this volume, pp. 86–87, points out the speculative nature of the hypothetical attribution to Theopompus of *Mir.* 80–81, 104–105, and 118.

30 On this item, see below, p. 29.

31 Schorn, in this volume, pp. 44–50, rather advocates for an indirect relationship between *Mir.* and Timaeus.

Item	Place	Content	Source
83	Crete	No wolves, bears or snakes are born there, because this is Zeus' birthplace	Theophrastus?
84	Desert island outside the Pillars of Heracles (Madeira?)	Richness of the island, control of the Carthaginians	Timaeus? (Etruscan source[32]?)
85	The Heraclean Road, from Italy to Iberia	Protection to Greek and local travelers	Timaeus? (Ephorus?)
86	Land of the Celts. (start of periegetic section up to 98.)	Poison and its antidote	Timaeus? (perhaps through Posidonius?)
87	Iberia	Wealth in silver	Timaeus? (perhaps through Posidonius?)
88	Gymnesiae Islands	Size of the islands; oil obtained from the terebinth; customs of the Iberian soldiers, mercenaries of Carthage	Timaeus[33]
89	Massalia, near Liguria	Lake, phenomena associated with it. Plentiful fish	Timaeus? (perhaps through Posidonius?)
90	Liguria	The Ligurians' skill with the slingshot	Timaeus? (perhaps through Posidonius?)
91	Liguria	Customs of the Ligurians (Couvade)	Timaeus? (perhaps through Posidonius?)
92	Liguria	River: water rises and prevents one seeing the other shore	Timaeus? (perhaps through Posidonius?)
93	Island of Aethaleia (Elba).	Wealth in copper and iron, exploited by Tyrrhenians	Timaeus? (perhaps through Posidonius?)[34]
94	Tyrrhenia; city of Oenaria	Governed by slaves in order to avoid tyranny	Unknown. Perhaps Timaeus, maybe through Posidonius?
95	Cumae	Sibyl; river †Capan†, whose water petrifies objects	Timaeus?

(Continued)

32 On this possibility, see Vanotti (2007) 170, with bibliography. Attribution to Timaeus is practically unanimous (cf. *FGrHist* 566, fr. 164: Diod. Sic. 5.19–20). However, cf. Schorn's remarks on *Mir.* 84 in this volume, pp. 64–67.
33 Cf. Str. 14.2.10, Diod. Sic. 5.17.1–4: *FGrHist* 566 F 65 and fr. 164. On the Timaean origin of (at least part of) this item, see Schorn in this volume, pp. 35–36.
34 However, Schorn, in this volume, pp. 50–54, appears rather skeptical about our real chances of identifying the source of *Mir.* 93.

(Continued)

Item	Place	Content	Source
96	Lacinan Promontory	The cloak of Alcisthenes the Sybarite	Timaeus?
97	Iapygian Promontory	Location of the battle between Heracles and the Giants; source of ichor	Unknown
98	Iapygian Promontory	Large stone moved by Heracles with one finger	Unknown. Probably same as 97
99	Orchomenos in Boeotia	Cave; supernatural facts	Unknown
100	Sardinia	Buildings built by Iolaus; etymology of the name of Ichnoussa; fertility of the land; presence of Aristaeus; policy of the Carthaginian power	Unknown. Perhaps Timaeus[35] or maybe another Sicilian author
101	Islands of Aeolus: Lipara	Supernatural phenomena	Timaeus? Theophrastus?
102	Cumae, lake Aornos	Description of its nature; criticism of its alleged lack of birds	Timaeus (possibly polemic against Heraclides Ponticus)[36]
103	Sirenusae islands	Precise location. Cult to the Sirens	Timaeus?
104	Mountain called Delphian between Mentorice and Istriane	Market where products from the Greek islands are sold to tradesmen from the Black Sea	Theopompus? Timaeus? Lycus?
105	Ister, Cyaneans, Planktai, Aethalia, Symplegades, Strait of Sicily	Discussion of the itinerary of Jason	Timaeus? Lycus (maybe through Timaeus)? Theopompus?
106	Tarentum	Cult to the heroes of Troy; place names and their origins	Timaeus? Lycus? Antiochus of Syracuse?
107	Sybaris, Crotone	Heroic cult to Philoctetes	Timaeus? Lycus?
108	Lagaria(?), near Metapontum	Temple of Athena where the tools used to build the Trojan Horse are displayed	Timaeus? Lycus?

35 For criticism of the arguments in favor of the alleged Timaean origin, see Schorn, in this volume, pp. 54–59.
36 Cf. *FGrHist* 566 F 57, ap. Antigonus Paradoxographus *Mir.* 152.

Item	Place	Content	Source
109	Daunia	Temple of Athena where the axes of the companions of Diomedes are displayed	Timaeus? Lycus?
110	Land of the Peucetii	Necklace dedicated to Artemis by Diomedes. Later dedication to Zeus of the same necklace by Agathocles, king of Sicily	Probably not Timaeus. Lycus? Theopompus? Duris of Samos?
111	Sicily, Pelorian Promontory	Extraordinary abundance of crocus	Timaeus or more probably Lycus
112	Sicily	Lake that changes its size if anyone gets in it	Polycritus, author of *Sikelika*. Perhaps through Lycus or Timaeus?
113	Sicily? Area under Carthaginian control	Smell of flowers close to Mount Uranium; oil spring	Very probably Polycritus, perhaps through Lycus
114	Sicily?	Spring of cold water and fire	Ditto?
115 (old 130)	Sicily	The Strait of Sicily and the violence of its currents	Ditto (οὗτος, apparently Polycritus). (Lycus?)
116 (old 131)	Eleusis	Discovery of the tomb of Deiope	Unknown. Lycus? Timaeus?
117 (old 132)	Aeolian Islands: Phoenicodes	Presence of date palms and explanation of the island's name; explanation of the Phoenicians' name	Timaeus? Lycus?
118 (old 133)	Aeniac country, Hypate	Discovery of an inscription in strange characters deciphered as containing an explanation of the name of Erytheia. Questioning of its western location	Lycus? Timaeus?[37]
119 (old 134)	Utica	Fossil salt	*Phoenician history* (through Timaeus?)

(Continued)

37 See also Vanotti (2007) 204 regarding a possible use of Theopompus as source; for criticism of this opinion, see Zaccaria in this volume, pp. 86 n. 2 and 136 n. 210.

(Continued)

Item	Place	Content	Source
120 (old 135)	Tartessos	Phoenician trade and wealth in silver	*Phoenician history*? (through Timaeus?)
121 (old 136)	Four days' sailing beyond Gadeira	A location rich in tunny fish	*Phoenician history*? (through Timaeus?

As can be observed, items 79, 81–84, 88, 93, 100–101, 103–105, 110–115, 117–118 and 121 – that is, at least 21 of the 44 items of interest – transmit information related to various islands of the Mediterranean Sea. Almost half of the anecdotes belonging to this section of *Mir.* share this interest in islands and the strange phenomena that happen on them, a focus that is not visible in the other sections of the work. Indeed, in the "peripatetic section" (1–77), only 18 *paradoxa* are connected to islands,[38] that is, less than a quarter, and only 4 in the "other" Peripatetic section (139–151),[39] one third, deal with islands. In the so-called Appendix, the proportion is even lower: only three[40] items between chapters 152 and 178 transmit *paradoxa* relating to islands. Finally, in the section containing material from Theopompus, encompassing chapters 122–138, islands are completely absent.

Moreover, some chapters that do not at first glance seem to focus on islands arguably should be added to the list. This is the case of chapter 78: even if it deals with a mountain on the Campanian coast (the Circaean Mount), there is a clear association with the figure of Circe, who has been presented as living on an island since the earliest testimonies about her, in the Homeric poems; indeed, in the ancient sources, information abounds reporting that this mountain was originally an island, and was later transformed into a sort of promontory or peninsula by certain geological processes.[41]

The insistence on islands is, in my opinion, essential to construe the original plan of the paradoxographer and his approach to his sources. Indeed, in *Mir.* 78–121 anecdotes that do not deal with islands seem to be concentrated in two specific sections: 85–98 and 106–110. If each one of these subsections is considered separately, one observes that both have a specific internal coherence and, perhaps, an independent origin.

4. *Mir.* 85–98: the periegetic sequence

Let us focus now on the first series. As scholars have rightly observed, chapters 86–98 can be read as a periegetic sequence that, despite some minor inconsistencies, starts

---

38 See *Mir.* 4 (Crete), 9 (Cephalonia), 16 (Melos, Cnidus), 25 (Gyaros), 26 (a small island near the coast of the Chalybes), 33 (Tenos), 34 (Lipara), 37 (Lipara and Pithecousae), 38 (Lipara and Sicily), 40 (Sicily), 43 (Cyprus), 44 (Melos), 55–57 (Sicily), 58–59 (Demonesus), 70 (Seriphus).
39 *Mir.* 140 (Naxus), 142 (Cyprus), 143 (Ceus), 148 (Sicily).
40 *Mir.* 154 (Etna, Sicily), 170 (Euboea), 172 (Sicily).
41 See Thphr. *HP* 5.8.3, Strab. 5.3.6. Cf. Eust. *in Dionys. Per.* 692.

at the western end of the Mediterranean and ends in southern Italy. Roughly speaking, the series of *paradoxa* follows the route of the Heraclean Road, the ὁδὸς Ἡράκλεια or *Via Herculea*, which is even explicitly mentioned in chapter 85, but in the opposite direction: according to *Mir.* 85, the Road[42] ran from Italy to Iberia, whereas this series of items starts in the Iberian Peninsula and arrives in the heel of Italy, in Puglia.

The *Via Herculea*, which in the times of Augustus became a well-established road (then called the *Via Augusta*) and whose path archaeologists are gradually reconstructing, thanks to the evidence of mile posts and other material traces, has its ultimate roots in the memories of the mythical itinerary followed by Heracles in his tenth Labor,[43] which consisted of bringing the cattle of Geryon from the Island of Erytheia, in the outer Ocean, to the Peloponnese.[44] According to the testimonies of Diodorus, Dionysius of Halicarnassus, and Apollodorus, after setting his famous Pillars at the Strait of Gibraltar,[45] the hero defeated the Iberian king Chrysaor, father of Geryon and owner of said cows, and drove the animals along the coast of the Iberian Peninsula[46] to the territory of the Celts,[47] from which he opened the path through the Alps to enter Italy,[48] then traveled through Liguria,[49] Tyrrhenia,[50] Rome,[51] and the Phlegraean plain, in Campania,[52] where authors such as Diodorus place the combat between Heracles and the Giants, attributing the information to Timaeus. The next stops on his trip are lake Avernus and various

---

42 On the *Via Herculea* and literary and archaeological–historical testimonies of it, see Knapp (1986). See also Plácido (1993) 75–80 and (2002) 129, which point out the importance of ancient routes of transhumance in the consolidation process of that road. On the diverse stages of Heracles' journey bringing the cattle of Geryon from Iberia to Greece, see Cordano (2014).

43 On the extant variants of this myth, see Gantz (1993) 408–409; on the particular development of Heracles' *nostos* of this Labor, and its value in the building of a conceptual map of the "far west" among the Greeks, see Finglass (2021) 142–146.

44 On Heracles' journey as the mythical precedent of the Roman road that started on the shores of Ocean and ran toward the east, see Rathmann (2006).

45 According to Diodorus, the hero reached the Ocean via Libya, along the southern shore of the Mediterranean. Apollodorus, however, gives him a route mainly along the European shore of the sea, sometimes briefly entering Libya (Apollod. *Bibl.* 2.5.10 [107]).

46 Diod. Sic. 4.18.2–3; Dion. Hal. 1.41.2; Apollod. *Bibl.* 2.5.10 (109).

47 Diod. Sic. 4.19.1–2.

48 Diod. Sic. 4.19.3–4; Apollod. *Bibl.* 2.5.10 (109).

49 Diod. Sic. 4.20; Dion. Hal. 1.41.3; Apollod. *Bibl.* 2.5.10 (109). On Heracles in Liguria, see also Aesch. F 199 Radt (ap. Strabo. 4.1.7, from the *Prometheus Lyomenos*), explaining that Heracles ran out of arrows and needed the help of Zeus, who sent him a shower of stones that the hero used to defend himself from the hostile and bellicose Ligurians. Apollodorus, however, attests a different story and refers to the hero's clash with Ialebion and Dercynus, two sons of Poseidon who tried to steal the cows (Apollod. *Bibl.* 2.5.10 [109]).

50 Diod. Sic. 4.21.1.

51 Diod. Sic. 4.21.1–4. Greek sources offer versions of the story in which the hero's adventures in the area of the future city of Rome are not particularly developed. Cf., however, the "Roman" versions of Dion. Hal. *Ant. Rom.* 1.39 and Verg. *Aen.* 8.190–272, in which the figure of Cacus as antagonist of the hero has crucial importance. See Dion. Hal. *Ant. Rom.* 1.39.2; Verg. *Aen.* 8.190–272, Prop. 4.9.1–20; Ov. *Fast.* 1.543–582.

52 Diod. Sic. 4.21.5–7. On Heracles in Campania, see also Dion. Hal. *Ant. Rom.* 1.44.1, about the foundation of Herculaneum.

places in Calabria[53] (Posidonia, Rhegium,[54] Locri). Some authors also included a stay in Sicily,[55] from which Heracles returns to Calabria. The final stages of the journey are not described in detail in the extant sources. According to Diodorus,[56] he traveled to Greece describing an arc all along the shore of the Adriatic Sea; in Apollodorus' version, however, he drove the cows to the Ionian Sea, to Thrace, and on to the Peloponnese.[57]

In their extant order, the *paradoxa* in our list of *mirabilia* correspond to the Land of the Celts (86), Iberia (87), the Balearic Islands (88), Massalia (89), Liguria (90–92), Tyrrhenia and its coast (93,[58] 94), Campania (95), and Calabria (96), in almost full agreement with the hero's path. However, the course recorded by the pseudo-Aristotle diverges from the standard versions of the heroic journey, in particular that in Diodorus, in two important respects: first, this subsection of *Mir.* does not include Sicily as part of the itinerary, and second, it includes two chapters set in Puglia (*Mir.* 97–98, the Iapygian Promontory), explicitly identifying it as the location of the battle with the Giants, attesting to a version of the myth that diverges from the tradition represented by the other extant authors.[59]

Only Strabo mentions a similar story. In his sixth book,[60] he talks about a spring (πηγή) of foul-smelling water in the area of Leuca, that is, at the southernmost tip of Puglia, whose origin he connects to Heracles' combat against the Giants. The geographer agrees with the paradoxographer in attributing the fetid smell to the *ichor* that poured from the wounds of the defeated monsters. However, Strabo's version still differs from the paradoxographer's: in his text the Giants arrive in Puglia after fleeing the battle with Heracles, which takes place in the Phlegraean plain, as in the standard account.

As mentioned above in the account of the attempts to explain *Mir.* 78–121 as a geographical series, scholars, particularly Geffcken, construed this periegetic subsection as reflecting the sequence of the content of Timaeus' original work, and therefore regarded it as evidence of the presence of Timaean material in the work. Indeed, as Stefan Schorn notes in his contribution to this volume, points of intersection with named fragments of Timaeus can be observed in this series of anecdotes. In particular, chapter 88, dealing with the Balearic Islands and their size, ranked immediately after the seven largest islands of the Mediterranean, seems to reveal a close correspondence in both content and wording, between Strabo 14.2.10,

---

53 Diod. Sic. 4.22.3–4.

54 Cf. Apollod. *Bibl.* 2.5.10 (110).

55 In particular, see Diod. Sic. 4.22.6–24.6. Also Dion. Hal. *Ant. Rom.* 1.44.2, Apollod. *Bibl.* 2.5.10 (110–111), Paus. 3.16.4.

56 Diod. Sic. 4.25.1.

57 Apollod. *Bibl.* 2.5.10 (112). At this point, the story is not clear. Hera sends a gadfly and disperses the cows across Thrace and the Hellespont. Once he has gathered them together again, the hero takes them to Mycenae.

58 *Mir.* 93: the island of Aethaleia, that is, Elba.

59 See Vanotti (2007) 179.

60 Strab. 6.3.5. Flashar (1972) 116, however, sees Timaeus as the likely source of both Strabo and the pseudo-Aristotle.

which explicitly mentions Timaeus,[61] and Diod. Sic. 5.17.1.[62] Nevertheless, if the sequence is examined as a whole, important details point in a different direction: in the first place, the absence of Sicily – Timaeus' homeland – from the geographical sequence, and second, the placing of Heracles' battle against the Giants and the resulting foul water on the Iapygian promontory, not on the Phlegraean plain, a detail the sources specifically identify as present in Timaeus' text.

Admittedly, it is likely that materials whose ultimate origins can be traced back to Timaeus' work are present in this section of our paradoxographical list, but the paradoxographer probably did not obtain them directly from Timaeus; rather, he might have used a work, perhaps a text specifically on Heracles' Labors, that contained astonishing news about the places visited by the hero, or even an earlier compilation of curiosities, arranged to follow the route of the *Via Herculea*.

Scholars have sometimes described *Mir.* 86–98 as arranged "in the opposite direction" with respect to the route of the *Via Herculea* introduced in chapter 85. Nevertheless, the sequence of chapters 86–98 in reality follows the journey of the hero with only minor deviations. Therefore, perhaps the reading of chapters 86–98 as running "in the opposite direction" to the *Via Herculea* as it is described in chapter 85 is stressing the wrong point, and maybe the right question should be why *Mir.* 85 takes the ὁδὸς Ἡράκλεια to run *from* Italy *to* Iberia, instead of respecting the original direction of the mythical journey.[63] However, this question is far beyond the scope of this chapter.

Be that as it may, *Mir.* 85–98 presents a series of items with an internal logic and coherence. Furthermore, it is precisely in this section that some scholars, in particular Flashar, have seen the possibility that some information comes from the work of Posidonius. This interpretation is by no means certain and, of course, the possibility remains that both pseudo-Aristotle and Posidonius obtained their material from the same source (Timaeus or perhaps some other author or authors). Still, the internal coherence of this set of anecdotes allows one to consider it as to some extent independent of what precedes and follows.

5. *Mir.* 106–110: the Greek heroes of Troy in Southern Italy

Let us now concentrate on the second subseries in which islands do not predominate. Chapter 105 is devoted to the route of Jason's travels in the Mediterranean, on both his outward and return journeys. The paradoxographer includes in the homeward journey anecdotes regarding islands in the Adriatic Sea, as proof that the hero traveled down the river Istrus,[64] as well as curiosities found on the Tyrrhenian islands (notably Aethalia), which are presented as results of the Argonauts' passage. The chapter finishes with a brief description of volcanic phenomena in Sicily,

---

61 *FGrHist* 566 F 65, ap. Strab. 14.2.10.

62 *FGrHist* 566 F 164, ap. Diod. Sic. 5.2.1–23.5.

63 It is noteworthy that, for instance, the itinerary *from* Gades *to* Rome, which appears engraved on the Goblets of Vicarello (1st–2nd c. CE) runs in the same direction as our sequence of pseudo-Aristotelian *mirabilia*. See *CIL* XI 3281–3284.

64 In particular, he mentions an island with a sanctuary of Artemis built by Medea herself.

also connected to the route of Jason and his companions. Next, chapters 106–110 abandon the natural *paradoxa* of Sicily, a topic that will be resumed in chapter 111. This new series thus apparently interrupts another coherent series of items, a subsection devoted to Sicily, which for its part was also connected with the items that preceded it, dealing with the other islands of the Tyrrhenian Sea (besides the mention of Aethalia in *Mir.* 105, see *Mir.* 100 on Sardinia, *Mir.* 101 on Lipara, and *Mir.* 103 on the Sirenusae).

Furthermore, in this case too it is possible to observe a situation similar to that of chapters 85–98: again, an internal coherence is visible in the subsection, which manifests in two different respects. First, chapters 106–110 focus on a very specific geographical area, that of eastern Magna Graecia: Tarentum, Sybaris, Croton, Metapontum, Daunia, and the land of the Peucetii (probably near Brindisi).[65] In this case, however, the series also shows a thematic coherence, given that all its chapters focus on a specific topic: offerings kept in temples, place names, and other similar material traces of the presence of the heroes of Troy in the area, as well as the heroic cults associated with them.

This is, then, another series of items in which islands are not the guiding thread that has an internal coherence and logic that suggests a possible independent origin. This time, however, a curious detail is to be observed, which perhaps helps us appreciate the way in which the paradoxographer worked and organized his collection: chapter 110, the last item of the Magna Graecia/Trojan subsection, has a special significance within the series. It maintains a thematic consonance with the rest of the chapters, in that it relates to Magna Graecia, but, importantly, it presents a sanctuary of Artemis among the Peucetians, that is, in central Puglia, which housed a necklace on which an inscription read "Diomedes to Artemis." This necklace, the paradoxographer explains, was once placed by the hero around (περιθεῖναι) the neck of a deer, and it remained fixed upon it (περιφῦναι).[66] Thus, it coincides with the rest of the sequence, both with respect to geography and in talking about the material traces of the Trojan heroes' *Nostoi* in southern Italy. Further, *Mir.* 110 is not only connected to Magna Graecia, it also has a connection to Sicily: at the center of the anecdote is not the necklace housed in the Peucetian sanctuary of Artemis, but the fact that it was found there by Agathocles, king of Syracuse (between 316 and 288 BCE),[67] exactly where the hero had placed it. Agathocles then offers it again, this time in a sanctuary of Zeus. Presumably, this second dedication was made in Sicily, though the compiler is not explicit on this point. In any case, the *paradoxon* starts by talking about a treasure housed in a sanctuary of Artemis in Puglia, and ends with its offering by a Sicilian tyrant.

Diomedes is indeed a hero with strong links with the Adriatic coast of Italy, with sanctuaries and local legends connecting him to Veneto and Daunia. The pseudo-Aristotelian compiler referred to him already in chapter 79, dealing with

65 See Vanotti (2007) 194, with bibliography.
66 Regarding the meaning of this verb, see Giacomelli (2023) 291–292.
67 On Agathocles, see Meister (2006).

the so-called "island of Diomedes" and the well-known story of the metamorphosis of his companions into birds that are friendly toward Greek sailors, but receive all the non-Greek visitors with violence and hostility. Beyond the areas of Veneto and Daunia, the presence of Diomedes' *vestigia* in southern Italy is also notable, as scholars have frequently acknowledged.[68] Indeed, Agathocles' interest in dedicating the necklace that once belonged to the hero has been interpreted by scholars as a political move: the Syracusan king is using a mythical figure of particular importance in eastern Magna Graecia to promote his own ambitions, to consolidate his own power over southern Italy.

There is little doubt that the original context of the anecdote in *Mir.* 110 was a discourse about Agathocles' political relations with the Peucetians. However, what is interesting to us here is the fact that *Mir.* 110 is immediately followed by a long series of *paradoxa* on natural phenomena of Sicily. Thus, the topic that was abandoned after *Mir.* 105 is here resumed. I believe that this reveals, on the one hand, the special character and, perhaps, the independent origin of the set of chapters between *Mir.* 106 and 110 and, on the other hand, the deep connection between the content that precedes and follows it, in *Mir.* 105 and 111–115.

It is not possible to glimpse a plan or an internal order in *Mir.* 106–110. There is no appreciable geographical or periegetic structure, in my opinion, nor any other kind of ordering criteria (e.g., alphabetical order). But is the organization of this section random? The paradoxographer's placement of chapter 110 appears by no means to be a matter of chance. On the contrary, the fact that the last item of this subseries deals with both Puglia and Sicily can be read as indicative of the will and the effort of the compiler to incorporate the series of anecdotes about heroic cults in a way that maintains the internal logic guiding the collection of curiosities, placing at its end an item that can serve as a sort of link between the series of anecdotes regarding heroic cults and the section dedicated to the curiosities of Sicily into which it is inserted.

In the case of the series *Mir.* 85–98, there is possibly a similar but less obvious effort to incorporate it into the series of islands in a way that respects the overall coherence of the final product. It should be noted that the sequence following the *Via Herculea*, in principle a terrestrial itinerary, includes two anecdotes connected to Mediterranean islands, the Balearics (88) and Aethalia (93). Moreover, these islands seem to fit the series of Tyrrhenian islands that follows (Sardinia, the Aeolians, the Sirenusae, etc.). Perhaps chapters 88 and 93 originally belonged to the series of islands and the paradoxographer moved them to better integrate the periegetic series in its new context. Of course, every statement in this sense must remain a mere hypothesis. Still, it does not seem to be mere chance that the periegetic series appears immediately after a chapter (84) dedicated to an island, perhaps Madeira, located outside the Pillars of Heracles. Indeed, the starting point of Heracles' homeward journey was precisely an island beyond the Strait of Gibraltar, Erytheia.

---

68 On Diomedes' traces in southeastern Italy, see Malkin (1998) 234–240; see also the map at ibid. 236 (partly based on Mastrocinque [1987] 90).

Going back to chapters 106–110, in view of their thematic and geographical consistency, scholars have wondered if they can all be attributed to a single source, and if so, what that source might be. Müllenhoff, for instance, saw the hand of Lycus in this section. In particular, he regarded Timaeus as unlikely in view of the mention of Agathocles, given Timaeus' clear hostility toward him, as he had been responsible for Agathocles' exile.[69] Others, however, have considered the sequence to contain material from Timaeus, despite Müllenhoff's objections: Geffcken thought this was certain, as does Flashar.[70] Later studies, however, have rehabilitated and developed Müllenhoff's position, regarding Lycus as the likely source of *Mir.* 106–110.[71]

Again, therefore, the sources used by our paradoxographer are a matter of debate among scholars, though they are all convinced that the compiler both (1) directly consulted the historiographical works that are the ultimate sources of the anecdotes, and (2) stitched together series of anecdotes extracted from one single author. Nevertheless, as we have seen, the elaboration process of the historiographical section of the *Mir.* appears to have been more complicated than the mere juxtaposition of anecdotes obtained from sources directly consulted and mechanically excerpted by the pseudo-Aristotelian compiler. What we can say, for now, is that we have rather a collection of *paradoxa* regarding islands, which seems to act as a sort of general framework, interrupted on two occasions, where the periegetic section and the chapters about hero cults in Magna Graecia were inserted. Further, there are visible traces of an effort on the part of the compiler to create a final product held together by a general coherence and logic – a way of working that is very different from the casual addition of clusters of anecdotes or the systematic epitomizing of Timaeus, arranging excerpts "in a motley order." In this sense, perhaps it would be better to imagine a compiler who, instead of directly using the works of Timaeus or Lycus, or other similar texts, could have used as his main source a work of the *Περὶ νήσων*, or *On Islands* type, whose content provides as a sort of backbone for the whole geographical/ethnographical section of *Mir.* To this main framework pseudo-Aristotle[72] added two further series of anecdotes: a collection of material arranged as a periegesis that followed the *Via Herculea*, perhaps elaborated with paradoxographical anecdotes from a mythographical work on Heracles, and a collection of information on hero cults in the east of Magna Graecia.

6. *Nēsiotika* and the *Mirabiles auscultationes*

Islands are clearly a central element in the geographical-historiographical section of the *Mirabiles auscultationes*, to such an extent that they seem to constitute the main thread of this part of the collection and the general framework into which the series *Mir.* 85–98 and *Mir.* 106–110 are inserted. This leads us to wonder what role

---

69 Müllenhoff (1890) 434–436, esp. 435.
70 Flashar (1972) 124–125. See also Giannini (1965) 275.
71 Nafissi (1992) passim, esp. 418–419. See Vanotti (2007) 189.
72 Or maybe a later paradoxographer, if the attribution of chapters 85–98 to Posidonius is accepted.

literature of the *On Islands* genre (Περὶ νήσων, Νησιωτικά) could have played in the process of composing this part of the work and, consequently, if it might be possible to propose a hypothesis about the identity of the specific *On Islands* our paradoxographer relied on. In order to answer these questions, one must be familiar with the features of this genre of literature, beyond its obvious interest in islands, and with the authors who cultivated it and the extant testimonies and fragments of their works, to compare them with our collection of anecdotes.

We do not know much about such works, unfortunately, apart from the fact that they apparently share with paradoxography a taste for the extraordinary and the curious and, perhaps, composition through compilation, at least in some cases.[73] We owe to Paola Ceccarelli a paper entitled "I Nesiotika,"[74] the most complete general survey of this genre to date, as well as a collection of the fragments of works that circulated under titles that indicate a focus on islands. Ceccarelli observes that *paradoxa* were a significant feature in these works, together with considerations regarding the names of various islands and their etymology, as well as mythological stories connected to them.[75] All these elements are represented in *Mir.* 78–121. The historiographical section of pseudo-Aristotle's *Mirabilia* thus appears to be consistent with the interests of the ancient literature on islands, and perhaps it can be read as inspired by a work of this kind, and therefore, as a possible further source of information about this literary genre we know so little about.

That there are intersections and coincidences between paradoxography and literature on islands is not surprising. Many studies have highlighted the significance of islands as spaces apt to host curious phenomena, even to become the perfect location for utopias.[76] Isolation and independence from the mainland, conceived as the normal sphere of everyday human life, unleashed the imagination of writers and at the same time created the idea that islands are "different," special, and full of marvels. It is no wonder, then, that the author of a collection of *mirabilia* should draw on a work of the *On Islands* genre as a source of inspiration.

Regarding our chances of identifying, even if only tentatively, the specific *On Islands* in the background of *Mir.* 78–121, some hypotheses can be put forward.[77] First, it is necessary to define the time span in which it could have been written. A *terminus ante quem* is probably easier to establish: chapter 100 of our paradoxographical collection makes reference to Sardinia as still being under Carthaginian control, which scholars have construed as indicative of a date before 238 BCE,

---

73 In many of the cases, the extant fragments do not allow us to identify the origins of the information or the authors' approach. Nevertheless, dependence on earlier sources and compilatory work can be glimpsed in the fragments of Philostephanus (see, e.g., *FGrHist* 1751 F5, 21), and, perhaps, Conon or Semus (on the problems of attribution of the reference in *Schol. Ap. Rhod.* 1.1165, see Ceccarelli [1989] 924–928).

74 Ceccarelli (1989).

75 Ceccarelli (1989) 931–935.

76 See, for instance, Ceccarelli (1989) 935; Ceccarelli (2009); De Vido (2009); Sulimani (2017).

77 On this point, I am deeply indebted to Stefan Schorn, who kindly has shared with me his thoughts regarding the possible authorship of the anonymous work Περὶ νήσων behind *Mir.* 78–121.

when the Roman power conquered the island.[78] The *terminus post quem*, however, is more problematic: the use of Timaeus, in principle, points to a period when his work was already circulating. Admittedly, coincidences with Diodorus book 5 indicate that probably some of the material of *Mir.* 78–121 goes back to Timaeus. However, it should be noted that the presence of Timaean material in *Mir.* 78–121 is certain only in two cases: *Mir.* 102 and (in part) *Mir.* 88,[79] neither of which can be attributed to the original catalogue of islands with any certainty. Indeed, chapter 102, even though it is surrounded by anecdotes about the islands of the Tyrrhenian Sea (the Aeolians and the Sirenusae), does not refer to *paradoxa* connected to islands, but to the peculiar nature of lake Aornos, in the Campanian mainland. In the case of *Mir.* 88, coincidence with Timaean materials corresponds to the place of the Balearics after the seven largest islands of the Mediterranean. In this case, therefore, the anecdote does fit the interest in islands. Nevertheless, in principle, it does not belong to the general series connected to Περὶ νήσων works, but it appears to be part of one of the series where islands do not predominate, namely, the periegetic sequence *Mir.* 85–98, for which we should probably assume a different origin.[80]

Perhaps chapter 78, on the poisons that grow on the Circaean mountain, can provide a more solid indication (if I am right that, despite appearances, this does belong with the discussion of islands; see p. 20). The second part of the anecdote mentions that Aulus "Peucestius"[81] and Gaius, probably Romans or allies of Rome, but otherwise unknown, used the poisonous substances of that place when they tried to assassinate "Cleonymus the Spartan," identified as the Spartan king of this name who visited Tarentum in 304 BCE.[82]

The year 304 BCE, then, apparently offers us a clear end point. That leaves us a time window between the late 4th or early 3rd century BCE and the year 238 BCE to date our work Περὶ νήσων, that is, between Cleonymus' stay in southern Italy, which roughly corresponds to the time when the works of Timaeus, Lycus, Duris, and other authors that scholars have suggested as possible sources of *Mir.* 78–121 started circulating, and the end of the Carthaginian control over Sardinia. Therefore, the only possible authors, among those considered by Ceccarelli, are Philostephanus of Cyrene (*On Islands*), Callimachus the Younger (*Nesias*), and Semus of Delos (*On Islands*).[83] A possible fourth candidate, Xenagoras (*On Islands*), is probably a bit too late.[84]

---

78 On the history of Sardinia, see Meloni (2006). On the value of this piece of information for dating this section of *Mir.*, or at least its source, see Flashar (1972) 119; Vanotti (2007) 181.

79 See Schorn's chapter in this volume, pp. 35–38.

80 Nevertheless, see above on the possibility that the compiler could have inserted not only the periegetic series into the series of islands, but maybe also some islands in the periegetic series, in order to provide the final work with overall coherence.

81 On this name, see Giacomelli (2023) 259.

82 See Flashar (1972) 105–106 and Vanotti (2007) 163 (with bibliography).

83 Philostephanus, *FGrHist* 1751 F 4–6; Callimachus iunior, *FGrHist* 1755 T 1; Semus of Delos, *FGrHist* 1756 F 1.

84 *FGrHist* 1757. The date of Xenagoras is, in any case, very uncertain.

None of these names, however, offers us a neat identification of the collection of islands behind *Mir.* 78–121: in principle, the fact that Callimachus' work was in verse[85] seems to exclude him, though the possibility that the paradoxographer paraphrased a poetic text in prose cannot be totally ruled out (this applies also in the case of Philostephanus, see below). Moreover, the fragments of these authors show that their works included both islands located in the eastern and the western Mediterranean, whereas the focus of our collection is almost exclusively on the western half of the sea, in some cases extending even beyond the Pillars of Heracles (Gadeira, Erytheia, perhaps Madeira). The Greek islands of the Aegean are almost completely absent,[86] as is Cyprus, which clearly were subjects of interest to Semus, Callimachus, and Philostephanus.

Still, the possibility remains that the paradoxographer could have selected only anecdotes about western islands from an original collection that included both western and eastern locations, and Philostephanus' fragments show a certain concordance with our paradoxographer: among the fragments without book title, one (F 22) deals with the presence of Daedalus in Sicily, after his flight from Crete,[87] which also appears in *Mir.* 81; another fragment (F 30)[88] deals with the area of the Eridanus, again a theme of chapter 81. Finally, F 34[89] of Philostephanus describes in elegiac couplets a small pond in Sicily which expulses those who go into it, a description that arguably bears a resemblance to *Mir.* 112, which is also about a Sicilian pond whose diameter expands to receive up to 50 men, and then expels the swimmers one by one.

Among the known authors of works in the *On Islands* genre, then, Philostephanus is the leading candidate for the source of *Mir.* 78–121. However, the extant evidence cannot support a firm attribution and it is very likely that there were other works *On Islands* that have totally disappeared from the tradition and that the pseudo-Aristotle may have used.

For the moment, then, it is not possible to identify the author of the Περὶ νήσων work that formed the basis for *Mir.* 78–121, or the precise nature of this work, beyond the fact that it included islands of the western Mediterranean. If the series of islands is considered in isolation from the subseries *Mir.* 85–98 and 106–110, a rough periegetic order from east to west can perhaps be glimpsed, specifically a sequence Adriatic Sea[90] – Tyrrhenian Sea[91] – Sicily[92] – Erytheia/beyond the Pillars

---

85 See *FGrHist* 1755 T 1, ap. *Sud.* K 227, s.v. Καλλίμαχος: ὁ νέος Καλλίμαχος, ὁ γράψας περὶ νήσων δι'ἐπῶν.

86 Only *Mir.* 83 deals with Crete, while the islands of the Aegean Sea are only briefly mentioned in chapter 104, in a statement that merchants from the Pontic area sell products from Chios, Lesbos, and Thasos, whereas merchants from the Adriatic sell amphorae from Corcyra.

87 *FGrHist* 1751 F 22; *FHG* III 34 F 36; F 18 Capel Badino, ap. *Schol. Hom. Il.* 2.145. See Ceccarelli (1989) 919–920 F 19.

88 *FGrHist* 1751 F 30; *FHG* III 32 F 22; F 29 Capel Badino, ap. *Schol. Dion. Perieg.* 289.

89 *FGrHist* 1751 F 34; *FHG* III 31 F 17; F 34 Capel Badino, ap. Tz. *Chil.* 7, 144, 662–667. See Ceccarelli (1989) 920–921, F 21.

90 *Mir.* 79, 81.

91 *Mir.* 100, 101, 103.

92 *Mir.* 110–115.

of Heracles.[93] However, this sequence is by no means perfect and it leaves "out of their order" chapters 78 (the Circean Mount), 82 (Sicily), and 83 (Crete),[94] 84 (Madeira?),[95] 104 (Greek islands, Corcyra), and 117 (Aeolian Islands).

Despite the lack of confidence in identifying the *On Islands* source, in my opinion, the traditional perspective that regarded this central section of the Aristotelian *mirabilia* as a mere sequence of excerpted sources consulted directly by the compiler and stitched together, or as a juxtaposition of geographical sections divided into western and eastern curiosities, should certainly be abandoned in favor of a new interpretation. The extant form of *Mir.* 78–121 is most probably the result of complex compilation by a writer who relied on earlier sources of diverse natures, perhaps even earlier compilations, and sought to create a collection of anecdotes united in geographical and thematic coherence, among other things, by astute use of the *On Islands* literature, a genre that enjoyed a golden age in the early Hellenistic period and whose points of contact with paradoxography still need to be explored.

Works cited

Capel Badino, R. 2010. *Filostefano di Cirene: Testimonianze e frammenti.* Milan: LED.
Ceccarelli, P. 1989. "I 'Nesiotika'." *Annali della Scuola Normale Superiore di Pisa: Classe di Lettere e Filosofia* 3(19): 903–935.
Ceccarelli, P. 2009. "Isole e terraferma: la percezione della terra abitata in Grecia arcaica e classica." In Ampolo, C. ed. *Immagine e immagini della Sicilia e di altre isole del Mediterraneo antico.* Pisa: Edizioni della Scuola Normale Superiore. 31–50.
Coppola, A. 1988. "Siracusa e il Diomede adriatico." *Prometheus* 14: 221–226.
Cordano, F. 2014 "Un periplo del Mediterraneo con le vacche di Gerione." In Breglia, L. and Moletti, A. eds. *Hespería: tradizioni, rotte, paesaggi.* Tekmeria 16. Paestum: Pandemos. 137–145.
De Vido, S. 2009. "Insularità, etnografia, utopie. Il caso di Diodoro." In Ampolo, C. ed. *Immagine e immagini della Sicilia e di altre isole del Mediterraneo antico.* Pisa: Edizioni della Scuola Normale Superiore. 113–124.
Finglass, P.J. 2021. "Labor X: The Cattle of Geryon and the Return from Tartessus." In Ogden, D. ed. *The Oxford Handbook of Heracles.* Oxford: Oxford University Press. 135–148.
Flashar, H. 1972. *Aristoteles: Mirabilia. Aristoteles Werke in deutscher Übersetzung. Band 18. Opuscula.* Teil II. Berlin: Wiley-VCH Verlag.
Gantz, T. 1993. *Early Greek Myth: A Guide to Literary and Artistic Sources.* Baltimore; London: Johns Hopkins University Press.
Geffcken, J. 1892. *Timaios' Geographie des Westens.* Berlin: Weidmannsche Buchhandlung.
Giacomelli, C. 2021. *Ps.-Aristotele De Mirabilibus Auscultationibus: Indagini sulla storia della tradizione e ricezione del testo.* Berlin: De Gruyter.
Giacomelli, C. 2023. *Pseudo-Aristotele De Mirabilibus Auscultationibus: Edizione critica, traduzione e commento filologico.* Rome: Accademia Nazionale dei Lincei.

93 *Mir.* 118.
94 Nevertheless, Sicily and Crete are mentioned in this order in chapter 81.
95 However, it should be noted that the item appears immediately before the series reflecting the route of Heracles, whose starting point was precisely the area beyond the Strait of Gibraltar.

Giannini, A. 1964. "Studi sulla paradossografia greca. II: Da Callimaco all'età imperiale: la letteratura paradossografica." *Acme* 17: 99–140.

Giannini, A. 1965. *Paradoxographorum Graecorum Reliquiae*. Milan: Istituto Editoriale Italiano.

Knapp, R.C. 1986. "La Via Heraclea en el occidente: mito, arqueología, propaganda, historia." *Emerita* 54(1): 103–122.

Malkin, I. 1998. *The Returns of Odysseus: Colonization and Ethnicity*. Berkeley; Los Angeles: University of California Press.

Mastrocinque, A. 1987. *Santuari e divinità dei paleoveneti*. Padova: La Linea Editrice.

Meister, K. 2006. "Agathocles [2]." In *Brill's New Pauly*. Leiden: Brill. Consulted online on 6 October 2022. <http://dx.doi.org/10.1163/1574-9347_bnp_e107240>

Meloni, P. 2006. "Sardinia." In *Brill's New Pauly*. Leiden: Brill. Consulted online on 6 October 2022. <http://dx.doi.org/10.1163/1574-9347_bnp_e1101420>

Müllenhoff, K. 1890². *Deutsche Altertumskunde. I. Neuer vermehrter Abdruck besorgt durch M. Roediger, mit einer Karte von H. Kiepert*. Berlin: Weidmann.

Müller, K. 1841–1873. *Fragmenta Historicorum Graecorum*. 5 vols. Paris: Didot.

Nafissi, M. 1992. "Atridi, Eacidi, Agamennonidi e Achille: Religione e politica fra Taranto e i Molossi. (Lico in Ps. Arist., *Mir*. 106)." *Athenaeum* 80: 401–420.

Pearson, L. 1987. *The Greek Historians of the West: Timaeus and His Predecessors*. Atlanta: Scholars Press.

Plácido, D. 1993. "Le vie di Ercole in occidente." In Mastrocinque, A. ed. *Ercole in occidente*. Trento: Dipartimento di Scienze Filologiche e Storiche. 63–80.

Plácido, D. 2002. "La Península Ibérica: arqueología e imagen mítica." *Archivo Español de Arqueología* 75: 123–136.

Rathmann, M. 2006. "E. Iberian Peninsula." In Lohmann, H., Wiesehöfer, J. and Rathmann, M. eds. *"Roads." Brill's New Pauly*. Leiden: Brill. Consulted online on 7 October 2022. <http://dx.doi.org/10.1163/1574-9347_bnp_e12225290>

Rose, V. 1854. *De Aristotelis librorum ordine et auctoritate commentatio*. Berlin: Georg Reimer.

Rose, V. 1863. *Aristoteles pseudepigraphus*. Leipzig: Teubner.

Schorn, S. ed. 2022. *Felix Jacoby: Die Fragmente der Griechischen Historiker Continued. Part IV. Biography and Antiquarian Literature. E. Paradoxography and Antiquities. Fasc. 2. Paradoxographers of Imperial Times and Undated Authors [Nos. 1667–1693]*. Leiden; Boston: Brill.

Schrader, H. 1868. "Über die Quellen der pseudoaristotelischen Schrift περὶ θαυμασίων ἀκουσμάτων." *Jahrbücher für Classische Philologie* 14 (= *Neue Jahrbücher für Philologie und Paedagogik* 97: 217–232).

Sulimani, I. 2017. "Imaginary Islands in the Hellenistic Era: Utopia on the Geographical Map." In Hawes, G. ed. *Myths on the Map: The Storied Landscapes of Ancient Greece*. Oxford: Oxford University Press. 221–242.

Vanotti, G. 2007. *Aristotele: Racconti meravigliosi*. Milan: Bompiani.

von Gutschmid, A. 1871. "Müllenhoff, Karl, *deutsche Alterthumskunde. Erster Band. Mit 1 Karte von Heinrrich Kiepert*. Berlin, 1870. Weidmannsche Buchhdlg. (XII, 501 S. 8.)." *Literarisches Centralblatt für Deutschland* 21: 521–529.

2 Timaeus in pseudo-Aristotle's *De mirabilibus auscultationibus*

Stefan Schorn

1. Timaeus as a source author of *Mir.*: history of scholarship

Modern source analysis of *Mir.* starts, to my knowledge, with Valentin Rose who was the first to split up the work into strings of excerpts taken from single authors with occasional insertions from other sources.[1] Basing himself on the results of his analysis of the sources of *Mir.* 1–77, where he had identified strings from Aristotle and Theophrastus, Rose assumed the same manner of composition for the following parts and regarded Timaeus as *Mir.*'s source of 78–121 and Theopompus of 122–138. He motivated the identification of Timaeus as the source only very briefly by referring to the agreement of content between *Mir.* 103 and Timaeus, *FGrHist* 566 F 57 and *Mir.* 88 and F 65, simply adding with "cf." a list of fragment numbers of Timaeus where, in most cases, already Karl and Theodor Müller in their collection of the Timaeus fragments had pointed to parallel reports in *Mir.* in their notes.[2]

As a reaction to Rose, Hermann Schrader was much more cautious and only admitted Timaeus as a source of those chapters for which (part of) their information is attested explicitly for this historian by other sources, that is, *Mir.* 88, 102, and 109.[3] In addition, he doubted direct use by the compiler.

We owe the most detailed discussions of Timaeus as a source of *Mir.* to Karl Müllenhoff and Johannes Geffcken.[4] They both treat *Mir.* as part of their reconstructions of Timaeus' description of the West. They do not only base their attributions of single chapters to our historian on parallels in named fragments of Timaeus but mainly on parallels in sources where Timaeus' name does not appear but have been assigned to him on the basis of *Quellenforschung*. These are in particular passages from Diodorus' 4th and 5th books, Lycophron's *Alexandra* and the scholia on

---

1 Rose (1854) 54–56 and (1863) 279–281 with a few minor differences in the details.
2 That is, *FGrHist* 566 F 85 ~ *Mir.* 105b; F 55 ~ *Mir.* 109b; F 63 ~ *Mir.* 100a; F 64 (here, Müller does not refer to *Mir.*; Rose will have thought of *Mir.* 100); F 73 ~ *Mir.* 55; F 68 ~ *Mir.* 81; F 46 ~ *Mir.* 169. The edition of Timaeus in *FHG* in vol. 2 pp. 193–233.
3 Schrader (1868) 226–231. He explicitly contests that *Mir.* 81, 100, and 105 stem from Timaeus because, according to him, Timaeus F 68 does not refer to the content of *Mir.* 81, F 63 contradicts *Mir.* 100, and *Mir.* 105 is rather taken from Theopompus.
4 Müllenhoff (²1890) 426–497 (to a large extent identical with ¹1870, 426–497); Geffcken (1892) esp. 83–99.

DOI: 10.4324/9781003437819-3

this poem, Dionysius Periegetes and the scholia on his *Periegesis*, Strabo, Pliny the Elder, and various other texts. The same is true for Alexander Enmann who dedicated an appendix in his book on the sources of Pompeius Trogus to the sources of *Mir.*[5] Müllenhoff claims *Mir.* 82, 84–98, and 100–103 for Timaeus, Enmann 78–82, 84–98, and 100–115,[6] and Geffcken 78–82, 84–98, and 100–121. Although Geffcken's general approach has been criticized in many subsequent studies as too far-going and speculative,[7] many of his and Müllenhoff's results are still accepted in Hellmut Flashar's commentary which is the standard treatment of the question today.[8] In his table of sources (p. 41), he assigns 19 chapters to Timaeus (82, 84, 88, 96, 100–103, 106–110, 117–121, 172), while in his commentary he is sometimes more cautious as for the certainty with which this source is to be identified. Five chapters he attributes to Timaeus, although he regards his authorship as "not certain" (85–86, 95, 97–98), four chapters may stem from Timaeus or Lycus (78–81), and seven may go back to Posidonius or the sources used by him, among which may have been Timaeus (87, 89–94).[9] Thus, for Flashar, too, Timaeus is the main source of the section 78–121.

Gabriella Vanotti is very skeptical about the results of source criticism which she calls "questionable" and "fruitless."[10] In her commentary, she usually only reports the various hypotheses and only twice considers Timaeus as *Mir.*'s direct or indirect source (82, 102).[11]

Faced with such discrepant results, one might be inclined to refrain from searching for Timaean material within *Mir.* But this would be very unsatisfactory. The fact that Timaeus was the most widely read historian of the Greek West makes it likely that some, perhaps even much, information from his *Sikelika* found its way, directly or indirectly, into our collection. Since it is certain in the cases of Aristotle, Theophrastus, and Theopompus that the compiler, for other series of excerpts, relied on one single author or work, it seems likely that he did not only excerpt one or a few scattered reports from Timaeus, if it can be shown that he did indeed use him somewhere. Even if *Mir.* 78–121 are lifted from a later work in which also Timaean material was used, the question still remains to what extent Timaeus was used *there* as a source, if it can be shown that he was among its sources, although

---

5 Enmann (1880) 193–203.

6 Enmann (1880) is not explicit about *Mir.* 111, which he mentions on p. 201. But considering that he accepts the idea of strings of quotations from the same author, he seems to have claimed it for Timaeus as well.

7 See, for example, the review of his book by Niese (1893); Jacoby on Timaeus, *FGrHist* 566 in various places. Cf. Flashar (1972) 45.

8 Flashar (1972) 41, 45–46.

9 See Flashar (1972) 110 on *Mir.* 87 and esp. 112 on *Mir.* 89. On p. 46, he is less explicit as for the source and the wording in his table on p. 41 regarding *Mir.* 87 and 89–94 is misleading or a slip of the pen: "Poseidonios (über Timaios?)". This sounds like Posidonius via Timaeus which is of course chronologically impossible.

10 Vanotti (2007) 44: "In definitiva la discussione sulla dipendenza del singolo capitolo dall'una piuttosto che dall'altra fonte, come è stata condotta dal buona parte della critica ottocentesca, porta a conclusioni opinabili e appare quindi sterile." Cf. 46.

11 Vanotti (2007) 169, 182.

then no *strings* of excerpts from Timaeus should be expected in *Mir.* At the same time, Geffcken's overoptimism is implausible as well because it has been shown by later scholars that his attributions are in many cases based on insufficient evidence or doubtful assumptions.

However, confronted with such a wide range of contradicting interpretations, one may ask whether source criticism is possible at all in a text such as *Mir.* and in addition one may ask if it is still something scholars should spend their time on today. I think that both questions deserve an affirmative answer in the case of *Mir.*, although this kind of research has become unpopular because of the excesses of earlier scholarship and Geffcken's study may be blamed for being such a negative example. Another reason for its unpopularity may be that it is an extremely complex and time-consuming endeavor with usually no definitive or absolutely compelling results. But if we see the small amount of texts we have of an author such as Timaeus compared to the original size of his *Sikelika* and to the popularity it had for centuries, it seems tempting to try to enlarge the material of the named fragments in a cautious and methodologically sound way by source criticism, even if we will reach only a more or less high degree of probability of Timaean authorship. For *Mir.* itself, it is important to know its sources if we want to learn more about the working method of the compiler, and if we can identify Timaeus as the unnamed source of information outside *Mir.*, it might even be possible to get a better understanding of how our compiler dealt with his sources and which types of information he excerpted in the "periegetic" or "ethnographic" middle section of his work. Thus, I think it is still worthwhile to look into the sources of *Mir.* and to ask which role Timaeus played in this regard. In addition, I think that such an endeavor may serve as a good case study of how such a problem may be addressed today and which alternatives of interpretation there are depending on how far one is willing to go with hypotheses. What I thus want to do in the following is to discuss, first, the chapters for which parallels in named Timaean fragments have been identified by scholars to see if they allow us to find in him their indirect or direct source and what he did with the information he found there. In a second step, I will have a look at the main text in which *Quellenforschung* has seen Timaeus as the source of passages containing parallel reports to *Mir.*, the fifth book of Diodorus' *Library*. Although it would be desirable to treat the whole question anew and study all chapters of *Mir.* in which Timaeus has been regarded as the source on the basis of *Quellenforschung*, this is not feasible here because it would require a book-length study. Nevertheless, I will deal with some of the other sources within my discussions as far as this is necessary. What needs to be established and what has to my knowledge not been done in recent literature is to check whether the fundamental assumptions of 19th and early 20th centuries source criticism, which constitute the basis of what is usually said today about *Mir.*'s sources, are still valid and, if not, what this means for establishing the sources of *Mir.* The main problem here is that source criticism of all the authors mentioned above is interrelated and it is a common feature of such treatments that material in, for example, *Mir.* is assigned to Timaeus because it appears in Diodorus in a "Timaean passage" and the same passage is, in a discussion of Diodorus' sources claimed for Timaeus because it appears in a group of chapters in *Mir.* presumably going back to Timaeus, and the same is true for Lycophron's *Alexandra*, the scholia on the *Alexandra* and other texts. What we need to try to find is some solid ground on which we can build further hypotheses, or at least we need to show the degree of probability of Timaean authorship of such passages and then decide whether it is high enough to support further attributions.

2. Chapters that overlap named fragments of Timaeus

2.1 The seven biggest islands (Mir. 88 and Timaeus F 65)

Mir. 88 reports some ethnographic curiosities about the barbarian population of the Gymnasian Islands, that is, the Baleares. It is introduced by the words:

Ἐν ταῖς Γυμνησίαις ταῖς κειμέναις νήσοις κατὰ τὴν Ἰβηρίαν, ἃς μετὰ τὰς λεγομένας ἑπτὰ μεγίστας λέγουσιν εἶναι, φασὶν.

In the Gymnasian islands, which lie off Iberia, which they say are the greatest after the so-called Seven, they say. (trans. Hett)[12]

It has generally been noted that this introductory remark has a clear, even literal parallel in a fragment of Timaeus preserved by Strabo (14.2.10.654 = Timaeus F 65) and Timaeus is therefore generally regarded as the source of this chapter.[13] This fragment is transmitted in Strabo's discussion of Rhodian colonization activity in which he reports that according to some sources the Gymnasian Islands were colonized (κτισθῆναι) by Rhodians and he adds:

ὧν τὴν μείζω φησὶ Τίμαιος μεγίστην εἶναι μετὰ τὰς ἑπτά, Σαρδὼ Σικελίαν Κύπρον Κρήτην Εὔβοιαν Κύρνον Λέσβον, οὐ τἀληθῆ λέγων· πολὺ γὰρ ἄλλαι μείζους.

The bigger one of the two Timaeus claims is the biggest after the Seven, i.e. Sardinia, Sicily, Cyprus, Crete, Euboea, Corsica, Lesbos. But he is wrong because others are much bigger.

Very close to Timaeus-Strabo's wording comes also Diodorus (5.17.1) in his chapter on the same islands:

τούτων δ' ἡ μείζων μεγίστη πασῶν ἐστι μετὰ τὰς ἑπτὰ νήσους, Σικελίαν, Σαρδώ, Κύπρον, Κρήτην, Εὔβοιαν, Κύρνον, Λέσβον.

The bigger one of the two is the biggest of all (islands) after the Seven Islands, i.e. Sicily, Sardinia, Cyprus, Crete, Euboea, Corsica, Lesbos.

---

12 Unless otherwise indicated, I use Hett's *LCL* translation of *Mir.* (1936), sometimes in a modified form.

13 See, for example, Müllenhoff (²1890) 463–464; Geffcken (1892) 66, 90; Flashar (1972) 111. Even Vanotti (2007) 172–173 comes close to admitting it.

This description is found in a passage which is generally claimed for Timaeus on the basis of source criticism, and the coincidence with F 65 is one of the main arguments for Timaean authorship. Diodorus' ethnographical notes on the population which follow show some agreement with *Mir.* and this has been considered further evidence for this source of *Mir.* I will come back to this below when I discuss the parallels in Diodorus.

If we limit ourselves, for now, to the piece of information attested explicitly for Timaeus, the case for him as a source of *Mir.* 88 is good but not absolutely conclusive. While Timaeus and Diodorus speak of the biggest of the two Balearic islands, that is, Mallorca, *Mir.* does so of *both* islands together. But this may be an inaccuracy due to epitomization. However, it is more important to note that the idea of the "Big Seven" was not a typically Timaean concept but reflects a widely accepted canon.[14] Most Greeks do not seem to have realized, or not to have cared, that this ranking is incorrect, even if we limit it to the Mediterranean. Mallorca is not the biggest island after the Big Seven, it *is* number seven and by far bigger than Lesbos which is part of the standard group (3,640 vs. 1,636 km^2). It is astonishing that Timaeus was not aware of this, but it seems that Mallorca, as an island inhabited by barbarians, was largely out of the sight of the Greeks. The point I want to make here is that the incorrect list of Seven is not an exclusively Timaean feature. Yet it is the striking similarity of the wording in the sources and the explicit claim that Mallorca is number eight after the Seven that are a strong argument for Timaeus as a (direct or indirect) source of *Mir.* 88.[15]

2.2 The Lake Avernus (Mir. 102 and Timaeus F 57)

Mir. 102 contains a clear parallel to Timaeus F 57, transmitted by pseudo-Antigonus.[16] Here *Mir.*'s source is obvious and there is general agreement about it.[17] The chapter deals with Lake Avernus (Lago d'Averno) near Cumae to which various *mirabilia* were related in antiquity. The most frequent *topos* in ancient literature is that birds cannot fly over it, which is often explained by poisonous vapor emanating

---

14　See Alexis F 270 Kassel-Austin (with their note); Trapanis (1960) 69 no. v with note on p. 73 = *PHI Chios* no. 291; App. *Prooem.* 17; Ps.-Aristot. *Mund.* 3.393a12–14; cf. Müllenhoff (²1890) 463–464; Lorimer (1925) 79–80; Arnott (1996) 756–757; Zucca (1998) 15; cf. Marasco (2004) 164–165.

15　Thus, most of the interpreters; see, for example, Müllenhoff (²1890) 463–464 and below for further discussion.

16　Ps.-Antig. *Mir.* 152 Musso/Giannini: (a) Τὴν δὲ ἐν τοῖς Σαρμάταις λίμνην Ἡρακλείδην (Heraclides Ponticus F 128b Wehrli = 137B Schütrumpf) γράφειν, ὅτι οὐδὲν τῶν ὀρνέων ὑπεραίρειν, τὸ δὲ προσελθὸν ὑπὸ τῆς ὀσμῆς τελευτᾶν. (b) (1) Ὁ δὴ καὶ περὶ τὴν Ἄορνίν τι δοκεῖ γίγνεσθαι καὶ κατίσχυκεν ἡ φήμη παρὰ τοῖς πλείστοις. (2) Ὁ δὲ Τίμαιος τοῦτο μὲν ψεῦδος ἡγεῖται εἶναι· τὰ πλεῖστα γὰρ κατατυχεῖν τῶν εἰθισμένων παρ' αὐτῇ διαιτᾶσθαι· ἐκεῖνο μέντοι λέγει, διότι συνδένδρων τόπων ἐπικειμένων αὐτῇ καὶ πολλῶν κλάδων καὶ φύλλων διὰ πνεύματα τῶν μὲν κατακλωμένων, τῶν δὲ ἀποσειομένων, οὐθέν ἐστιν ἰδεῖν ἐπ' αὐτῇ ἐφεστηκός, ἀλλὰ διαμένειν καθαράν.

17　See Schrader (1868) 228; Geffcken (1892) 91; Jacoby on Timaeus, *FGrHist* 566 F 57; Flashar (1972) 120. Even Vanotti (2007) 182 call this "assai probabile."

from it.[18] Both versions contain elements not present in the other and are thus independent of each other. In pseudo-Antigonus, the chapter is part of his section lifted from Callimachus' paradoxographical work. It first reports, with reference to Heraclides (Ponticus), about a birdless lake in Sarmatia over which birds cannot fly, but die from the smell, and he then adds that the same is told about the Aornis (Giannini, Musso, Aorneitis cod. P, Aornos Geffcken). But while the report from Heraclides is given in indirect speech, showing that it has been taken from Callimachus, the critique by Timaeus is in direct speech. Unfortunately, the information on Aornis stands in a relative clause so that it is uncertain if it is still part of the Heraclides fragment. Pfeiffer sees it as an addition by pseudo-Antigonus who would have taken this *thaumasion*, including the critique of it from Timaeus,[19] while Jacoby claims the whole lemma for Timaeus (via Callimachus), probably because Timaeus elsewhere criticizes Heraclides for being a "narrator of marvels" (παραδοξολόγος).[20] In this case, Heraclides would have told about both lakes. As *Mir.* seems to summarize a description of the Italian region around Lake Avernus, one is wondering how the Sarmatian lake could have been part of it and why Timaeus should have included it in his polemics against Heraclides, especially as he does not seem to have doubted the existence of this birdless lake. Thus, Pfeiffer's reading seems to be preferable, but some uncertainty remains.

Some literal agreement and especially the similar structure show that both texts summarize the same passage of Timaeus in which he denied the traditional idea of the "birdlessness" of Lake Avernus and claimed instead that the marvel it presents is that no leaves (and twigs) are found on its surface, although it is surrounded by trees.[21] The content of the chapters in *Mir.* and pseudo-Antigonus only shows a few, but not substantial differences. The most interesting one we find in the sentence with which they deny the "birdlessness":

Mir.	*Antigonus*
ὅτι δὲ οὐδὲν διίπταται ὄρνεον αὐτήν, ψεῦδος· οἱ γὰρ παραγενόμενοι λέγουσι πλῆθός τι κύκνων ἐν αὐτῇ γίγνεσθαι. "It is false that no bird flies over it; for those who have been there say that there are a quantity of swans on it."	ὁ δὲ Τίμαιος τοῦτο μὲν ψεῦδος ἡγεῖται εἶναι· τὰ πλεῖστα γὰρ κατατυχεῖν (Bentley, κατὰ τύχην cod. P) τῶν εἰθισμένων παρ' αὐτῇ διαιτᾶσθαι· "Timaeus regards this as false. For most (of the birds) customarily living on it, are able to do so."

---

18 See the *apparatus similium* of Giacomelli's edition (2023).
19 Callim. F 407 xxiv p. 334 Pfeiffer (19): "quae sequitur ex Timaeo . . . add. Antigonus."
20 Jacoby on Timaeus, *FGrHist* 566 F 57. The critique of Heraclides is in F 6. The same is assumed by Wehrli (²1969) 103 and Schütrumpf *et al.* (2008) 246. Cf. also Shannon-Henderson (in this volume), 153–157; 165.
21 See the table in Müllenhoff (²1890) 439.

Antigonus seems to imply that not all birds are able to fly over it, and one may reconcile *Mir.*'s remark about the swans with such a claim. So did Timaeus not completely deny that the lake had an effect on (some) birds? The presence of many swans on it in *Mir.* raises the question why leaves (*Mir.* and pseudo-Antigon.) and twigs (pseudo-Antigon.) are not seen on its surface. It cannot have been explained by the idea, not uncommon in paradoxography, that the water is so light that everything immediately sinks to the ground, which is also claimed about our lake by Coelius Antipater.[22] Thus, it seems that some important information has been lost in the process of epitomization. In addition, we cannot tell whether Timaeus was used in *Mir.* directly or indirectly (as in pseudo-Antigonus). Diodorus (4.22.1–2) and Strabo (5.4.5.244) contain some elements of the tradition found in *Mir.*/pseudo-Antigonus, but they do not help reconstructing Timaeus' account, so it cannot be assessed how reliable or not our chapter represents what Timaeus had written.

2.3 Diomedes in the West (Mir. 109 and 79 and Timaeus F 55)

Λέγεται περὶ <Λουκερίαν> τὸν ὀνομαζόμενον τῆς Δαυνίας τόπον ἱερὸν εἶναι Ἀθηνᾶς Ἀχαΐας καλούμενον, ἐν ᾧ δὴ πελέκεις χαλκοῦς καὶ ὅπλα τῶν Διομήδους ἑταίρων καὶ αὐτοῦ ἀνακεῖσθαι. Ἐν τούτῳ τῷ τόπῳ φασὶν εἶναι κύνας οἳ τοὺς ἀφικνουμένους τῶν Ἑλλήνων οὐκ ἀδικοῦσιν, ἀλλὰ σαίνουσιν ὥσπερ τοὺς συνηθεστάτους. Πάντες δὲ οἱ Δαύνιοι καὶ οἱ πλησιόχωροι αὐτοῖς μελανειμονοῦσι, καὶ ἄνδρες καὶ γυναῖκες, διὰ ταύτην, ὡς ἔοικε, τὴν αἰτίαν. τὰς γὰρ Τρῳάδας τὰς ληφθείσας αἰχμαλώτους καὶ εἰς ἐκείνους τοὺς τόπους ἀφικομένας, εὐλαβηθείσας μὴ πικρᾶς δουλείας τύχωσιν ὑπὸ τῶν ἐν ταῖς πατρίσι προϋπαρχουσῶν τοῖς Ἀχαιοῖς γυναικῶν, λέγεται τὰς ναῦς αὐτῶν ἐμπρῆσαι, ἵν' ἅμα μὲν τὴν προσδοκωμένην δουλείαν ἐκφύγωσιν, ἅμα δ' ὅπως {μετ'} ἐκείνων μένειν ἀναγκασθέντων συναρμοσθεῖσαι κατάσχωσιν αὐτοὺς ἄνδρας. πάνυ δὲ καὶ τῷ ποιητῇ καλῶς πέφρασται περὶ αὐτῶν· ἑλκεσιπέπλους γὰρ καὶ βαθυκόλπους κἀκείνας, ὡς ἔοικεν, ἰδεῖν ἔστιν. (*Mir.* 109)

In the Daunian region called Luceria, there is said to be a temple of Athena called Achaean, in which are dedicated the bronze axes and the arms of Diomedes' companions and his own. In this place they say there are dogs that do no harm to any Greeks who come there, but fawn on them as though they were their dear friends. But all the Daunians and their neighbours dress in black, both men and women, apparently for the following reason. The Trojan women who were taken prisoners and came to that region, in their anxiety to avoid bitter slavery at the hands of the women who belonged to the Greeks before in their own country, burned their ships according to the story, that they might at the same time escape the slavery which they expected, and that, joined with them as husbands, as they were compelled to remain, they might

---

22   Plin. *HN* 31.21 = Coel. Antip. F 51 Peter, *HRR* I = 11 F 33 Beck-Walter = 15 F 52 Cornell, *FRHist.* Flashar (1972) 120 has misunderstood pseudo-Antigonus; none of the two authors gives a (rationalistic) explanation of the absence of leaves.

keep them. A very good account of them is given by the poet; for one can see that they were "with trailing robes" and "deep-bosomed."

The chapter clearly consists of two different parts which are only loosely connected in *Mir.*'s compressed resumé: the description of the Daunian sanctuary, and the *aition* that explains the black clothing of the Daunians and their neighbors. It is the second part that has a parallel in a fragment of Timaeus (F 55):

ὁ δὲ Τίμαιός φησιν, ὅτι Ἕλληνες ἐπειδὰν ἀπαντήσωσι ταῖς Δαυνίαις ὑπεσταλμέναις μὲν ἐσθῆτα φαιάν, ἐζωσμέναις δὲ ταινίαις πλατείαις, ὑποδεδεμέναις δὲ τὰ κοῖλα τῶν ὑποδημάτων, ἐχούσαις δὲ ἐν ταῖς χερσὶ ῥάβδον, ὑπαληλιμμέναις δὲ τὸ πρόσωπον καθάπερ πυρρῶι τινι χρώματι, τῶν Ποινῶν ἔννοιαν λαμβάνουσι τῶν τραγικῶν.

Timaeus says that the Greeks, when they come across the Daunian women who wear grey clothing, broad belts and curved (?) shoes and carry a staff in their hands and have their face colored red, are reminded of the Goddesses of Vengeance of tragedy.

This fragment has often been seen as evidence for Timaean provenance of *Mir.* 109.[23] If our chapter goes back to this description of Timaeus, its text has been thoroughly reworked: Timaeus only speaks of Daunian women, as does Lycophron whose verses are commented on by the scholiast who quotes Timaeus' description.[24] In *Mir.* this is extended to all Daunians and their neighbors but at the end of the chapter the reference to Homer only refers to the women's dress which may be an indication that also *Mir.*'s source contained a separate description of Daunian women. Timaeus describes the clothing as grey, while it is black in *Mir.* That the women wear broad belts in Timaeus refers to the same type of Homeric female dress as βαθυκόλπους ("deep-bosomed") and ἑλκεσιπέπλους ("with trailing robes") in *Mir.*, for which our author refers to Homer.[25] It is thus possible that all these expressions were found in *Mir.*'s source. On the other side, the most spectacular element in Timaeus, the red faces of the women, as well as their particular shoes, are missing in *Mir.* This all has as a consequence that the persons described have a very different character: scaring, Erinyes-like women in Timaeus versus a whole population in mourning dresses that still commemorates the destruction of the ships and their being stuck in Daunia in early times. Thus, *Mir.* rather looks like an alternative story than a modified summary of Timaeus.[26] That Timaeus is, indeed, not the source of *Mir.* 109 seems to be confirmed by Timaeus F 53, also

23 See, for example, Geffcken (1892) 22; Flashar (1972) 126.

24 Lycoph. 1137–1140 (red face and staffs) with *schol.* on 1138 Leone.

25 On the meaning of βαθύκολπος, βαθύζωνος, ἑλκεσίπεπλος, and τανύπεπλος in Homer, which all refer to the same kind of garment, see van Wees (2005) 6–9 (with illustrations of the dress) and Nawratil (1959) (on the first two adjectives).

26 Jacoby, on Timaeus, *FGrHist* 566 F 55 points to the difference between Daunian women versus all Daunians and neighbors and warns of "complementing Timaeus from this collection." Flashar's (1972) 126 short comparison is not careful enough.

transmitted by the scholia on Lycophron.[27] There, Timaeus and Lycus (*FGrHist* 570 F 3) are referred to for various pieces of information on Diomedes. One is that Diomedes came home from Troy to Argos, but was cast out by his wife Aigialeia and thus came to Italy. If this comes from Timaeus (which is hard to say for sure in such "collective" references),[28] it contradicts the claim of *Mir.* that the Trojan women burned the ships because they did not want to be brought to the homes of the Greek warriors and serve as their wives' slaves. This story presupposes a version of the myth in which Diomedes came to Italy without returning home first, and such versions did, indeed, exist.[29] All in all, there is not sufficient evidence to see in Timaeus the source of *Mir.* 109, unless we want to postulate that his account was transformed into a totally new story by our compiler or by his source.

Although it is probable that *Mir.* 109 does not come from Timaeus, it seems worthwhile to discuss here briefly also *Mir.* 79 which shows some similarity with it and which has often been claimed for Timaeus too.[30] Its treatment in literature shows on how shaky ground many attributions of chapters to sources often stand. Flashar states in his commentary on *Mir.* 109[31]: "Man wird dieses Kapitel nicht von 79 trennen können, wo ebenfalls von den Spuren des Diomedes in Italien die Rede ist und die Philhellenie der Vögel derjenigen der Hunde hier entspricht." The similarity in the motive is indeed striking. In *Mir.* 79, the birds on the Diomedean Island[32] are said to be the former companions of Diomedes who have been transformed into birds after the murder of their leader. They attack and kill the local barbarians who enter the island while they keep quiet when Greeks arrive. They obviously do this because they want to take revenge for the murder.[33] In *Mir.* 109, it is said that the dogs in the sanctuary in Luceria do not attack Greeks and treat them as old friends which must imply that they do attack the non-Greek visitors. It is an obvious guess that they do this because they were presented, in the original account, as former Greeks as well. So what we have here is a competing legend to that in *Mir.* 79. In addition, *Mir.* 79 assumes that the Greeks coming with Diomedes to Daunia were all removed from the country in the end: Diomedes was killed and his companions were transformed into birds. This is not compatible with what we read in the second part of *Mir.* 109: all Daunians and neighboring peoples wear black cloths in commemoration of the ships burnt by the Trojan women. This implies that they were regarded as descendants of the Greeks (mixed with the locals and the Trojan women) who were forced to stay there by the lack of ships.[34] Thus, this legend belongs to the branch of the tradition that did not speak of a murder of Diomedes: he

---

27 *Schol. Lycoph.* 615a Leone.

28 Jacoby on Timaeus, *FGrHist* 566 F 53–56 leaves it undecided whether Timaeus is only quoted for the last item in the list (the statue of Diomedes) or also the other ones.

29 See Genovese (2009) 204. A different explanation of our chapter in Martinez Pinna (1996) 35–38.

30 See, for example, Enmann (1880) 198; Geffcken (1892) 7–8. Flashar (1972) 107 suspects contamination of Timaeus and Lycus, Štrmelj (2015) 68–70 thinks of Theopompus (without sufficient evidence).

31 Flashar (1972) 126.

32 Now with a high amount of probability identified with the Croatian island of Palagruža: Castiglioni (2008); cf. Štrmelj (2015) 52–55.

33 See Corssen (1913) 328; Flashar (1972) 107.

34 Cf. Genovese (2009) 254–256.

lived in Daunia, married the daughter of the local king, and ruled there until his death.[35] This explains the "Hellenization" of the region in *Mir.* 109. Although it cannot be excluded that one and the same author reported contradicting local legends,[36] they rather point to different sources.[37] The story in *Mir.* 79 is compatible with Lycus, *FGrHist* 570 F 6, so that he may be considered as its source.[38] But as this fragment is in all likelihood lacunose, this has to remain uncertain.[39]

These are the main parallels in the so-called "Timaean section" of *Mir.* to named fragments of Timaeus.[40] I will discuss a few more below when dealing with Diodorus' book 5. For those who do not "believe" in the possibility of assigning anonymous material in other sources to Timaeus through source criticism, the story ends here. In this case, there is no sufficient evidence for a Timaean section in *Mir.* and we cannot even tell whether our compiler had knowledge of the Timaean tradition directly or indirectly or how he treated this material.

3. Parallels to *Mir.* in Diodorus book 5

3.1 Diodorean Quellenforschung

There are some striking parallels to *Mir.* in the first chapters of book 5 of Diodorus' *Library*, the so-called "Island Book." Its first part contains a geographical and ethnographical treatment of Sicily, other islands of the western Mediterranean, and islands outside the Pillars of Heracles, also including their early, often mythological, history and some aspects of their later history. It is generally assumed that this section (5.2–23) is mainly based on Timaeus and the following one (5.24–40) dealing with Gauls, Iberians, and others on Posidonius. For 5.41–46 on Panchaea, Euhemerus of Messene, is certain as the source, while the source of the section on the eastern Greek islands (5.47–81) has not been establish yet.[41]

Before we have a look at the first section, it may be useful to familiarize ourselves with the main aspects of Diodorean source criticism because the approach one follows will have an impact on the chapters of *Mir.* one may claim for Timaeus.

---

35   On this version of the legend, see Bethe (1903) 821.21–65 (variant 3); cf. Preller and Robert (1926) 1493 n. 3, 1494–1495; Genovese (2009) 205.

36   This is how Enmann (1880) 200 and Geffcken (1892) 22 n. 2 try to save both accounts for the same source (= Timaeus in their view). Similarly, Müllenhoff (²1890) 434–435 who claims both fragments for Lycus.

37   On the two chapters, see also Greene (in this volume), 199–202.

38   Thus, Müllenhoff (²1890) 430–431. Jacoby on Lycus, *FGrHist* 570 F 6 is cautious.

39   See Jacoby on Lycus, *FGrHist* 570 F 6.

40   For the sake of completeness, it is to be mentioned that two chapters in the last section of *Mir.* (151–178), the late appendix, coincide with information in Timaeus: *Mir.* 169 and 172. They are not discussed in this contribution.

41   On the whole book 5, see, for example, Schwartz (1903) 678; Jacquemin in Casevitz and Jacquemin (2015) xxviii–xxxi. For attribution of the first section to Timaeus, see, for example, Müllenhoff (²1890) 442–748; Geffcken (1892); Jacoby, *FGrHist* 566 F 164 with commentary; Meister (1967) 31–36; Pearson (1987, 48–49, 1991, 17–22). Authors who also assume the use of other sources besides Timaeus include Brown (1958) 118 n. 16 and especially Dudziński (2016). Champion on *BNJ* 566 F 164 thinks that also elements from other sources are possible, but does not provide a discussion besides pointing to Diodorus' usual additions of information on later times.

Many interpreters since the second half of the 19th century have regarded Diodorus as a compiler summarizing the material of one source for extended passages.[42] This one-source theory became dominant in the late 19th and early 20th centuries and assumes that, for example, Ephorus was the source of books 11–16 and Hierony-mus of Cardia that of books 18–20 with only a few additions. Although it can still be found in contemporary research,[43] also other interpretations have been proposed in the meantime and it is often held that Diodorus worked differently in the vari-ous parts of the *Library*. Laqueur posited that Diodorus used Ephorus as his main source for the history of the Greek West. According to this scholar, he made a sum-mary of Ephorus' *History* and added longer and shorter supplements from Timaeus to this excerpt, additions that Laqueur thought to be able to distinguish on the basis of contradictions and inconsistencies.[44] But this approach has generally been rejected as arbitrary, and there is indeed no evidence that Ephorus was Diodorus' main source here.[45] Burton tried to show that book 1 relies on a variety of sources and others have assumed such a multi-sources theory for other parts of his work.[46] If correct, this would make source criticism almost impossible as far as information is concerned for which there is no independent evidence elsewhere. According to my assessment of the discussion, advocates of this interpretation have, however, not adduced sufficient evidence that Diodorus applied this method on a large scale and they generally ignore cases in which the use of one single source can be proven for long sections (Ctesias in book 2; Agatharchides in book 3; Polybius in books 28–32).[47] Yet, it has been made plausible that in some passages Diodorus combined two (or perhaps more) sources,[48] and there is now a tendency to assume for whole books or thematic units the use of one main source to which Diodorus added some limited material from one, or sometimes perhaps more, sources.[49] For the history of the Greek West, Timaeus is generally assumed to have been this main source.[50]

---

42 For sketches of the history of Diodorean source criticism and overviews of the state-of-the-art source criticism, see Seibert (1983) 27–36; Meister (1990) 176–181; Ambaglio (2008) 20–34; Hau (2009) 174–176; Scardino (2014) 670; Rathmann (2016) 156–162; Hau *et al.* (2018) 3–9; Meeus (2022) 11–25; Müllenhoff ([1]1871).

43 See especially Stylianou (1998) 49–84 on Ephorus; differently Parmeggiani (2011) 349–394; Meeus (2022) 11–25, 58–90. More modern exponents of the one-source theory in the works quoted in n. 42.

44 Laqueur (1936) 1082–1083 for the principle, and 1174–1187 for books 4–5.

45 See against Laqueur, Jacoby on Timaeus, *FGrHist* 566 F 69 (III b Noten, 334 n. 354); Meister (1967) 15–19; Ambaglio (2008) 25 n. 121; Hau (2009) 174; Rathmann (2016) 159 with n. 18 (all with further literature).

46 Burton (1972) 1–34; *contra* Stylianou (1998) 49 n. 139 (with literature); other advocates of the use of many sources include, for example, Brown (1952) (Sicilian history; books 17–20); Chamoux, in Chamoux *et al.* (2002) xxviii–xxix (book 1, following Burton), xxix–xxx (book 2: other sources besides Ctesias), xxx (books 13 and 15 on Dionysius I); Dudziński (2016) (Timaeus and other sources for the Greek west).

47 For criticism of this approach, see also Meister (1990) 178.

48 For example, for the Third Sacred War in book 16: Ambaglio (2008) 25 (with literature) and Hau (2009) 175; for the Sicilian expedition: *ibid.*

49 See, for example, Boncquet (1987) 14–15; Meister (1990) 178–179 and Ambaglio (2008) 26 (books 17–20) with references.

50 See especially Meister (1967); cf. Ambaglio (2008) 25. Differently Dudziński (2016).

Recently, Rathmann has proposed a new theory.[51] He argues that Diodorus epitomized compilations in which earlier research was included and that he did so for very long passages. In addition, he regards it as possible that Diodorus looked up the literature quoted there and added information from it to his excerpt as well as information from other sources and of his own. He denies that such material added to the original excerpt can usually be identified. He contests that references to predecessors in Diodorus prove the use of them as a source in the passage in question beyond the specific piece of information they are quoted for and maintains that sources can, if at all, be identified on the basis of the tendency of a passage. If correct, this would make source criticism an almost hopeless endeavor as well. But Rathmann has not, in my view, shown that Diodorus ever looked up a source quoted in the work he used directly, and his own source analysis of books 18–20 as being based on Agatharchides does not convince.

Thus, the use of one main source for one topic, supplemented to a limited extent by probably one other source plus Diodorus' own additions referring to the situation of later times, is the most likely scenario. But this need not apply for his whole work. As it has been shown that Timaeus was his main source for the later history of the Greek West, he is a priori a likely candidate for our passage. Yet, there is one caveat: additions to Diodorus' main sources have usually been identified because there is independent evidence for their provenance from other authors, because information obviously refers to a later date than that of the main source or because its content or tendency contradicts that of the main source. In other words: if Diodorus integrates material that is consistent with his main source skillfully, we will probably not be able to detect the addition.

One last thing is important, also regarding our comparison with *Mir.* Research of the last decades has shown how Diodorus dealt with the material in his sources. It has become clear that he considerably condensed their content and made uniform their language and style.[52] He made a selection which was determined by the topics he was interested in and which was fitting his didactic goals and he sometimes imposed his own political and philosophical interpretations upon it.[53] In addition, most researchers today agree with those of the past that Diodorus' text contains many mistakes, misunderstandings, and contradictions as a consequence of his not being able to deal sufficiently with the material he had at his disposal. All this we have to take into account when we compare material in Diodorus that was claimed for Timaeus with chapters in *Mir.* So let us have a look at the passage in question and evaluate the arguments for Timaeus as its source. Parallels to *Mir.* are found in the following chapters of the first part of book 5:

Diod.	5.2.4–3.3	5.13.1–2	5.15 (+ 4.29–30, 4.82)	5.17.1–4	5.19–20	5.23
Mir.	82	93	100	88	84	81

51 Rathmann (2016) 156–270.
52 Palm (1955) whose results have been widely accepted by later research.
53 See Sacks (1990); Sulimani (2011) 57–108; Scardino (2014) 670–672; Rathmann (2016); contributions in Hau *et al.* (2018). But see Hau (2009) who shows that some topoi that seem Diodorean only appear in passages from certain sources.

3.2 *Diod. Sic. 5.2–23: the islands of the west*

3.2.1 *The mythical history of Sicily (5.2–6 and* Mir. *82)*

Mir. 82 describes the peculiarities of the place where Hades abducted Kore. There are clear parallels to Diodorus 5.2.4 and 5.3.2–4. The second of these passages is the more important one for us, but here Diodorus' account is obviously confused and cannot truly represent what his source had written. There we read:

Γενέσθαι δὲ μυθολογοῦσι τὴν ἁρπαγὴν τῆς Κόρης ἐν τοῖς λειμῶσι τοῖς κατὰ τὴν Ἔνναν. Ἔστι δ' ὁ τόπος οὗτος πλησίον μὲν τῆς πόλεως, ἴοις δὲ καὶ τοῖς ἄλλοις ἄνθεσι παντοδαποῖς ἐκπρεπὴς καὶ τῆς θεᾶς ἄξιος. Διὰ δὲ τὴν ἀπὸ τῶν φυομένων ἀνθῶν εὐωδίαν λέγεται τοὺς κυνηγεῖν εἰωθότας κύνας μὴ δύνασθαι στιβεύειν, ἐμποδιζομένους τὴν φυσικὴν αἴσθησιν. Ἔστι δ' ὁ προειρημένος λειμὼν ἄνωθεν μὲν ὁμαλὸς καὶ παντελῶς εὔυδρος, κύκλῳ δ' ὑψηλὸς καὶ πανταχόθεν κρημνοῖς ἀπότομος. Δοκεῖ δ' ἐν μέσῳ κεῖσθαι τῆς ὅλης νήσου, διὸ καὶ Σικελίας ὀμφαλὸς ὑπό τινων προσαγορεύεται. (3) Ἔχει δὲ καὶ πλησίον ἄλση καὶ λειμῶνας καὶ περὶ ταῦτα ἕλη, καὶ σπήλαιον εὐμέγεθες, ἔχον χάσμα κατάγειον πρὸς τὴν ἄρκτον νενευκός, δι' οὗ μυθολογοῦσι τὸν Πλούτωνα μεθ' ἅρματος ἐπελθόντα ποιήσασθαι τὴν ἁρπαγὴν τῆς Κόρης. Τὰ δὲ ἴα καὶ τῶν ἄλλων ἀνθῶν τὰ παρεχόμενα τὴν εὐωδίαν παραδόξως δι' ὅλου τοῦ ἐνιαυτοῦ παραμένειν θάλλοντα καὶ τὴν ὅλην πρόσοψιν ἀνθηρὰν καὶ ἐπιτερπῆ παρεχόμενα.

And the Rape of Corê, the myth relates, took place in the meadows in the territory of Enna. The spot lies near the city, a place of striking beauty for its violets and every other kind of flower and worthy of the goddess. And the story is told that, because of the sweet odour of the flowers growing there, trained hunting dogs are unable to hold the trail, because their natural sense of smell is balked. And the meadow we have mentioned is level at the centre and well watered throughout, but on its periphery it rises high and falls off with precipitous cliffs on every side. And it is conceived of as lying in the very centre of the island, which is the reason why certain writers call it the navel of Sicily. Near to it also are sacred groves and meadows, surrounded by marshy flats, and a huge grotto which contains a chasm which leads down into the earth and opens to the north, and through it, the myth relates, Pluton, coming out with his chariot, effected the Rape of Corê. And the violets, we are told, and the rest of the flowers which supply the sweet odour continue to bloom, to one's amazement, throughout the entire year, and so the whole aspect of the place is one of flowers and delight. (trans. Oldfather, *LCL*, modified).[54]

---

54 Unless otherwise indicated, I use Oldfather's *LCL* translation of Diodorus, in some cases modified.

Cicero follows the same source in *Verr.* 4.2.106–107. His report shows what went wrong when Diodorus made his summary[55]: Diodorus transferred the description of Enna, located on a mountain (ἄνωθεν . . . ἀπότομος), to the meadows where Kore was kidnapped,[56] which were outside the city in the plain according to all sources providing information on this. This transferral has as a consequence that he states in 3.3 that "near to them (scil. the meadows) also are groves and meadows (!), surrounded by marshy flats, and a huge grotto." Cicero and the double mention of the meadows in Diodorus prove, however, that the subject of ἔχει was originally Enna.[57] It is hard to believe that Diodorus created such a mess with the consequence that he gives a completely wrong description of a locality less than 40 km away from his birthplace, which he must have known. So one might be tempted to consider manuscript corruption and write ἔστι δ' ὁ προειρημένος τόπος (scil. Enna) etc. However, it is more likely that Cicero calls the meadows where Persephone was abducted and not Enna "the navel of Sicily."[58] Writing τόπος instead of λειμών, Enna would become the navel. So it seems that Diodorus has, in the sentence ἔστι

55 *Vetus est haec opinio, iudices, quae constat ex antiquissimis Graecorum litteris ac monumentis, insulam Siciliam totam esse Cereri et Liberae consecratam. Hoc cum ceterae gentes sic arbitrantur, tum ipsis Siculis ita persuasum est ut in animis eorum insitum atque innatum esse videatur. Nam et natas esse has in iis locis deas et fruges in ea terra primum repertas esse arbitrantur et raptam esse Liberam, quam eandem Proserpinam vocant, ex Hennensium nemore, qui locus, quod in media est insula situs, umbilicus Siciliae nominatur. Quam cum investigare et conquirere Ceres vellet, dicitur inflammasse taedas iis ignibus qui ex Aetnae vertice erumpunt. Quas sibi cum ipsa praeferret, orbem omnem peragrasse terrarum. (107) Henna autem, ubi ea quae dico gesta esse memorantur, est loco perexcelso atque edito, quo in summo est aequata agri planities et aquae perennes, tota vero ab omni aditu circumcisa atque directa est. Quam circa lacus lucique sunt plurimi atque laetissimi flores omni tempore anni, locus ut ipse raptum illum virginis, quem iam a pueris accepimus, declarare videatur. Etenim prope est spelunca quaedam conversa ad aquilonem infinita altitudine, qua Ditem patrem ferunt repente cum curru exstitisse abreptamque ex eo loco virginem secum asportasse et subito non longe a Syracusis penetrasse sub terras, lacumque in eo loco repente exstitisse, ubi usque ad hoc tempus Syracusani festos dies anniversarios agunt celeberrimo virorum mulierumque conventu.*

56 See already Holm (1870) 367 who rewrites Diodorus on the basis of Cicero and makes his description fit the real topography. But he thinks that in Cicero, Enna is the navel of Sicily. Cf. Ziegler (1912) 285; Rossbach (1912) 15 n. 7; Heinze (1919) 8 n. 1 (mistake by Diodorus or a scribe); Herter (1941) 245 n. 20. Laqueur's (1936) 1182–1184 analysis of the sources of these chapters is not convincing. The confusion is not mentioned in the commentary of Casevitz and Jacquemin (2015). Translators try to solve the problem by rendering ἄνωθεν μὲν ὁμαλός in 3.2 by "level in the centre" Oldfather (1933) 103; "in ihrer Mitte eben" Veh in Wirth *et al.* (1993) 434. See also Casevitz in Casevitz and Jacquemin (2015) 5: "(la prairie) . . . forme au centre un plan"; "(il prato) . . . è piano all'interno" Cordiano and Zorat (2014) 279. But that is not what ἄνωθεν means, which is "on its top."

57 It is thus methodologically wrong to delete καὶ λειμῶνας as do Vogel (1890) 6 and Oldfather (1933) 102. Correct is Casevitz in Casevitz and Jacquemin (2015) 6 who does not mention this conjecture anymore. Malten (1910) 526 accepts Diodorus' description of the meadow.

58 Cic. *Verr.* 4.2.106 (see n. 55). It seems to be generally acknowledged that Enna was the navel of Sicily. Of course, the position of Enna on the Monte San Giuliano was a landmark, but the two passages that use the expression seem to point to the meadow. One might argue that a navel is a hollow, not an elevation, but Delphi as the navel of the world was also located on a hill. In some sources, the

δ' ὁ προειρημένος . . . πορσαγορεύεται, mixed up information on the meadows and on the city. In addition, we are confronted with the problem that Diodorus reports things twice: he introduces the location of the meadows as close to Enna in 3.2, followed by a description of them which includes the *thaumasion* that hunting dogs cannot hold the trail because of the smell of the flowers. Then he introduces the location again at the beginning of 3.3 adding a second wonder (παραδόξως), which is that the flowers bloom all year. One has the impression that Diodorus has either summarized the same passage twice, picking out different features each time, or added one description from a different source to the excerpt from his main source. In the latter case, the second description would come from this main source because it contains the characterization of the lawn we also read in Cicero (*omni tempore anni* ~ δι' ὅλου τοῦ ἐνιαυτοῦ), who does not mention the effect on dogs.

Let us now have a look at *Mir.* 82 and compare it with Diodorus-Cicero:

Ἐν τῇ Σικελίᾳ περὶ τὴν καλουμένην Αἴτνην, σπήλαιόν τι λέγεται εἶναι, περὶ ὃ κύκλῳ πεφυκέναι φασὶ τῶν τε ἄλλων ἀνθέων πλῆθος ἀνὰ πᾶσαν ὥραν, πολὺ δὲ μάλιστα τῶν ἴων ἀπέραντόν τινα τόπον συμπεπληρῶσθαι, ἃ τὴν σύνεγγυς χώραν εὐωδίας πληροῖ, ὥστε τοὺς κυνηγοῦντας, τῶν κυνῶν κρατουμένων ὑπὸ τῆς ὀδμῆς, ἐξαδυνατεῖν τοὺς λαγὼς ἰχνεύειν. Διὰ δὲ τούτου τοῦ χάσματος ἀσυμφανής ἐστιν ὑπόνομος, καθ' ὅν φασι τὴν ἁρπαγὴν ποιήσασθαι τὸν Πλούτωνα τῆς Κόρης. Εὑρίσκεσθαι δέ φασιν ἐν τούτῳ τῷ τόπῳ πυροὺς οὔτε τοῖς ἐγχωρίοις ὁμοίους οἷς χρῶνται οὔτε ἄλλοις ἐπεισάκτοις, ἀλλ' ἰδιότητά τινα μεγάλην ἔχοντας. Καὶ τούτῳ σημειοῦνται τὸ πρώτως παρ' αὐτοῖς φανῆναι πύρινον καρπόν. Ὅθεν καὶ τῆς Δήμητρος ἀντιποιοῦνται, φάμενοι παρ' αὐτοῖς τὴν θεὸν γεγονέναι.

In Sicily, around the so-called Aetna, there is said to be a cave, around which is an abundance of flowers at every season of the year, and particularly that a vast space is filled with violets, which fill the neigbourhood with sweet scent, so that hunters cannot chase hares, because the dogs are overcome by the scent. Through this cave there is an invisible underground passage, by means of which Pluto is said to have made the rape of Core. They say that wheat is found in this place unlike any that they have or that is imported, but having great peculiarities. And from this it is concluded that wheat appeared first among them. Therefore, they also claim Demeter, saying that the goddess was born among them.

Most modern editors write περὶ τὴν καλουμένην Ἔνναν, thus locating the *thaumasia* in Enna.[59] But the reading Ἔνναν is only attested as a marginal note in G and P, which read Αἴτνην *in textu*, and in R. These three manuscripts go back to the lost cod. x. The majority of the manuscripts, that is, those of the group ψ as well as H

---

lawn is located by the Lago di Pergusa, which is only 5 km away from Enna (Ov. *Met.* 5.385–390; more passages in Bömer [1976] 326). So this area may well be called the navel of the island.

59 See, for example, Westermann (1839); Flashar (1972) with commentary on p. 109; Vanotti (2007).

which is, together with x, a descendant of the lost γ, read ἐν τῇ καλουμένῃ Αἴτνῃ.[60] So this reading is much better attested and should be accepted.[61] There were contradicting traditions about the place where Kore was kidnapped by Hades, and Enna was among the most popular ones which explains why a learned reader may have added this name to x as a correction.[62] But it is even more often attested that the abduction took place at mount Aetna, a tradition first found in Moschus (3.121).[63] The lawn of flowers[64] and the cave used by Hades[65] are located there as well, and Plutarch (*Quaest. Nat.* 23.917e-f) not only locates the lawn of mountain violets there but even speaks of the paradox that hunting dogs are misled by the smell of the flowers and that this happens all year. The main reason why scholars prefer Enna to Aetna in *Mir.* is that they assume that *Mir.* and Diodorus-Cicero go back to the same source, Timaeus. Yet, that is what needs to be proven, and this is not the only point where *Mir.* differs from them. In *Mir.* the cave is at the center of interest and also geographically, with the flowers growing around it, while in Diodorus-Cicero the cave is said to be close to the meadow. In addition, one has the impression that in *Mir.* Hades returns to his realm with the girl through the same cave. In Diodorus-Cicero and in Ovid, who goes back to the same tradition,[66] Hades races over land on his chariot and enters earth near Syracuse. Thus, commentators often speak of an inaccuracy of the author of *Mir.*[67] But that need not be the case. It is obvious that Diodorus' source author tries to reconcile different stories about the place of abduction. He has Hades exit the underworld at Enna and enter it at Syracuse[68] and he even takes the Aetna version into account to some extent by stating that Demeter lit her torches with the fire of the volcano after she had not found her daughter in Sicily and started out to search the rest of the world.[69] It is perhaps not by chance that in the versions of the story in which the abduction takes place at Mt. Aetna, it is never claimed that Hades returned to his reign elsewhere, so this may explain why this is the case in *Mir.* as well.

60 See Giacomelli's note on *Mir.* 82 and his stemma in Giacomelli (2021) 56.

61 It is now accepted by Giacomelli (2023) 260.

62 See, besides Diodorus and Cicero, Ov. *Met.* 5.385ff.; *Fast.* 4.419ff.; Sil. *Pun.* 7.689–690, 14.242–247, etc. The various traditions are collected in Bloch (1890–1894) 1310–1311, 1313–1315, and Bräuninger (1937) 951–952, 966; cf. Bömer (1957–1958) II 245; Jacquemin in Casevitz and Jacquemin (2015) 121.

63 See the collection of passages in Bräuninger (1937) 952, Bloch (1890–1894) 1310, and add Lydus *Mens.* 4.137 Wünsch. The sheer number of attestations excludes that Aetna is a mistake for Henna, for example, Malten (1910) 526 n. 0 holds.

64 Hyg. *Fab.* 146; Plut. *Quaest. nat.* 23.917e-f; Auson. *Epist.* 14.49; *Mythographus Vaticanus* 2.115 Kulcsár; *Schol. Pind. Nem.* 1.20.

65 Philagrius *ad Verg. Ecl.* 3.105 (I). Cf. the cave at Ov. *Fast.* 4.495–498, which may be a reminiscence of such a version located at mount Aetna.

66 Ov. *Met.* 5.385–437. In the version in *Fast.* 4.417–454, the place of reentrance is not mentioned. On Timaeus as Ovid's (ultimate) source, see Zinzow (1906) 15–37; Malten (1910) 521–540; Herter (1941) 240–248; Bömer (1957–1958) II 245; (1976) 325.

67 Flashar (1972) 109 with reference to Müllenhoff (²1890) 447.

68 See Jacquemin in Casevitz and Jacquemin (2015) 126.

69 Diod. Sic. 5.4.3. This is a common motive: see the references in Bloch (1890–1894) 1315.

Despite these differences, it is clear that the traditions in Diodorus-Cicero and in *Mir.* are related, and that *Mir.* helps answer the question whether Diodorus combined two sources or badly worked together two excerpts from the same source. As for its content, *Mir.* corresponds in the main to 3.2, but it also has information only found in 3.3, namely, the bloom of the flowers during all year. Furthermore, there are verbal reminiscences with 3.2 and especially with 3.3:

3.2	3.3	Mir.
ἴοις δὲ καὶ τοῖς ἄλλοις ἄνθεσι παντοδαποῖς	τὰ δὲ ἴα καὶ τῶν ἄλλων ἀνθῶν τὰ παρεχόμενα	τῶν τε ἄλλων ἀνθέων πλῆθος . . . πολὺ δὲ μάλιστα τῶν ἴων
εὐωδίαν	εὐωδίαν τὸν Πλούτωνα . . . ποιήσασθαι τὴν ἁρπαγὴν τῆς Κόρης	εὐωδίας τὴν ἁρπαγὴν ποιήσασθαι τὸν Πλούτωνα τῆς Κόρης

Thus, the easiest way of explaining Diodorus' account is to assume that he only summarized one source in the form of various excerpts which he finally combined in a clumsy and in part misleading way.

After the abduction story *Mir.* speaks about a type of wheat with a very particular appearance (not specified further) only found in this place.[70] This fact is seen as evidence (σημειοῦνται) by (probably) the locals that this fruit had appeared there first and it explains why they claim Demeter for themselves. This passage becomes clearer by the parallel tradition in Diodorus (5.2.3–5) and Cicero (*Verr.* 4.2.106). It shows that these sentences are part of the lively discussion about the place where Demeter came first to men and where grain was cultivated first, making the case for the Sicilian position.[71] In Diod. Sic. 5.2.3, we read shortly before the passage quoted above that Demeter and Kore had appeared (φανῆναι) first in Sicily and that this island produced grain first because of the excellence of its soil. After citing verses of the *Odyssey* (9.109–111) claiming that the land of the Cyclopes (here, as often, identified with Sicily) brings forth fruits without sowing and plowing, the historian adds: ἔν τε γὰρ τῷ Λεοντίνῳ πεδίῳ καὶ κατὰ πολλοὺς ἄλλους τόπους τῆς Σικελίας μέχρι τοῦ νῦν φύεσθαι τοὺς ἀγρίους ὀνομαζομένους πυρούς ("Indeed, in the plain of Leontini, we are told, and throughout many other parts of Sicily the wheat men call 'wild' grows even to this day."). The argument, as we can see, is similar to that of *Mir.* but not identical. In both cases, a specific kind of wheat is used as evidence. But in Diodorus this wheat testifies to the quality of the Sicilian soil and confirms Homer's claim. In agreement with this, it is found in many places on the island. In *Mir.*, one special kind of wheat is found *only* in the Aetna region and serves as evidence that grain was first introduced *there* by Demeter. The idea

70 Cf. Müllenhoff (²1890) 337.

71 On this discussion, see Diod. Sic. 5.69.1–3 with the claims of the Egyptians, Athenians, and Sicilians.

seems to be that it is a type of wheat that the goddess only brought here, where she resided first, while the grain she gave to peoples who welcomed her during her quest for Kore was different.[72] Thus, what we have here are two different versions of the "grain motive." Neither excludes the other and they may both have been present in a common source that Diodorus and *Mir.* excerpted differently. But the difference may also be explained by the use of different sources or the assumption that one version is the reworking of the other.

There can be little doubt that Timaeus was Diodorus' main or only source in the chapter on Sicily and this has been the *communis opinio* for a long time.[73] One argument for this assumption is the parallel in Cicero who knew Timaeus well and held him in high esteem.[74] Another comes from Diodorus himself. In 5.6.1–2, he declares that the origin of the Sicani was disputed and that Philistus claimed they were immigrants from Iberia.[75] Timaeus is said to have criticized this and to have shown, with many arguments, that they were autochthonous.[76] The same position is found in 5.2.4 with reference to "the best historians" (νομιμώτατοι τῶν συγγραφέων) and it is added that according to them, Demeter and Kore appeared first on this island. The same authorities, that is, Timaeus, are still the subject of the verbs λέγουσιν and μυθολογοῦσιν in 5.3.1–2 (and probably of the verbs in 5.3.3–4 and 5.4.1 and 2) which shows that everything narrated there is in all likelihood taken from Timaeus. Therefore, probably everything related to the introduction of grain and to Demeter and Kore stems from this source. That Diodorus' source obviously tries to reconcile, in 5.4.3–4, the Sicilian and Athenian claims to priority in the cultivation of grain, assigning Sicily priority in time, but making Athens second and, in addition, the city that gave the fruit to others, perfectly fits the Western Greek historian who lived for decades and wrote his *Sicilian History* in Athens.

Timaeus F 37, however, transmitted by a scholion to Ap. Rhod. 4.965, is problematic. It explains the expression Θρινακίης λειμῶνα in the poem. It states: "Timaeus says that Sicily was called Thrinakia (Θρινακίαν) because it has three capes (ἄκρας)." This contradicts Diod. Sic. 5.2.1 where it is stated that Sicily was once called Trinakria (Τρινακρία) because of its shape. But Jacoby may be right when he argues that there is a mistake in the text of the scholion because the explanation of the name there presupposes that Timaeus spoke of Trinakria, and the

---

72 On the idea of Demeter bringing grain to mankind during her quest, see Diod. Sic. 5.4.3–4.

73 See, for example, Jacoby on Timaeus, *FGrHist* 566 F 164; Meister (1967) 31–32; but not all of his arguments, sometimes based on source criticism of other texts or the assumed tendency of a passage, are compelling. Especially his references to Diodorus' book 4 are often problematic. On the other hand, there is no need to regard 5.4.4 (thus Jacoby) or 5.5.2–3 (thus Meister and Jacoby n. 597 as a possibility) as additions from a different source.

74 For comparisons of the two reports and the identity of the common source, see Schwabe (1870) 337–341; Müllenhoff (²1890) 444–447; cf. also Meister (1967) 31. A somewhat superficial comparison of the two texts in Romano (1980). Laqueur (1936) 1182–1184 assumes that Cicero and Diodorus' source (i.e., Ephorus) both follow an oral local tradition. This is excluded by the structural and verbal similarities between the two texts. Cf. Meister (1967) 31 with n. 10 on p. 173.

75 Philistus, *FGrHist* 556 F 43.

76 Timaeus, *FGrHist* 566 F 38.

mistake is easily explained by the fact that Thrinakia is the form used in Apollonius' poem.[77]

It needs to be asked, however, whether Timaeus was used here by Diodorus directly or indirectly. Both are possible and the main argument for direct use is perhaps the parallel in Cicero of whom we know that he was familiar with this author. If we are willing to admit that Cicero used a different source, then indirect use is possible as well.

To conclude: As many similarities in structure, narrative elements, and single expressions as there are between *Mir.* and the source of Diodorus-Cicero, there are also significant differences. Although they may be explained in more than one way, they show that the two traditions are closely related. It is possible that *Mir.* depends on the report used and reworked by Diodorus' source or that *Mir.* is a report that further develops and modifies what Diodorus' source had written, adding elements from other traditions such as the location at mount Aetna. In this case, the reworking may be due to *Mir.*'s source or the author of *Mir.* himself.

3.2.2 *Elba and Populonia (5.13–1–2 and* Mir. *93)*

Mir. 93 describes a marvel on the island of Aethalia (Elba):

Ἐν δὲ τῇ Τυρρηνίᾳ λέγεταί τις νῆσος Αἰθάλεια ὀνομαζομένη, ἐν ᾗ ἐκ τοῦ αὐτοῦ μετάλλου πρότερον μὲν χαλκὸς ὠρύσσετο, ἐξ οὗ φασὶ πάντα κεχαλκευμένα παρ' αὐτοῖς εἶναι, ἔπειτα μηκέτι εὑρίσκεσθαι, χρόνου δὲ διελθόντος πολλοῦ φανῆναι ἐκ τοῦ αὐτοῦ μετάλλου σίδηρον, ᾧ νῦν ἔτι χρῶνται οἱ Τυρρηνοὶ τὸ καλούμενον Ποπλώνιον οἰκοῦντες.

In Tyrrhenia there is said to be an island called Aethaleia, in which in the old days copper was dug from a mine, from which everything made of copper by them comes; after that it was no longer found, but then, after the lapse of considerable time, iron appeared from the same mine, which the Tyrrhenians who live in the region called Poplonium still use.

Mir. distinguishes three different phases[78]: (1) at some point in the past, copper was extracted from a mine on Elba, (2) after everything was extracted and much time had passed without mining, (3) iron was found in the same place. We have here thus the *topos*, not only popular in paradoxography, that metal can grow.[79] In our case, it appears with the modification that a different kind of material grew after the previous one was exhausted. It seems to be implied that the quantity of the extracted copper was very high, while it is not stated where it was worked into all kinds of tools. As for the iron, we do not get information about its quantity, but we

---

77   See Jacoby on Timaeus, *FGrHist* 566 F 37 with nn. 247 and 248. He shows in n. 247 how hopelessly confused the transmission of the various names of the island is. Cf. Meister (1967) 31 with n. 7. Dudziński (2016) 56 is imprecise.

78   Cf. Corretti (2004) 270–273 (with the attempt of a historical reconstruction); Vanotti (2007) 174.

79   Radt (2007) 42; Corretti (2004) 281–284.

learn that it is still "used" today by the Etruscans in Populonium (today Populonia) on the mainland right across from Elba. It is unclear what this exactly means, but the linking with the copper seems to suggest the production of tools as well.

Diod. 5.13.1–2 also reports about the iron mines on Elba, but as we shall see, there is not much thematic overlap with *Mir.* He introduces the island with words that seem to echo the beginning and the end of *Mir.* 93[80]:

Τῆς γὰρ Τυρρηνίας κατὰ τὴν ὀνομαζομένην πόλιν Ποπλώνιον νῆσός ἐστιν, ἣν ὀνομάζουσιν Αἰθάλειαν.

Off the city of Tyrrhenia known as Poplonium there is an island which men call Aethaleia.

Diodorus then explains the name Aithaleia, which he derives from *aithalos*, smoke, and gives a description of the mining of the iron rock and its first treatment:

Οἱ γὰρ ταῖς ἐργασίαις προσεδρεύοντες κόπτουσι τὴν πέτραν καὶ τοὺς τμηθέντας λίθους κάουσιν ἔν τισι φιλοτέχνοις καμίνοις· ἐν δὲ ταύταις τῷ πλήθει τοῦ πυρὸς τήκοντες τοὺς λίθους καταμερίζουσιν εἰς μεγέθη σύμμετρα, παραπλήσια ταῖς ἰδέαις μεγάλοις σπόγγοις.

For those who are engaged in the working of the ore crush the rock and burn the lumps by means of a great fire and form them into pieces of moderate size which are in their appearance like large sponges.

He then reports that these sponges are shipped to Dicaearchia (Puteoli; Pozzuoli) and other trading places where all kinds of tools are manufactured and exported to all parts of the world.

With some goodwill, one may see in both texts excerpts from the same source with different focalizations, *Mir.* with interest in the *thaumasion* and Diodorus with interest in the economic aspects.[81] But if we do so, we have to assume that *both* authors have somewhat misleadingly summarized or modified their common source: reading Diodorus' description, one gets the impression that the sponges are produced on Elba only, but in the common source the iron must have been melted in Elba *and* Populonia. On the other side, *Mir.*'s author gives the impression that iron tools were produced in Populonia while only the first step of the working of the metal took place there.

It has been claimed that the role played by Dicaearchia in Diodorus' chapter presupposes the amplification of its harbor by the Romans after the Second Punic War (199 BC) and the foundation of a *colonia* there in 194 BC.[82] If this interpretation

<hr>

80 Cf. Müllenhoff (²1890) 452; Geffcken (1892) 65.

81 Cf. Müllenhoff (²1890) 452.

82 See Corretti (2004) 277. In n. 6, he refers to predecessors for such an interpretation. In Corretti (2009) 135, he, however, argues that Diod. Sic. 5.13 corresponds to the situation of the last years of the war against Hannibal and the production of weapons at that time. In other words, without saying so explicitly, he presupposes that the Roman harbor is not a prerequisite of Dicaearchia's situation as described by Diodorus. I do not agree with Jacquemin in Casevitz and Jacquemin (2015) 149 who

is correct, this would seriously invalidate the general assumption that the first part of Diodorus' book 5 goes back to Timaeus. But there was a harbor in the old Dicaearchia as well and we do not know what role exactly it played in Diodorus' source.[83] Thus, Diodorus' words may, it seems, be reconciled with an earlier date.

Things are made more complicated by a passage on Populonia in Strabo (5.2.6.306–307) and by the archaeological evidence. Strabo describes Populonia on the basis of his visit to that place (in the 40s or 30s?).[84] At that time, it was in decline. The city itself on the hill was almost deserted, while the harbor was better off. He reports that the iron is no longer melted out on Elba but in Populonia and he adds two marvels:

Εἴδομεν δὲ καὶ τοὺς ἐργαζομένους τὸν σίδηρον τὸν ἐκ τῆς Αἰθαλίας κομιζόμενον· οὐ γὰρ δύναται συλλιπαίνεσθαι καμινευόμενος ἐν τῇ νήσῳ· κομίζεται δ᾽ εὐθὺς ἐκ τῶν μετάλλων εἰς τὴν ἤπειρον. Τοῦτό τε δὴ παράδοξον ἡ νῆσος ἔχει καὶ τὸ τὰ ὀρύγματα ἀναπληροῦσθαι πάλιν τῷ χρόνῳ τὰ μεταλλευθέντα, καθάπερ τοὺς πλαταμῶνάς φασι τοὺς ἐν Ῥόδῳ καὶ τὴν ἐν Πάρῳ πέτραν τὴν μάρμαρον καὶ τὰς ἐν Ἰνδοῖς ἅλας, ἅς φησι Κλείταρχος.

And I also saw the people who work the iron that is brought over from Aethalia; for it cannot be brought into complete coalescence by heating in the furnaces on the island; and it is brought over immediately from the mines to the mainland. However, this is not the only remarkable thing about the island; there is also the fact that the diggings which have been mined are in time filled up again, as is said to be the case with the ledges of rocks in Rhodes, the marble-rock in Paros, and, according to Cleitarchus, the salt-rock in India. (trans. Jones, *LCL*)

The amount of iron melted in Populonia at that time does not become clear from Strabo's description, but considering the run-down state of the city, it need not have been high. Such an assumption is confirmed by the archaeological evidence. Until the mid-20th century, enormous heaps of iron slags were seen near the shore of the Baratti Bay, before most of them were removed for re-smelting. On the basis of what remains, the end of extensive iron production in Populonia is usually dated by archaeologists to the turn of the 2nd/1st century BC. This may be related to a *senatus consultum* ordering the closing of all mines in Italy that Pliny the Elder labels "old." It cannot be dated exactly, and it is a matter of discussion whether it applied to the Elba mines at all as they were not located *in* Italy.[85] Camilli now even argues

concludes from the use of the old Greek name Dicaearchia instead of Puteoli that Diodorus' source must predate the founding of Puteoli. The Greek name remained in use after that time, especially in Greek authors; see the collection of sources in Hülsen (1903) 546.

83 An earlier date is considered by Aranguren *et al.* (2004) 337–338 and by the scholars quoted by Corretti (2004) 227 n. 22.

84 On the possible date of this trip, see Honigmann (1931) 82.

85 On the end date of intensive iron production in Populonia, see Acconcia and Giuffré (2009) 155–157: "La fase intensiva della produzione del ferro sembra esaurirsi, con modalità che appaiono abbastanza repentine, tra la fine del II e gli inizi del I secolo a.C. Lo stesso fenomeno si registra per

for the end of the iron production in Populonia 100 years earlier.[86] If this is correct, the mention of Dicaearchia in Diodorus must refer to the Greek harbor. Before the issue can be settled, more archaeological research is necessary, and as long as that is not the case, we lack a firm basis for interpreting the claims in Diodorus and *Mir.* Be that as it may, an end date around 100 or 200 BC shows that Diodorus' claim of mass production and export to Dicaearchia and other places cannot be regarded as one of his many updates in which he adds information about his own time to the report of his source but that it must stem from his source.[87]

So what is the source of *Mir.*? Flashar opts for Posidonius, because he sees in him Strabo's source of the passage quoted above.[88] But the evidence for such an assumption is weak, although Theiler even goes so far as to include part of the Strabo passage as F 37 in his edition of the Posidonius fragments.[89] But even if we could be sure that he was Strabo's source here, we would have to admit that the miracle of the growing metal is slightly different in *Mir.* and Strabo[90]: Strabo seems to assume that in *all* mines of the island iron grows again, while *Mir.* speaks of *one* mine in which iron grew after copper had been found there before. There is thus not much evidence to postulate a common source for Strabo and *Mir.* (but see below, n. 186).[91] Therefore, Geffcken's argument for Timaeus as *Mir.*'s source

---

parte dei siti metallurgici di età repubblicana noti nell'area posta lungo la costa tra il golfo di Follonica e San Vincenzo e per quelli dell'Isola d'Elba." (156) This points to the *SC* as the reason for the end of the production. It is mentioned in Plin. *HN* 3.138 (*vetere consulto patrum*); 33.78 (*vetere interdicto patrum*); 37.202 (allusion to the end of mining in Italy). Cf. Corretti (2004) 282–286; Corretti *et al.* (2014) 184–185. On iron smelting places on Elba, see Corretti *et al.* (2014) 183–184. These authors date the production "end of the 3rd to 1st century BC."

86 Camilli (2016, 51–55, (2019), 125.
87 Diodorus was working on his *Library* ca. 60–30 BC: see Meister (1990) 172.
88 Flashar (1972) 113–114.
89 For the arguments, see his commentary in Theiler (1982) II 51–52: an example from Rhodes, Cleitarchus as Posidonius' source elsewhere, and asphalt that grows again in another Posidonian fragment (F 46 Theiler = 235 Edelstein-Kidd = A 153 Vimercati). See already Aly (1957) 244.
90 This is emphasized by Corretti (2004) 272–273.
91 Aithaleia is also mentioned in *Mir.* 105 (the conjecture is certain), but this does not help us identify the source of *Mir.* 93 because we do not know whether both chapters go back, in final analysis, to the same source (even if Pajón Leyra's hypothesis, presented below on p. 75 is correct: the work *On Islands* that she assumes as the source of *Mir.* 93 and 105 may have assembled different traditions). In *Mir.* 105, Timaeus is excluded as the source because he assumed a different route of the Argonauts from Colchis to the western Mediterranean than our compiler; see Flashar (1971) 122–124 (with earlier literature) against Geffcken (1892) 91–94 who tried to claim the chapter for Timaeus. As Geffcken also Enmann (1880) 194–196. Timaeus' route is summarized by Diod. Sic. 4.56 (=Timaeus, *FGrHist* 566 F 85) and in this context he mentions the foundation of the harbor of Argoos on Elba among the signs left by the Argonauts in this part of the Mediterranean. It is interesting to see that Strabo also mentions this event in the lines following those on the production of iron in Populonia and that he adds the marvel of the colored pebbles which also appears in *Mir.* 105. It suffices to note that here too the *thaumasion* is slightly different: while in *Mir.* 105, the pebbles at the shore are *colored by the oil* the Argonauts scraped off their bodies, in Strabo the colored pebbles *consist of the hardened oil* of the Argonauts. Cf. Dini *et al.* (2007) on these pebbles, which can indeed be found on the beach of Capo Bianco on Elba. It is obvious that Strabo's source, which is again often identified with Posidonius without hard evidence (see, e.g., Flashar (1972) 124), in the chapter on Elba and Populonia uses the same material that we find in the two chapters of *Mir.* and

is problematic as well. He starts from the assumption that *Mir*. 85–87 and 89–94 show many points of contact with Posidonius and he concludes that these are due to Posidonius' frequent use of Timaeus as a source.[92] But that cannot be regarded a strong enough argument. Thus, the source of *Mir*. 93 cannot be established at the moment, and neither can that of Diod. Sic. 5.13.1–2. Depending on the end date of the iron production in Populonia and the interpretation of Dicaearchia's role as a trading center, the latter's source can be dated to either before ca. 100 BC or to before ca. 200 BC.[93] *Mir*. might go back to the same source but that is not certain and requires the assumptions of mistakes or modifications in both Diodorus and *Mir*. If there was such a common source, it could be Timaeus, but proof is lacking.

3.2.3 Sardinia (5.15 and Mir. 100)

Ἐν τῇ Σαρδοῖ τῇ νήσῳ κατασκευάσματά φασιν εἶναι εἰς τὸν Ἑλληνικὸν τρόπον διακείμενα τὸν ἀρχαῖον ἄλλα τε πολλὰ καὶ καλὰ καὶ θόλους περισσοῖς τοῖς ῥυθμοῖς κατειργασμένους· τούτους δὲ ὑπὸ Ἰολάου τοῦ Ἰφικλέους κατασκευασθῆναι, ὅτε τοὺς Θεσπιάδας, τοὺς ἐξ Ἡρακλέους, παραλαβὼν ἔπλευσεν εἰς ἐκείνους τοὺς τόπους ἐποικήσων, ὡς κατὰ συγγένειαν αὐτῷ τὴν Ἡρακλέους προσήκουσαν, διὰ τὸ πάσης τῆς πρὸς ἑσπέραν κύριον Ἡρακλέα γενέσθαι. Αὕτη δὲ ἡ νῆσος, ὡς ἔοικε, ἐκαλεῖτο μὲν πρότερον Ἰχνοῦσσα, διὰ τὸ ἐσχηματίσθαι τῇ περιμέτρῳ ὁμοιότατα ἀνθρώπων ἴχνει. Εὐδαίμων δὲ καὶ πάμφορος ἔμπροσθεν λέγεται εἶναι· τὸν γὰρ Ἀρισταῖον, ὅν φασι γεωργικώτατον εἶναι ἐπὶ τῶν ἀρχαίων, τοῦτον αὐτῆς ἄρξαι μυθολογοῦσιν, ὑπὸ μεγάλων ὀρνέων ἔμπροσθεν καὶ πολλῶν κατεχομένης. Νῦν μὲν οὖν οὐκέτι φέρει τοιοῦτον οὐδέν, διὰ τὸ κυριευθεῖσαν ὑπὸ Καρχηδονίων ἐκκοπῆναι πάντας τοὺς χρησίμους εἰς προσφορὰν καρπούς, καὶ θάνατον τὴν ζημίαν τοῖς ἐγχωρίοις τετάχθαι, ἐάν τις τῶν τοιούτων τι ἀναφυτεύῃ. (*Mir*. 100)

In the island of Sardinia they say that there are many beautiful buildings arranged in the ancient Greek style, and among others domed buildings, carved with many shapes; these are said to have been built by Iolaus the son of Iphicles, when he took the Thespiads, descended from Heracles, and sailed to those parts to colonize them, on the grounds that they belonged to him by his kinship with Heracles, because Heracles was master of all the country towards the west. Apparently the island was originally called Ichnussa because its circumference made a shape like a man's footstep, and it is said before this time to have been prosperous and fruitful; for the legend was that Aristaeus, who, they say, was the most efficient husbandman in ancient times, ruled them, in a district previously full of many great birds. Now the island no longer bears anything, because the Carthaginians who got possession of it cut down all the

---

that seems to have been part of the standard repertoire of the (mythical) history of Elba (the pebbles and Argoos also in Ap. Rhod. 4.654–658; an allusion to the pebbles in Lycoph. *Alex*. 874–875). But as we often note in such traditions, authors consciously modify the traditional material to show their individuality or to adapt it to their literary agenda. So the only thing that is sure is that Timaeus F 85, Apollonius of Rhodes, the source(s) of *Mir*. 93 and 105 on Elba, and Strabo are somehow related to each other or to a common pool of information, but how exactly we cannot tell.

92 Geffcken (1892) 65, 95–96; for Timaeus also Corretti (2004) 272–273.

93 Corretti (2004) 278–281 argues for Polybius, an author often used by Diodorus in later books. But there is no hard evidence for this assumption.

fruits useful for food, and prescribed the penalty of death to the inhabitants, if any of them replanted them.

The interest of the paradoxographer was triggered by the presence of buildings on Sardinia "in the ancient Greek manner," although the island was inhabited by barbarians. With these buildings, the so-called *nuraghi* are meant characteristic buildings of the indigenous Sardinian culture, which indeed resemble Mycenean architecture.[94] As the island is presented as under Carthaginian rule, this describes a situation before 238 BC, when it became Roman, which is a valuable indication for the date of *Mir.*'s source. Unfortunately, we cannot say whether it also means that this part of our collection predates that year or that our author, although writing in later times, did not care about updating the information of his source.[95]

Diodorus dedicates 5.15 to the history and culture of Sardinia. Although significant verbal accordance is lacking, the presence of a number of identical topics and some structural agreement show that both reports are related to one another. The differences in the order of the elements are explained by the different starting points of both authors, the *nuraghi* in *Mir.* and the colonization by Iolaus and the Thespiads in Diodorus.[96]

It will be useful to have a look at the single narrative elements in *Mir.* and compare them to the corresponding ones in Diodorus. This will show that nothing hinders us from assuming that both go back to the same source.

1. In Diodorus, Iolaus founds cities, divides the land and, what interests us here, constructs buildings, temples, and gymnasia, which are still preserved. It is to be noted that the *tholoi* are not mentioned (5.15.2).
2. The foundation of the colony is in both texts a joint enterprise by Iolaus and the Thespiads, with Iolaus as the leader (see Diod. Sic. 15.1–2). That his position was due to his family relation with Heracles is reflected in Diodorus by the remark that he was the son of Heracles' sister. Noteworthy additional information here is that many other Greeks and barbarians participated.
3. The old name of the island, Ichnoussa, and Aristaeus as the one who first cultivated the island previously inhabited by birds, are not mentioned in Diodorus, but the Carthaginians are and here as well their domination of the island is the reason why the Iolaeans put an end to agriculture (Diod. 5.15.4–6). Diodorus speaks of an oracle announcing that those participating in the *apoikia* will be free forever and adds that this has "miraculously" (παραδόξως) been the case up to his day. In addition, there are some striking differences from *Mir.*: when the Carthaginians brought the island under their control, Diodorus narrates, the Iolaeans fled into the mountains, built subterranean abodes, and became nomads, feeding on milk, cheese, and meat. It is added that they left the plains

---

94 Cf., for example, Flashar (1972) 117–118; Vanotti (2007) 180; Neri (2002) 35–37; Mastino (2004) 21–22.
95 This is generally acknowledged. See, for example, Flashar (1972) 119. Cf. Müllenhoff (²1890) 455.
96 Cf. Flashar (1972) 118–119. On this chapter of *Mir.*, see also Greene (in this volume) 196–198.

and stopped farming which is interpreted positively in two regards: it freed them from hard work (κακοπάθεια), and the Carthaginians (and later the Romans) were never able to subdue them because they are living in inaccessible places. This account may appear as contradictory to *Mir.*, but it may as well be seen as complementary if we assume that the Iolaeans fled to the mountains and the Carthaginians, in order to prevent them from returning, destroyed the plantings, and forbade agriculture. Yet it is remarkable that as in the case of the "Paradise Island" in the Atlantic Ocean in *Mir.* 84 (see below), our compiler speaks of a Carthaginian interdiction with death as punishment for violators, while Diodorus (5.20.4) in both cases lacks this element, although his general view on the Carthaginians is negative and there is no reason why he should have left this aside. This may either indicate a special interest of our compiler or a type of addition he introduced to his source, or it may point to the use of an intermediary source in which this element had been introduced.

4. Diodorus has some additional pieces of information, for example, on the fate of Iolaus and the Thespiads (15.5.6), which do not have anything extraordinary, so it need not surprise us that a paradoxographer did not select them if they were present in his source. Diodorus states, in addition, that the remaining population became barbarians, which is also implied in *Mir.* where the marvel consists in the fact that Greeks seem to have lived on the island in early times while it is now inhabited by barbarians.

Two more passages in Diodorus deal with the same facts described in 5.15 and in *Mir.* and need to be taken into account.[97] The first is 4.29–30 which is part of Diodorus' long account of the deeds of Heracles. In these chapters, the same events as in 5.15 are described, with the difference that some pieces of information are left aside and others are added or narrated with more detail. It is obvious from verbal and structural similarities that both texts are somehow related. One group of scholars argues that both passages summarize the same source, generally identified as Timaeus, in a different manner and with a different degree of elaboration.[98] Another group argues that 4.29–30 is a reworked version of Timaeus' account and that the latter is the direct source in book 5. In this reworking, especially Athenian claims would have been taken into account.[99] Both interpretations are possible. If we want to favor one of them, it basically comes down to how much intervention by Diodorus

---

97 Strab. 5.2.7.225 is not related to the traditions in Diodorus, although this is sometimes claimed, for example, by Bernardini (2004) 55. See the various interpretations listed by Anello (2014) 1–2 n. 4.

98 Thus (with Timaeus as the source): Pais (1881) 98–112; Geffcken (1892) 52–59; Meister (1967) 23–24, 34; Sanna (2004) 101; Didu (2005) 55–56; Mariotta in Mariotta and Magnelli (2012) 117 (he does not speak about 4.81–82); Jacquemin in Casevitz and Jacquemin (2015) 156–158 (as it seems); differently Müllenhoff (²1890) 455 (the source is Dionysius Scytobrachion); a doxography of older research in Mastino (2004) 16–17 and esp. in Anello (2014); literature also in Vanotti (2007) 180.

99 The representatives of this group discuss agreements and differences: Breglia Pulci Doria (1981, § 27–30, (2005), 68–73); Bernardini (2004) 46–55; Chiai (2004) 120–122; Galvagno (2004) 33–36; Anello (2014) 15–16. On Athenian elements, see esp. Coppola (1995) 77–85. The differences are sometimes overemphasized in these studies and too many Athenian elements are seen in 4.29–30. It is possible that Timaeus took the Athenian claims into account as he did in the case of Demeter, who brings grain to Sicily and then to Athens (see p. 49).

in rewriting his sources we are willing to assume. If we accept a high amount of it, we may explain the differences and contradictions on this basis. But all in all, the second alternative seems to be preferable. It is especially the function of Iolaus which is different in books 4 and 5. In book 4, it is Iolaus who cultivates the land, plants fruit trees first (29.6), and then constructs buildings, which are described in a similar way in book 5. In the same report, Diodorus adds that Iolaus had Daedalus come from Sicily and build the so-called *daidaleia* on his behalf.[100] This is in plain contradiction to what Diodorus narrates a few chapters later, in 82.4,[101] about Aristaeus who is said to have "set out plantings in it (= Sardinia) and brought it under cultivation, whereas formerly it had lain waste." This chapter is the second one of book 4 that needs to be accounted for in interpreting the tradition. Most scholars who see in Timaeus the source of 5.15 claim it for this historian, because it may be combined with 5.15 and *Mir.*, as can be seen in the following table.[102]

	4.29–30	4.82.4	5.15	Mir. *100*
Land uncultivated in the beginning	(Yes)	Yes		Yes
Introduction of cultivated plants	Iolaus	Aristaeus	–	Aristaeus
Introduction of polis elements	Iolaus (+ Daedalus)	–	Iolaus	Iolaus

It is clear that 4.29–30 contains a version in which Aristaeus has been replaced by Iolaus and that 4.82 and 5.15 complement each other. That they belong together may then be seen to be confirmed by *Mir.* if one assumes that it goes back to the same source as 5.15. It is indeed possible that the source of 5.15 originally presented Aristaeus as the one who cultivated that island, but I would not go so far as to consider 4.82.4 an excerpt from that source.[103] This piece of information is part of a longer chapter on the life or Aristaeus (4.81–82) and it is very unlikely that Diodorus' source for the Island Book narrated these things with such detail.

It remains to be determined whether the source used by *Mir.* and Diod. Sic. 5.15 was the same, and if it may have been Timaeus. Not much is known about Timaeus' treatment of Sardinia. *FGrHist* 566 F 64 explains the "Sardonian laughter" but does not help us understand to what extent he discussed Sardinian culture and early history. It has been claimed that Diodorus' statement that Sardinia is of similar size as Sicily agrees with F 65, which lists Sardinia and Sicily (in this order) among the

---

100 It is only these buildings that seem to be executed by Daedalus in this version of the story and not also all the others. See on this Neri (2002) 26–27.

101 Geffcken's (1892) 56 n. 1 attempt at harmonizing the traditions fails to convince.

102 4.81–82 as Timaeus: Flashar (1972) 118–119; Breglia Pulci Doria (1981) § 24–26; Bernardini (2004) 45 n. 16. Differently Galvagno (2004) 33–36: 4.29–30 and 81–82 are from Timaeus, 5.15 is from Ephorus.

103 See also Jacoby on Timaeus, *FGrHist* 566 F 63–64 who claims Diod. Sic. 5.15 for Timaeus but none of the passages of book 4 (with polemic against Geffcken). He is hesitant to regard *Mir.* as Timaeus as it combines the Iolaus and the Aristaeus stories. As has been shown above, this need not be seen as an argument against assuming a common source.

biggest islands (see pp. 73–75), and that this may point to Timeaus as Diodorus' source.[104] But this is a very weak argument. Any observer of both islands could state what we read in Diodorus. The only evidence for Sardinia in Timaeus is in F 63 (= Plin. *HN* 3.85) where we read:

> *Sardiniam ipsam Timaeus Sandaliotim appellavit ab effigie soleae, Myrsilus Ichnusam a similitudine vestigii.*
> Timaios has called Sardinia itself Sandaliotis from its similarity to the sole of a shoe, and Myrsilos calls it Ichnousa, from its resemblance to a footprint.
> (trans. Champion, *BNJ*)

If this correctly describes what Timaeus claimed, *Mir.* cannot represent Timaeus' description.[105] However, Müllenhoff has argued that Pliny confused the names of his authorities and that it was Timaeus who spoke of Ichnoussa, and he has been followed by many scholars.[106] His main argument is that it is unlikely that a denomination coined or propagated by an authority such as Timaeus should have left almost no trace in later tradition,[107] while the variant of the "obscure" Myrsilus became famous in antiquity. In addition, he regards Sandaliotis as a secondary formation fitting Myrsilus, an author later than Timaeus, and he finds Ichnoussa attested in texts that he regards as going back to Timaeus. A closer look at this argumentation shows that it is weak. The last argument can be disregarded because Timaean authorship can be substantiated in none of these cases with any certainty. In addition, we cannot date Myrsilus and he may have been older or a contemporary of Timaeus;[108] further, Müllenhoff's first argument is far from conclusive. If Sandaliotis was Timaeus' name by which he wanted to "correct" a perhaps well-established tradition, it need not surprise us that he was not successful. The only reason scholars seem to stick to Müllenhoff's theory is that they want to save *Mir.* 100 and Diod. Sic. 5.15 for Timaeus. But this is a *petitio principii*.[109] If we accept the transmission, Diod. Sic. 5.15 can still be Timaeus, but *Mir.* 100 cannot in its

---

104 For example, Geffcken (1892) 56; Breglia Pulci Doria (1981) § 23.

105 Champion on BNJ 566 F 63 states that "Timaios may have given both the names Sandaliotis and Ichnousa, perhaps also adding the name Sardo as a third alternative."

106 Müllenhoff (²1890) 455–456. This interpretation is accepted by Geffcken (1892) 56 n. 3; Jacoby on Timaeus, *FGrHist* 566 F 64–64 and on Myrsilus, *FGrHist* 477 F 11; Flashar (1972) 118; Sanna (2004) 105–106.

107 The only other attestation of Sandaliotis is in Hsch. σ 162: Σανδαλιῶτις (Salmasius, σανδαλώπη H)· ἡ Σαρδὼ πάλαι.

108 The *terminus ante quem* is Strabo who refers to him (*FGrHist* 477 F 16). See A. Belousov, Introduction to Myrsilus, *FGrHist* 1654. The date of A. Kandellis (Biographical Essay to Myrsilus, *BNJ* 477) is based on the assumption that the paradoxographer Antigonus who refers to Myrsilus four times (Ps.-Antigon. *Mir.* 5, 15c, 117 and 118 = Myrs., *FGrHist* 477 F 2, 5, 6 and 1b) is the Carystian biographer which is very unlikely.

109 The transmission is accepted as correct by Meloni (1942–1944) 63 (but with reference to Solinus who depends on Pliny); Brown (1958) 41; Nicosia (1981) 435; Bernardini (2004) 41; Vanotti (2007) 181; Breglia Pulci Doria (see n. 111).

totality.[110] However, the closeness between the texts still shows that they belong together. It is thus possible that the compiler used the same text as Diodorus and added this piece of information himself. This need not have been from a different source. Assuming his source author was Timaeus, Timaeus may have polemicized against the name Ichnoussa and have propagated his own variant Sandaliotis. It is then possible that the author of *Mir.*, however, followed this other tradition and created a new account by combining Timaeus' version with the one the latter criticized. Another possibility is that the paradoxographer summarized a text in which the tradition of Diodorus' source had been reworked. This is what Breglia Pulci Doria proposes and she sees in Myrsilus *Mir.*'s source, whom she assumes to have reworked Timaeus' account.[111] This is not impossible because a paradoxographical work *Historical Paradoxes* (Ἱστορικὰ παράδοξα) is attested for him[112] and he is quoted four times in ps.-Antigonus' paradoxographical compilation. However, the chapters in Antigonus all seem to stem from his *Lesbiaka* and the name Ichnoussa was obviously of widespread use so that it is more prudent not to put a name tag on *Mir.* 100 at the moment. But it is possible to see in Timaeus the source of Diodorus and of *Mir.*, if we assume that the latter has modified Timaeus' text.[113] Alternatively, Timaeus may be the source of Diodorus and *Mir.* may rely on a source in which the Timaeus tradition had already been modified. If we do not regard Timaeus as Diodorus' source, it is not even necessary to assume that the author of *Mir.* or an intermediary source made a change here because this unknown author may have spoken of Ichnoussa.

It goes too far to see in the mention of Heracles' "rule of the West" an argument for the use of the same source in *Mir.* 97 and 98. The topic was too widely treated in Greek literature as to allow such a conclusion.[114]

3.2.4 Baleares (5.17 and Mir. 88)

Ἐν ταῖς Γυμνησίαις ταῖς κειμέναις νήσοις κατὰ τὴν Ἰβηρίαν, ἃς μετὰ τὰς λεγομένας ἑπτὰ μεγίστας λέγουσιν εἶναι, φασὶν ἔλαιον μὴ γίνεσθαι ἐξ ἐλαιῶν, ἐκ δὲ τῆς τερμίνθου κομιδῇ πολὺ καὶ εἰς πάντα ἁρμόττον. Λέγουσι δὲ οὕτω τοὺς οἰκοῦντας αὐτὰς Ἴβηρας καταγύνους εἶναι ὥστε ἀντὶ ἑνὸς σώματος θηλυκοῦ διδόναι τοῖς ἐμπόροις τέτταρα καὶ πέντε σώματα ἄρρενα. Στρατευόμενοι δὲ παρὰ Καρχηδονίοις τοὺς μισθοὺς ὅταν λάβωσιν, ἄλλο μὲν οὐδέν, ὡς ἔοικεν, ἀγοράζουσι, γυναῖκας δέ· οὐ γὰρ χρυσίον οὐδὲ

110 The argument of Chiai (2004) 121–122 against Timaeus as the source of Diod. Sic. 5.15 is not compelling. Meloni (1942–1944) 61 doubts that the source of 5.15 can be identified.

111 Breglia Pulci Doria (1981) § 13 and n. 26, § 22–23 with reference to La Penna (1968) 305 n. 183. In (2005) 70 she is more cautious: "*Mir. ausc.* 100, in cui Timeo è certamente presente, forse accanto a qualche altra fonte." Bernardini (2004) 55–56 sees in Timaeus the source of *Mir.* Nicosia (1981) 435–436 argues for a Sicilian author, writing between the 5th and the mid-3rd century who was not Timaeus. Against a confusion in Pliny, also Didu (2003) 13–14 with n. 19 (literature).

112 *FGrHist* 477 F 4 = *FGrHist* 1654 F 1.

113 Müllenhoff (²1890) 455 doubts this and regards 5.15 as a shorter version of 4.29–30 where he thinks Diodorus relies on Dionysius of Mytilene; cf. also Jacoby on Timaeus, *FGrHist* 566 F 164 n. 597.

114 *Contra* Flashar (1972) 118 (source is Timaeus).

ἀργύριον ἔξεστι παρ' αὐτοῖς οὐδένα ἔχειν. Ἐπιλέγεται δέ τι τοιοῦτον ἐπὶ τῷ κωλύειν χρήματα εἰσάγειν αὐτούς, ὅτι τὴν στρατείαν Ἡρακλῆς ἐποιήσατο ἐπὶ τὴν Ἰβηρίαν διὰ τοὺς τῶν ἐνοικούντων πλούτους. (*Mir.* 88)

In the Gymnasian islands, which lie off Iberia, which they say are the greatest after the so-called Seven, oil is said to come not from olives, but from the terebinth, in great quantity and suitable for everything. They also say that the Iberians who live there are so much given to women, that they give the merchants four or five male persons in exchange for one female. On service with the Carthaginians, when they receive their pay, they apparently buy nothing but women. None of them is allowed to possess any gold or silver. It is added that this is done with a view to preventing them from bringing in gold, because Heracles made an expedition against Iberia because of the wealth of the inhabitants.

Diodorus' description of the Gymnesian Islands (5.17), that is, the Baleares, is more detailed, but all the narrative elements of *Mir.* appear there as well and, for the most part, in the same order. Thus, although clear linguistic agreement is lacking, we can be sure that both texts are related and in this case it seems clear that *Mir.* summarizes and reworks Diodorus' source or that it uses a source in which this had been done.[115] As we have seen above (p. 36), Diodorus claims that the larger of the two Baleares, Mallorca, is the biggest island after the Big Seven, while *Mir.* confusingly states this for both together. *Mir.* then skips the discussion of the alternative name Baliarides, the note on Menorca, and the claim that wine does not grow on both islands, so the inhabitants are crazy about it. Then also Diodorus claims that there is no olive oil on the islands and that they use the liquid of the mastix (σχῖνος; *Pistacia Lentiscus*), mixed with the fat of pigs, to anoint their bodies. *Mir.* names a different plant, the terebinth (τέρμινθος; *Pistacia Terebinthus*), but this is a minor difference because both plants belong to the same family and their resin is chemically identical.[116] Both grow in the Mediterranean and we cannot decide which text is closer to the common source. It is more important that *Mir.* does not mention the pig's fat and claims that the resin can be used for all purposes for which oil is used, that is, also for cooking, which is an obvious exaggeration of what Diodorus has. When Diodorus calls the inhabitants "of all men the most fond of women," this is illustrated by the fact that they pay three or four men as a ransom to pirates who have kidnapped a woman, which is enhanced in *Mir.* in two ways: they give four or five men for a woman, and they do so to buy women from merchants, which makes this a much more frequent and almost everyday practice.

After a few remarks about the abodes of the inhabitants, not present in *Mir.*, Diodorus mentions the prohibition to import silver and gold and explains it with Heracles' campaign against the rich Iberian king Geryones. Here again we note a slight exaggeration in *Mir.*: Heracles was attracted by the wealth of the *Iberians*, not only by that of their king. When Diodorus then speaks of the use the Balearic

---

115   The comparison by Müllenhoff (²1890) 464–465 remains fundamental; however, he does not discuss most of the differences.

116   https://de.wikipedia.org/wiki/Mastix. Müllenhoff (²1890) 464 prefers Diodorus' tradition.

mercenaries made of the payment they received in the past from the Carthaginians, this is presented by *Mir.* as something in the present. The present tense in *Mir.* has always been used to date *Mir.*'s source of this chapter.[117] It indeed points to a date before the end of the 3rd century BC. The Balearic slingers were famous in antiquity and they are attested in the Carthaginian army from 406 (or perhaps 409) to the battle of Zama (202) in the Second Punic War.[118] With Roman hegemony in this region after the end of this war (201), such recruitment seems to have come to an end and the islands were finally conquered by the Romans in 123/2.[119] Here, obviously, *Mir.* is closer than Diodorus to the common source. But here as well we detect a manipulation because in Diodorus the mercenaries buy women *and* wine, while in *Mir.*, where their love of wine is left aside, it is only women.

Thus, I think that one should not speak of a "total agreement in all details" between both texts, as is usually done.[120] It is clear that the author of *Mir.* or his source has not only made a selection. He has reduced the two "addictions" of the inhabitants to one and has, as a consequence, adapted the text to fit this peculiarity. In addition, he has enhanced the claims of his source throughout and, for this goal, reworked the motives present there.

It is remarkable that he has not included what we read in the following chapter of Diodorus (5.18), the description of the marriage customs, labeled as "marvellous" (παράδοξον), of the funeral habits, called "peculiar" (ἴδιον), and of the Balearic slingers and the corresponding training of their children, which both were famous in antiquity.[121] Yet the prominence of this latter aspect may, indeed, have been the reason not to include it because the author of this version may have wanted to present to his readers something new.[122] This would then also explain why he did not mention the second name of the islands, Baliarides, while we find a discussion of it in Diodorus (5.17.1). According to him, it stems from βάλλειν, "to throw" (stones with a sling).

As we have seen above, the idea of Mallorca as the biggest island after the Big Seven (including the list) constitutes an argument for Timaean authorship of Diod. Sic. 5.17–18. The Timaeus fragment in which this idea is found is preserved by Strabo as an insertion into a Rhodian context, but in 5.5.1.167–168 Strabo dedicates a chapter to the Baleares. Much of what he reports dates from Roman times and he describes in detail the armament of the Balearians, among which are the slings (and also other equipment). He has the men carry the three slings worn around

---

117 See Flashar (1972) 111 with reference to Müllenhoff (²1890) 465; Vanotti (2007) 173.

118 See Schulten (²1974) 251 with n. 110; Tovar (1989) 260–264; Ameling (1993) 220–221; Zucca (1998) 60–62, 69–91; Marasco (2004) 171–173; Panzram (2013) 9.

119 See Morgan (1969) 226–232; Zucca (1998) 91–96; Panzram (2013) 11.

120 Flashar (1972) 111 "hier (scil. in Diodorus) finden sich die genauen Entsprechungen zu *Mir.* 88 in allen Einzelheiten, nur ist Diodor (= Timaios) ausführlicher;" cf. Vanotti (2007) 172: "Quest'ultimo, senz'altro attingendo a Timeo, riferisce in termini più precisi e più ampi le stesse notizie presenti nel testo ps.aristotelico." Also, Jacquemin in Casevitz and Jacquemin (2015) 162–163 points to the similarities but does not note the differences.

121 For the sources, see Lacroix (1969) 394 with nn. 2, 3, and 7; cf. Curchin (2007); cf. Marasco (2004) 169–174.

122 Cf. Greene (in this volume) 179 with n. 32 with reference to a possibly similar procedure *in Paradoxographus Florentinus*.

their heads, while in Diodorus they have one around the head, another around the stomach, and the other in the hand, and he adds other details not found in Diodorus. However, he narrates the same anecdote as Diodorus about the Balearic children who only get bread to eat after they have hit it with a stone thrown by a sling. The situation is thus similar to what we have encountered above in the case of Elba-Populonia: there are some similar elements in Strabo and Diodorus, but also differences and later information. The interpretation usually found in the literature is the same here as there and may in the present case indeed hit the mark: Strabo may have visited the Baleares (as he visited Populonia), but he used Posidonius as a source who, on his part, used Timaeus.[123]

There is, however, another fragment of Timaeus on the Baleares we have to take into account, F 66, preserved by the scholia on Lycophron's *Alexandra* (633b Leone). In vv. 633–647 of this poem, Cassandra prophesizes that some Boeotians returning from Troy will be cast to the shores of the Gymnasian Islands and live there "without cloak and without shoes." Lycophron then mentions their three slings and adds the anecdote about the training of the children. The scholium *ad loc.* assembles references to the island from different authors, one of which is Timaeus:

Τίμαιος δέ φησιν εἰς ταύτας τὰς νήσους ἐλθεῖν τινας τῶν Βοιωτῶν, ἄστινας νήσους Χοιράδας εἶπεν.

And Timaios says that some of the Boiotians came to these islands, which islands he (scil. Lycophron) calls Choiradai. (trans. Champion, *BNJ*)[124]

This claim may be reconciled with Diodorus' description if we assume that his barbarians were originally these Boeotians who became barbarians after their shipwreck. The fact that they still anoint their bodies might be a relic of such a provenance.[125] The scheme of explanation would then be the same or similar to that of the Greeks in Sardinia.

Because of its similarities to Diodorus' account, this passage of Lycophron has often been seen as stemming from Timaeus as well, and in general parallels between *Mir.* and Lycophron have been regarded as a sign of common use of Timaeus. The latter is dangerous, however, and one should not apply such a general rule,[126] because there is no certainty about Lycophron's sources in most of his passages on the West (see the Excursus below). In the passage under discussion,

---

123 See, for example, Müllenhoff (^{2}1890) 452, 467; Morr (1926) 118–123; Jacoby on Timaeus, *FGrHist* 566 F 65–66 n. 339 (*FGrHist* III b Noten, 333); Morgan (1969) 219–221; Marasco (2004) 164, 170; Timaeus and Posidonius used independently by Strabo (which is unlikely): Jacquemin in Casevitz and Jacquemin (2015) 166.

124 That εἶπεν refers to Lycophron, not to Timaeus, was seen by Jacoby (*in textu*); Curchin (2010) 156 n. 9; somewhat doubtfully Hornblower (2015) 270. Differently, for example, Schulten (^{2}1974) 252 (Timaeus as the subject).

125 See Marasco (2004) 167–168. The same may be the case with their love of wine; cf. Jacquemin in Casevitz and Jacquemin (2015) 163 on the Greeks' love of wine.

126 The main exponent of this kind of interpretation is Geffcken (1892) 3 for the "rule" and *passim* for its application.

nothing excludes Timaeus as his source.[127] The fact that the Greeks on the islands are Boeotians only in Timaeus and Lycophron, while they are Rhodians in most other sources,[128] may be seen as an additional argument for such a dependence. As for Diodorus, everything points to Timaeus as the source, and this is what is generally assumed.[129]

Excursus: Timaeus and Lycophron, Alexandra

The question of Lycophron's sources on the West and those used on that topic in the scholia on his poem is very complex and can only be touched upon here. It would need a discussion of single passages similar to those in Diodorus, but their number would be much higher. Wilamowitz claimed that Lycophron completely relies on Timaeus for information on the West; his student Geffcken followed him in this regard, and even Hornblower in his recent commentary accepts Timaeus as the source, or possible source, behind many passages, the one under discussion here as well.[130] Indeed, there are some parallels between Lycophron and named fragments of Timaeus, but most of the parallels adduced in literature to argue for Timaeus as the source of a passage in the *Alexandra* stem from passages in which Timaean authorship is based on *Quellenforschung*. An alternative authority for at least part of the Western material is Lycus of Rhegium, author of a Sicilian history, who is cited together with Timaeus as a parallel for Lycophrons' account of Diomedes in the West in *Schol.* 615a Leone (see p. 40).[131] The role the latter may have played depends mainly on the identity of the author of the *Alexandra*. According to an entry of the *Suda*, he is identical with the tragedian of Euboean Chalcis who was the adoptive son of Lycus of Rhegium[132]; according to another entry and to the

---

127 Jacoby on Timaeus, *FGrHist* 566 F 65–66 p. 568.15–19 with n. 337 points to a difference between Diodorus and Lycophron, but this is not decisive. Geffcken's (1892) 4 explanation does not convince; see Jacoby's n. 337. Morr (1926) 118–123 argues that Diodorus has combined Timaeus with Posidonius; but *contra* see Morgan (1969) 219–221. Lycophron shows that it is not necessary to postulate a second source. For Timaeus as Lycophron's source, see, for example, Günther (1889) 34; Müllenhoff (²1890) 466; Geffcken (1892) 4; von Holzinger (1895) 266; Lacroix (1969) 396–397; Marasco (2004) 165–166 with n. 23 (literature); Curchin (2007) 84–85; Hornblower (2015) 22 (cautiously), 269 ("an obvious inference").

128 Apollod. *Bibl.* epit. 6.15b; Strab. (*loc. cit.*); Sil. *Pun.* 3.364–365; cf. Curchin (2007) 84 with n. 7; Hornblower (2015) 269–270. The Rhodian descendance has probably been constructed because the Rhodians were famous slingers too; see Lacroix (1969) 395–396; Hornblower (2015) 269–270. The Greeks are called "comrades of Odysseus" in the anonymous *P. Oxy.* 2688.26–32. Curchin's (2007) 85–88 idea that there was a link between the colonization of Sardinia by the Boeotian Aristaeus and that of the Baleares is not convincing.

129 Starting with Müllenhoff (²1890) 461–467. See, for example, Brown (1958) 38 (with some more elements suggesting Timaean authorship); Meister (1967) 34–35; Flashar (1972) 111; Marasco (2004); Jacquemin in Casevitz and Jacquemin (2015) 162; and the many scholars who regard the whole first part of book 5 as Timaean (a list in Marasco p. 163 n. 2).

130 Von Wilamowitz-Moellendorff (1883) 11; Geffcken (1892) 1–51; Hornblower (2015) 21–24 with the caveat at the end (24) of his discussion: "Fraser . . . observed that 'the Timaean material in the *Alexandra* may have been exaggerated at the expense of other and older sources'. This is right; and we can add that even where we think it likely that Lyk(ophron) used Timaios direct, we should bear in mind that Timaios himself had his sources."

131 Timaeus, *FGrHist* 566 F 53; Lycus, *FGrHist* 570 F 3.

132 *Suda* λ 872, s.v. Λυκόφρων = Lycus, *FGrHist* 570 T 2b.

anonymous *Life of Lycophron*, he was even his natural son.[133] The identity is still a matter of discussion. Hornblower rejects the biographical tradition on the tragedian and, in addition, considers the *Alexandra* a pseudonymous work written around 190 BC.[134] But this is far from certain and other scholars accept the biographical tradition and argue for a strong influence of Lycus' *Sicilian History* on the *Alexandra*.[135] Another problem regards the priority of Lycus and Timaeus. Jacoby, with due caution, saw in Lycus a predecessor and source of Timaeus as do others,[136] but hard proof is lacking. Be that as it may, the double reference to Timaeus and Lycus in the above-mentioned scholium suggests that both authors of Sicilian histories were somehow related and one was used by the other. This means that regardless of which of them is quoted for a piece of information, we cannot rule out the possibility that it was *also* present in the other in the same or a similar form. In addition, as rightly noted by Jacoby, Lycophron will have used more than one historical source in his poem.[137] The consequence of all this is that at the current state of research, parallels in Lycophron and authorless material in the scholia should not be used as an argument for Timaean authorship of a report.

3.2.5 The 'paradise island' in the Atlantic Ocean (5.19–20 and Mir. 84)

Ἐν τῇ θαλάσσῃ τῇ ἔξω Ἡρακλείων στηλῶν φασὶν ὑπὸ Καρχηδονίων νῆσον εὑρεθῆναι ἐρήμην, ἔχουσαν ὕλην τε παντοδαπὴν καὶ ποταμοὺς πλωτούς, καὶ τοῖς λοιποῖς καρποῖς θαυμαστήν, ἀπέχουσαν δὲ πλειόνων ἡμερῶν πλοῦν· ἐν ᾗ ἐπιμισγομένων τῶν Καρχηδονίων πολλάκις διὰ τὴν εὐδαιμονίαν, ἐνίων γε μὴν καὶ οἰκούντων, τοὺς προεστῶτας τῶν Καρχηδονίων ἀπείπασθαι θανάτῳ ζημιοῦν τοὺς εἰς αὐτὴν πλευσομένους, καὶ τοὺς ἐνοικοῦντας πάντας ἀφανίσαι, ἵνα μὴ διαγγέλλωσι, μηδὲ πλῆθος συστραφὲν ἐπ' αὐτῶν ἐπὶ τὴν νῆσον κυρίας τύχῃ καὶ τὴν τῶν Καρχηδονίων εὐδαιμονίαν ἀφέληται. (*Mir.* 84)

In the sea outside the Pillars of Heracles they say that a desert island was found by the Carthaginians, having woods of all kinds and navigable rivers, remarkable for all the other kinds of fruits, and a few days' voyage away; as the Carthaginians frequented it often owing to its prosperity, and some even lived there, the leaders of the Carthaginians announced that they would punish with death any who would sail there, and that they killed all the inhabitants, that they might not tell the story, and that a crowd

133 *Suda* λ 814, s.v. Λύκος; *Vit. Lycoph.* p. 4.25–26 Scheer = Lycus, *FGrHist* 570 T 2a.
134 Hornblower (2015) 36–41, 47–49. He is the last in a long list to plead for such a late date, see, for example, with a good sketch of the history of scholarship, Gigante Lanzara (2000) 5–21 (author a Lycophron iunior active at the beginning of the 2nd century BC).
135 Influence of the father on the (adoptive) son: Amiotti (1982); Ottone (2002) 430–431; cf. already Manni (1961/1990) who, in addition, challenges the use of Timaeus by Lycophron whom he regards as the older author. The biographical tradition is discussed and accepted by Hurst in Fusillo *et al.* (1991) 17–27; Meyer (2014) 90–99 with n. 264 (literature); Smith, Biographical Essay on *BNJ* 570; Rozokoki (2019).
136 Jacoby, *FGrHist* III b Text, Introduction to no. 570, p. 598–599; cf. p. 528 in the introduction to Timaeus; but see Ottone (2002) 414–415 on the uncertainty of such an assumption.
137 Jacoby, *FGrHist* III b Text 528–529, 562–563 (on F 53–56); cf. Manni (1961/1990) 539–541; Vattuone (1991) 271; Schade (1999) 96–98; Smith, Biographical Essay on Lyco, *BNJ* 570.

might not resort to the island against them, and get possession of it, and take away the prosperity of the Carthaginians.

Diodorus dedicates two and a half pages to the same island (5.19–20). Despite its length the description does not contain further elements that could make its identity certain, but there is now a strong tendency to identify it with Madeira.[138] Diodorus starts with a detailed description of the land, including fauna, flora, and climate, but also elements of human settlement in the form of villas and gardens, using the present tense. He then narrates about the commercial activities of the Phoenicians and how some of their explorers sailing down the African west coast were cast away by a storm and discovered it. Realizing its beauty, they told everybody about it. Then he continues (5.20.4):

Διὸ καὶ Τυρρηνῶν θαλαττοκρατούντων καὶ πέμπειν εἰς αὐτὴν ἀποικίαν ἐπιβαλλομένων, διεκώλυσαν αὐτοὺς Καρχηδόνιοι, ἅμα μὲν εὐλαβούμενοι μὴ διὰ τὴν ἀρετὴν τῆς νήσου πολλοὶ τῶν ἐκ τῆς Καρχηδόνος εἰς ἐκείνην μεταστῶσιν, ἅμα δὲ πρὸς τὰ παράλογα τῆς τύχης κατασκευαζόμενοι καταφυγήν, εἴ τι περὶ τὴν Καρχηδόνα ὁλοσχερὲς πταῖσμα συμβαίνοι· δυνήσεσθαι γὰρ αὐτοὺς θαλαττοκρατοῦντας ἀπᾶραι πανοικίους εἰς ἀγνοουμένην ὑπὸ τῶν ὑπερεχόντων νῆσον.

Consequently the Tyrrhenians, at the time when they were masters of the sea, purposed to dispatch a colony to it; but the Carthaginians prevented their doing so, partly out of concern lest many inhabitants of Carthage should remove there because of the excellence of the island, and partly in order to have ready in it a place in which to seek refuge against an incalculable turn of fortune, in case some total disaster should overtake Carthage. For it was their thought that, since they were masters of the sea, they would thus be able to move, households and all, to an island which was unknown to their conquerors.

Mir. consists of the same two parts as Diodorus' account, a descriptive and an historical one.[139] It is a very compressed and modified version of the same text that is summarized more extensively by Diodorus. Such a common source is proven

---

138 Hennig (²1944) 43–44; Huß (1985) 31, 66, 69–70; Amiotti (1988); Manfredi (²1996) 65–70; Roller (2006) 46; Vanotti (2007) 170. Janni (2016) 31–32, however, refrains from identifying the island. For other identifications proposed by scholars, see Jacquemin in Casevitz and Jacquemin (2015) 167.

139 There exist various comparisons of the two accounts with, for the most part, different explanations of the differences from those proposed above. Usually, scholars assume that both authors have excerpted the same source differently and that the author of *Mir.* has made a few mistakes by sloppily excerpting it. See, for example, Müllenhoff (²1890) 467–468; Geffcken (1892) 66–67 n. 4; Flashar (1972) 109–110; some differences are discussed by Rebuffat (1976) 887–891; Amiotti (1988) 170–172; Manfredi (²1996) 79–83. The best treatment is still that by Rohde (³1914) 231–232 n. 4 who has seen most of the differences and pointed to many problems of interpretation. He explains the differences by intermediary sources used by Diodorus and the author of *Mir.*, both of which had used Timaeus (or the use of such an intermediary source by only one of them; this

by some significant linguistic congruences in the first part.[140] But while Diodorus (and obviously his source) in this part emphasizes the paradise-like character of the island,[141] this aspect has been tuned down in *Mir.* where it only appears as a beautiful and wealthy island. The point its author makes is that it was uninhabited when it was discovered and that, despite of its quality, it was depopulated again by the Carthaginians out of fear. Therefore, all the elements of human presence have been deleted in *Mir.* from the description of the island in the first part of the entry. As the paragraph on the Phoenicians is also omitted, the author of *Mir.* or his source has to make the Carthaginians discover the island. In addition, the indication of its distance to the African coast has now become the distance to the Pillars of Heracles.

In Diodorus, the story about the Phoenician discovery and the Carthaginian measures to keep away settlers from the island is not very clear, probably due to Diodorus' summarizing his source.[142] Although the Etruscans are said to rule the sea, the Carthaginians are able to prevent them from founding a colony on the island. The participles εὐλαβούμενοι and κατασκευαζόμενοι are subordinated to διεκώλυσαν, which seems to mean that the Carthaginians feared that many of their own citizens would emigrate to an Etruscan colony. This is very unlikely.[143] In addition, the idea of having a safe refuge on the island presumes that the Carthaginians control the sea and this is what the text explicitly states, although a few lines earlier the same had been said about the Etruscans. It seems as if Diodorus has combined two events into one: the Carthaginians prevent the Etruscans from founding a colony when the latter were ruling the sea. Whether this was by military or diplomatic means, we do not learn.[144] Later, when the Carthaginians were the dominant naval power, they forbade their citizens to move to the island because they feared mass emigration and wanted the island to be forgotten so that they could use it in case of defeat. The two intentions of the Carthaginians are linked by ἅμα μέν and ἅμα δέ, but it is unclear whether both are related. The fact that they did not want "many" (πολλοί) Carthaginians to emigrate to the island may indicate that they were fearing that this would diminish the manpower of the mother city and weaken it. But since it was also their intention that the island be forgotten, *any* emigration needed to be stopped. Thus, what we read in *Mir.*, that they did not want Carthaginians to move there and spread knowledge of the island's existence or location, seems to be implied here too and to go back to the common source.

---

does not become clear). I have not indicated all aspects in the following where he has preceded my interpretation.

140 *Mir.*: ἀπέχουσαν δὲ πλειόνων ἡμερῶν πλοῦν ~ Diod. Sic. 5.19.1: ἀπέχει πλοῦν ἀπὸ τῆς Λιβύης ἡμερῶν πλειόνων; *Mir.*: ποταμὸς πλωτούς ~ Diod. Sic. 5.19.2: ποταμοῖς πλωτοῖς; *Mir.*: ὕλην τε παντοδαπήν ~ Diod. Sic. 5.19.3: δένδρα παντοδαπὰ καρποφόρα. Cf. Müllenhoff (²1890) 467–468.

141 On these aspects, see esp. Manfredi (²1996) 60–61.

142 Geffcken (1892) 67 n. 0 thinks of textual corruption and proposes a different explanation.

143 Cf. Manfredi (²1996) 64. He stresses that also the idea of the Etruscans founding a colony on this island is probably unhistorical (80–81).

144 Also Huß (1985) 66 is very cautious when he uses this passage for his reconstruction of the Etruscan–Carthaginian relations. Hypotheses on the historical situation of this conflict in Charles-Picard (1958) 173; Rebuffat (1976) 892–897; Antonelli (1990) 176; Manfredi (²1996) 62–63; Vanotti (2007) 170.

Whether the severe law in *Mir.* was already present in the original source is difficult to decide. In the case of Sardinia, a similar law was suspect, because it is not mentioned by Diodorus, who was not a friend of the Carthaginians, so one might doubt here as well that it was mentioned by his source. How the Carthaginians hoped the erasure of knowledge of the island was to happen after the Phoenicians "had told everybody" about the island is difficult to guess. Possibly, Diodorus' source (and Diodorus himself?) wanted to present them as naive in this regard. That the Carthaginians were afraid of mass emigration, as *Mir.* claims, because they feared that the multitude moving there would become independent and turn against Carthage[145] does not sound very realistic.[146] I think that what we have here is the same phenomenon we have encountered twice before in *Mir.* 84 and also earlier in this study: the author of *Mir.* or his source has used a narrative element from a different context in his source and referred it to the *thauma* he has selected. My suggestion is that it comes from the story about the Etruscan attempt to found a colony there. This may have been regarded as a serious threat to Carthaginian economic interests which was the reason why they did not let it happen. It is probably also an addition of *Mir.* or its source that the Carthaginians killed their own people on the island. It is contradicted by Diodorus' description of the island as inhabited in 5.19.[147] In addition, it is improbable that Diodorus would have left out such a measure. In 5.11, he narrated in detail the murder of the 6,000 mercenaries on the island of Ustica ordered by the Carthaginian government. So I think that this may be, together with the severe law, an invention of our author or of his source.

If the above interpretation is correct, the author of *Mir.* or an intermediary source has extremely condensed what Diodorus' source had narrated, making the emptiness of the island the central aspect of his report. For this reason, he has deleted all elements of inhabitation from the description and has exaggerated the Carthaginians' measures to make the existence or the location of the island forgotten by adding that they even killed their own settlers on the island. In his usual way of accumulating information from other contexts and relating it to his topic, he has made the Carthaginians the discoverers of the island, has arbitrarily changed the point of reference of the distance to the island, and has moved a narrative element here from a different context.

There is no specific evidence who was *Mir.*'s/Diodorus' source here. It can perfectly have been Timaeus, and the anti-Carthaginian bias would fit him. But that tendency is, of course, not limited to this historian.[148]

---

145 This is what the text seems to mean but ἐπ' αὐτῶν ἐπὶ τὴν νῆσον is odd and the text perhaps corrupt. αὐτῶν can only mean the Carthaginians at home, not those on the island who were killed.

146 Cf. Rebuffat (1976) 890.

147 I do not see any reason to assume that Diodorus has "updated" the text of his source because Carthage no longer existed, that is, that he presented as the state of his time what was in his source the description of the island before the Carthaginian law. It was an island that he could not identify and he had no further knowledge of. Cf. Amiotti (1988) 172.

148 Manfredi (²1996) 61, 70–72, 74–77 with reference to Amiotti (1988) 171 and others assumes a Carthaginian source transmitted by Posidonius. The idea (in Diodorus) to use the island as a safe place in case of a defeat has been seen as an element that presupposes the Carthaginian defeat

3.2.6 Basileia and the provenance of amber (5.23 and Mir. *81)*

The last chapter to be discussed here is Diod. Sic. 5.23 on the island of Basileia (unidentifiable)[149] in the Northern Ocean, where amber is washed ashore and from where it is exported to the Mediterranean. As part of this exposition also, the mythical story connecting the genesis of amber to the Phaethon myth and localizing it in the Po area is reported and criticized. It is such a mythical explanation that is found in *Mir.* 81: it deals with the Electrides Islands located in the Adriatic vis-à-vis the mouth of the Po River. It reports the myth of Phaethon who, stricken by Zeus' thunderbolt, fell into a lake near this river. There are many poplars in (ἐν) it (i.e. at its edge) from which the amber emanates. What our author has left out before this statement, probably because it was general knowledge, is that these poplars are the sisters of Phaethon, who had been mourning his death and were thus transformed into these trees. In addition, *Mir.* reports details about this lake (see below) and information about Daedalus in this region. Müllenhoff denied that both texts stem from the same source because they argue for different provenances of the amber. He claims Diodorus for Timaeus and suggests Lycus as the source of *Mir.* (78-)81.[150] As for Diodorus, he is followed by Jacoby who calls *Mir.* 81 "a wild account" in which Theopompus has possibly been combined with other sources, for example, Lycus,[151] and Flashar agrees with him and states that the problem of the sources of *Mir.* 81 cannot be resolved.[152]

However, Enmann and Geffcken argue that both followed the same source which they excerpted in different ways and they identify it with Timaeus.[153] According to them, the author of *Mir.* has excerpted the version of "many poets and prose authors" whose explanation is criticized by Diodorus. I think that such a reading is problematic.[154] It is not a problem that there are no striking linguistic parallels. The structure is very similar and what Diodorus says at the end about the locals in

---

in the Second Punic War or even its destruction which would exclude Timaeus as the source; thus, for example, Amiotti (1988) 171 n. 13 (Timaeus via Posidonius); Manfredi (²1996) 64–65, 76–77. Timaeus transmitted by Posidonius also in Borca (2000) 58 (without arguments). We cannot exclude that Diodorus added this as the intention *ex eventu*, but such an assumption is not compelling. The idea of moving the population of a whole city or region to a different place, also to an island, in times of military threat is not unparalleled and Timaeus may have claimed such an intention for the Carthaginians; attestations of the motive in Jacquemin in Casevitz and Jacquemin (2015) 170.

149 Jacquemin in Casevitz and Jacquemin (2015) 180 (with literature).
150 Müllenhoff (²1890) 429–431, 473–491.
151 Jacoby on Timaeus, *FGrHist* 566 F 68 with n. 352 (III b Noten, p. 334): "ein wilder Bericht."
152 Flashar (1972) 108. Štrmelj (2015) 68–70 suspects Theopompus as the source.
153 Enmann (1880) 197–199; Geffcken (1892) 91. With different arguments also Grilli (1983) 7 n. 4.
154 Against Geffcken, see Jacoby on Timaeus, *FGrHist* 566 F 68: "Das problem (scil. of the provenance of amber) hatte eine lange vorgeschichte; . . ., wenn die quellenkritik die vorgeschichte in rechnung gestellt hätte, würde sie den wilden bericht Θαυμ. ἀκ. 81 über die Ἠλεκτρίδες νῆσοι, αἱ κεῖνται ἐν μυχῶι τοῦ Ἀδρίου nicht wieder als T(imaios) behandelt haben." Further criticism of Geffcken in III b Noten, p. 334 n. 352.

the North may be seen as echoing what *Mir.* claims about those of the Po region.[155] However, we should not make too much of the structural analogy here because it corresponds to the standard structure of the Phaethon myth. The problem with this interpretation lies in the differences between the two accounts of the Phaethon myth. The main difference is the place where Phaethon fell. The Po-Eridanus[156] that we find in Diodorus is the standard place that we also encounter in other sources,[157] whereas in *Mir.* it is a lake (λίμνη) which is described in the passage preceding the myth as warm, smelly, and causing birds flying over it to fall dead from the sky; further, the author even gives its measures as 200 stadia in circumference and 10 stadia in perimeter. This lake appears elsewhere in literature in the same context, for example, in Apollonius of Rhodes (4.595–611, 619–626).[158] His account is very similar to that of *Mir.* It describes the toxic character of the lake (λίμνη) and gives the same explanation of the genesis of amber.[159] It is very likely that Apollonius follows the same source as *Mir.*[160] This shows that we have in *Mir.* not one of the usual modifications made by the paradoxographer or his source to the text he used to enhance its marvelous aspects. He rather follows a different tradition from that criticized by Diodorus' source. In addition, there is another problem with Enmann's interpretation. *Mir.* combines the myth of Phaethon and the genesis of amber with information about Daedalus' stay in the Adriatic region. In Diodorus, the Phaethon story is part of the historian's description of the islands of the Northern Ocean, of Britain (5.21–22) and Basileia (5.23), which clearly forms a unity.[161] While it is not surprising that Diodorus' source discusses in the context of Basileia the traditional idea of amber coming from the Po region, it can hardly be imagined that it had also given, in this context, an account of Daedalus in the Po region. In

---

155 Diod. Sic. 5.23.5: τὸ γὰρ ἤλεκτρον συνάγεται μὲν ἐν τῇ προειρημένῃ νήσῳ, κομίζεται δ' ὑπὸ τῶν ἐγχωρίων πρὸς τὴν ἀντιπέρας ἤπειρον, δι' ἧς φέρεται πρὸς τοὺς καθ' ἡμᾶς τόπους, καθότι προείρηται. ~ *Mir.* 81.5: καὶ συλλεγόμενον ὑπὸ τῶν ἐγχωρίων διαφέρεσθαι εἰς τοὺς Ἕλληνας.

156 On the problem of the identification of the Eridanus in ancient sources and modern studies, see Bianchetti (1990) 77–108.

157 See, for example, Hyg. *Fab.* 154; *Schol. Hom. Od.* 17.208 = Aesch. *Heliad.* p. 185 Radt, *TrGF*; Ov. *Met.* 2.324 etc.; for more attestations, see Knaack (1902–1909); Türk (1938).

158 Other attestations: *Paradoxographus Florentinus*, *FGrHist* 1680 ch. 31 (probably from *Mir.*); *Schol. Lycoph. Alex.* 704 Scheer (not in Leone) (perhaps from *Mir.*); [Paul. Sil.] *Therm. Pyth.* 116–23 = *Anth. Pal.* append. Epigr. exhort. et suppl. 75 Cougny; cf. Öhler (1913) 106–107; S. Sørensen on *FGrHist* 1680 ch. 31.

159 On the meaning of λίμνη in vv. 599 and 608 ("lake" not "lagune" which would mean the mound of the Po), see Livrea (1973) 183 on v. 599; cf. Hunter (2015) 163 on vv. 599–603. Knaack (1902–1909) 2187 and Vanotti (2007) 168 detect in *Mir.* a reference to the Abano Fountain and quote Claud. *Carm. min.* 26.27ff. Braccesi (²1977) 39 sees in the λίμνη a "palude:" "Il particolare della palude in cui sarebbe precipitato Fetone chiaramente rispecchia una delle caratteristiche morfologiche salienti della foce del Po, ricca di 'valli' e di acquitrini; il fatto poi che la palude avesse acqua ribollente e maleodorante potrebbe indurci a pensare a una sorgente sulfurea." Alternatively, he considers a reference to the Abano Fountain a possible explanation.

160 See the comparison by Bianchetti (1990) 92–96 (with further literature in n. 56) who convincingly explains the slight differences by Apollonius' combining various poetic traditions.

161 See Müllenhoff (²1890) 469–474.

contrast, *Mir.* clearly stems from a description of this area, not from an excursus within the description of a Northern island.

As for Diodorus' source for the chapters on Britain and Basileia, there is evidence that it was Timaeus who relied himself on Pytheas of Massilia. However, the tradition we owe to Pliny the Elder is quite confused.[162] In one passage he relates that according to Pytheas, the amber is washed ashore on the island of Abalus in spring time and he adds that Timaeus agrees with him, but calls the island Basilia.[163] According to another passage in which the same is narrated with very similar words, one gets the impression that this island was nameless in Timaeus, and in the same context it is stated that Pytheas spoke of an island named Basilia.[164] The question is made more complicated by other names attested in fragments of other authors and various suggestions have been made to explain the mess. If we accept the tradition of the first passage, Timaeus' version agrees with Diodorus' account.[165]

As for Britain, another fragment preserved by Pliny seems to contradict at first sight the view that 5.21–22 stem from Timaeus. There we read:

> The historian Timaios says there is an island Ictis lying six days' voyage from Britain where tin is found, and to which the Britons cross in boats of osier covered with stitched hides.[166]
>
> (trans. Champion, *BNJ*)

In Diodorus, the tin is found in Britain and brought to the offshore island of Ictis on wagons during the low tide from where it is transported to the continent.[167] However, the six-day journey in Pliny is possibly a mistake and refers to the distance

---

162 Interpretations of the tradition with differences in the details but all assuming that Timaeus called the island Basileia in Müllenhoff (²1890) 469–491; Jacoby on Timaeus, *FGrHist* 566 F 75; Grilli (1983) 13–15; Bianchetti (1996) 73–78; (1998) 195–204; Magnani (2012) 211–223. For Timaean authorship, see also, for example, Meister (1967) 35.

163 Plin. *HN* 37.35–36 = Timaeus, *FGrHist* 566 F 75b = Pytheas F 11a Mette = F 15 Bianchetti = F 21 Scott: *Pytheas Guionibus, Germaniae genti, accoli aestuarium oceani Metuonidis nomine spatio stadiorum sex milium; ab hoc diei navigatione abesse insulam Abalum; illo per ver fluctibus advehi (scil. the amber) et esse concreti maris purgamentum; incolas pro ligno ad ignem uti eo proximisque Teutonis vendere.* (36) *Huic et Timaeus credidit, sed insulam Basiliam vocavit.*

164 Plin. *HN* 4.94–5 = Timaeus, *FGrHist* 566 F 75a = Pytheas F 11b Mette = F 16 Bianchetti = F 17 Scott: *Transgressisque Ripaeos montes litus oceani septentrionalis in laeva, donec perveniatur Gadis, legendum. Insulae complures sine nominibus eo situ traduntur, ex quibus ante Scythiam quae appellatur Baunonia (?) unam abesse diei cursu, in quam veris tempore fluctibus electrum eiciatur, Timaeus prodidit. . . .* (95) *. . . Xenophon Lampsacenus a litore Scytharum tridui navigatione insulam esse inmensae magnitudinis Balciam tradit, eandem Pytheas Basiliam nominat.*

165 See Müllenhoff (²1890) 481; Mette (1952) 39–40; Jacoby on Timaeus, *FGrHist* 566 F 75; Bianchetti (1996) 77; Magnani (2012) 211–223. Dudziński (2016) 58 is somewhat superficial.

166 Plin. *HN* 4.104 = Timaeus, *FGrHist* 566 F 74: *Timaeus historicus a Britannia introrsus sex dierum navigatione abesse dicit insulam Ictim (Ictim Mayhoff, Ictin iam Salmasius, mictim codd., mictin excerpt. Crickl.), in qua candidum plumbum proveniat; ad eam Britannos vitilibus navigiis corio circumsutis navigare.*

167 Diod. Sic. 5.22.2.

between Britain and Thule about which Pliny speaks a little later,[168] both ways of transport do not exclude each other and may reflect transport during low tide and high tide, and the provenance of the tin from Ictis itself may be not much more than a minor inaccuracy.[169] So if one is willing to go that far, it is possible to see in Timaeus the source of Diodorus.[170]

The source of *Mir.* cannot be established with any certainty. Müllenhoff's candidate Lycus cannot be substantiated, and it does not help us either to know that Apollonius of Rhodes used the same source. But it is important, as Bianchetti has noted, that the "Adriatic part" of Apollonius' fourth book shows very little influence by Timaeus.[171] This may be seen as additional confirmation that the common source used by Apollonius and *Mir.* is different from the tradition in Diodorus, which seems to come from Timaeus. That it was Theopompus combined with other sources, as Jacoby suggested,[172] is possible, because he was one of the authors who located the Electrides in the Adria and who possibly told the story about the amber coming from the tears that fall into the Eridanus.[173] But as this is the most common explanation in ancient literature, it is not enough to see in him *Mir.*'s source.

3.3 The source of Diod. Sic. 5.2–23

The discussion above has shown that there are several parallels to named fragments of Timaeus in Diod. Sic. 5.2–23 and that explanations for the contradicting information in other such fragments may be found. It is especially clear that much of the chapter on Sicily is based on Timaeus, who is most probably meant by "the most famous historians," and the close parallel in Cicero, who knew Timaeus' work well, is further evidence for this. The problem of the slightly different name in F 37 (Thrinakia) may be explained as a minimal textual corruption or a mistake by the scholiast in a confused scholion. In the chapter on the Gymnasian Islands, there is even a literal parallel with F 65 on the Big Seven, and F 66 on the Boeotians is compatible with Diodorus' account. The name Basileia is confirmed by F 77b, but according to F 75a, Timaeus considered the amber island as nameless. However, the latter may be explained as a mistake by Pliny who is very confused in his reports on the amber island. It is clear that Pliny's claim must be wrong in one

---

168 Plin. *HN* 4.104.
169 See Müllenhoff (²1890) 472; Mette (1952) 3–4, 41–42; Jacoby on Timaeus, *FGrHist* 566 F 74; Magnani (2012) 147–148 (with literature).
170 Dudziński (2016) 57–58 emphasizes the differences. In addition, he tends to see it as a problem that Diodorus does not mention the impact of the high and the low tide on rivers in Gaul which is mentioned in Timaeus F 73. However, such an argument *ex silentio* is generally weak and Diodorus is in this part of his work concerned with islands and not with the continent.
171 Bianchetti (1990) 91.
172 See n. 151; also Bianchetti (1990) 93 n. 59 thinks of him.
173 Whether the latter is the case depends on where F 130 stops. It is transmitted by pseudo-Scymnus. Jacoby prints vv. 369–90 as the Theopompus fragment but notes in the commentary that probably also vv. 391–414 go back to him (with reference to *Mir.* 81 to back up this interpretation). This is the passage on the Eridanus and the amber. Differently, Bianchetti (1990) 99–108. Cf. also Zaccaria (in this volume), 87.

place. Also the many differences in F 74 on Ictis can be explained, but this is the most difficult case and requires some goodwill from the interpreter.

In the chapters of Diodorus 5.2–23 that were not discussed above, there is no clear additional evidence for or against Timaeus. It has, however, been argued that the description of Kyrnos/Corsica in Timaeus F 3 contradicts Diod. Sic. 5.14.1, but this is not a compelling conclusion:[174] In F 3, which is part of a polemical attack by Polybius on Timaeus' reliability as a witness on Corsica, we read that according to Timaeus there are many wild goats, sheep, bulls, deer, and other animals on the island and that the Corsicans spend all their time hunting them. Polybius polemicizes against such a view and denies the existence of these wild species on the island. He explains that the herders let all their animals graze unattended and call them by the signal of a trumpet, which may explain the erroneous view that these animals are wild. Diodorus relates that the Corsicans feed on milk, honey, and meat and he emphasizes their justice. One example for this claim is that their sheep are distinguished by brands and left unattended without being stolen. It is perfectly possible that Timaeus reported that the Corsicans were passionate hunters *and* bred animals and that Polybius overemphasized the one activity where he thought he had caught Timaeus on a mistake. It is a well-known fact that he misrepresents the reports of his adversaries, especially of Timaeus, in his polemics to undermine their reliability, so we should not overestimate the value of his testimony here.

Neither is there a contradiction between Timaeus F 67 and Diod. Sic. 5.20.2, as has been claimed.[175] The latter mentions the Phoenician foundation of Gadeira and of its temple of Heracles on an *peninsula*. Timaeus F 67, preserved by Pliny,[176] states that Timaeus named two *islands* on the Iberian coast on which in the course of time parts of Gadeira were located: Aphrodisias and Potinusa (or Kotinusa). As the mention of Gadeira is found in a short excursus within Diodorus' account of the Paradise Island, we need not expect from him more detailed information on this area such as the names of the location. From Pliny, on the other hand, who is only interested in the names of what *he* regards as islands, it does not follow that Timaeus spoke of an island instead of an peninsula on which the oldest part of Gadeira was located. Thus, both texts are compatible with each other.

If we look at the whole passage of Diod. Sic. 5.2–23, it seems very coherent, the chapters on the single islands are interrelated by various references (e.g. the distance from Sicily) and by common topics. Besides the usual information that Diodorus adds about the situation in Roman times, there is no passage that stands out as an addition from a different source. Further, Jacoby points to the "pre-Posidonian indication" of the size of Sicily in 5.2.2 and the use of Olympiad dating in 5.9.2, which was introduced by Timaeus, as indications pointing to this source.[177] Meister adds more elements that he regards as typically Timaean but these are not very

---

174 The contradiction is emphasized by Dudziński (2016) 56; but see Geffcken (1892) 66; Müllenhoff (²1890) 454–455; cf. Jacoby on Timaeus, *FGrHist* 566 F 3 on the possibly unreliable character of Polybius' report.
175 Dudziński (2016) 57.
176 Plin. *HN* 4.120.
177 Jacoby on Timaeus, *FGrHist* 566 F 164 n. 594; cf. Meister (1967) 33.

significant and may be characteristic of most histories of the Greek West.[178] The praise of Ephorus and the critique of Timaeus' excessive use of polemics in Diodorus' proem to book 5 (5.1.3–4) is neither an argument for nor against Timaeus' use in the following chapters. Thus, there is some evidence for Timaeus as the source of this part of book 5, and the fact that he was Diodorus' main source for the later history of this region also points in this direction.[179] Jacoby is convinced that these chapters of Diodorus are based on the first books of Timaeus' *Sicilian History* in which the historian described "colonization activities, foundations and relationships,"[180] and he follows Geffcken in the idea that these books contained a "geography of the West." Against his usual principle not to include fragments on the basis of *Quellenforschung*, he does so in this case and prints the whole passage in small letters as F 164 of Timaeus.

But some doubts remain, not only because of the errors in three or four[181] named fragments of Timaeus we have to assume to make them fit our passage, although most of them are admittedly tiny. To be sure, in each individual case, the explanations brought forward by scholars sound plausible, but taken together, very critical interpreters may start doubting whether it is legitimate to assume all these mistakes.[182] Also problematic may be that Jacoby assumes that in Timaeus the islands of the West were probably described "in the manner of old geography," which means in the course of the description of the mainland, they were close to.[183] If this idea of the introductory books of the *Sicilian History* is correct, Diodorus did not summarize a long and continuous passage from Timaeus, but went through several of his books and picked out here and there the information on the islands he was interested in. Although such a procedure is possible, it does not fit very well what we think we know about the way he usually worked. So one may consider that Jacoby's idea about the first part of Timaeus' work is wrong and that it may have contained a section on islands similar to the one in Diodorus too. Or one may hypothesize that Diodorus rather summarizes a work *On Islands* in which Timaeus was used as a source, and probably even as an important one, for the islands of the West. Such an assumption might find confirmation in the chapters on the Eastern Mediterranean in the same book (5.47–81). No possible source has yet been found for that part, but it may have been a book *On Islands*, which must then, however, have

---

178 Meister (1967) 31–35: Sicily as geographical point of reference, detailed information on Sicily, "abstruse stories," the *topos* of the peaceful barbarians, *eudaimonia* and *tryphe* as topics. One may add the anti-Carthaginian bias. Jacoby (on F 164) mentions the discussion of Homer's verses and the praise of Sicily in this regard.

179 As for the later history, Dudziński (2016) now tries to significantly limit Timaeus' role. I am doubtful about this.

180 *FGrHist* 566 T 7 with Jacoby's introduction to Timaeus on p. 542.

181 This depends on whether or not we count the error in Pliny on Basileia (F 75a/b): in one passage his version agrees with Diodorus, in another it contradicts it. The other mistakes counted here are Thrinak(r)ia in F 37, Itkis in F 74 and Corsica in F 3.

182 It is now especially Dudziński (2016) 56–58 who concludes on the basis of the differences that Timaeus was not Diodorus' only source in this passage.

183 See Jacoby on F 164. Jacoby (p. 593) notes this, but obviously does not see it as an obstacle to Timaean authorship.

been different from that used in the first part of the book, as is shown by the different topics and interests in both parts.[184] Such an "island book" would explain well the strong coherence of the passage, in which the islands are in various ways related to each other by common topics and information about distances. Such an interconnectedness is more easily realized in a work focusing on islands than in a geographical description of the West in which islands were only one, and not even the central, topic.

Before we consider which explanation may be more probable, it is advisable to summarize the way in which the information in Diodorus differs from *Mir.* because this will be relevant for such an assessment.

4. *Mir.*'s and Diodorus' sources

Looking back at the chapters studied above, we can now assess quite well what happened to information from the source of Diodorus' first chapters of book 5 in the course of the transmission, until it was included in *Mir.*[185] Regardless of whether Timaeus or a work *On Islands* was Diodorus' source, it is especially evident in the chapters on the Baleares (*Mir.* 88) and the Paradise Island (*Mir.* 84) that this source was used directly or indirectly in *Mir.* We state that our author or an intermediary source has extremely condensed the content of this source, picking out one or a few marvelous aspects, and that he has modified other information there to make it fit this selection. He has adapted the historical and ethnographical information to his goal to make his reports singular and to increase their sensational aspects. Therefore, he has exaggerated numbers and frequency, substituted the name of one people with that of another, transferred a distance between an island and one place to another place, and he has probably transferred information to a different historical situation. He even seems to have invented a drastic law and to have introduced the killing of the Carthaginian settlers. We get the impression that historical and ethnographical data are for him narrative elements that can be adapted and used freely if this is necessary for his own literary agenda.

If we apply these insights to the other chapters where it has been left undecided above whether Diodorus and *Mir.* used the same source or are in some other way related to each other, we see that most of the differences may now be explained easily as the consequence of such a reworking by the same author. In the chapter on the abduction of Kore (*Mir.* 82), the focalization on the cave as the spatial and narrative center becomes understandable and explains some discrepancies from Diodorus-Cicero, and the difference in the grain motive is likely to be understood in such a way that the author kept the motive but made it more spectacular and

---

184   An argument against such a source might be that Cicero must then have known it as well. But this is not an insurmountable obstacle. Ceccarelli (1989) 931–935 considers a work *On Islands* Diodorus' source in 5.47.81.

185   Of the named fragments of Timaeus not related to Diodorus, F 57 ~ *Mir.* 102 allows some comparison, but the only thing we can state for sure is that *Mir.* has left out some information that would have been necessary to fully understand the thaumasion this chapter describes.

relocated it. The transfer of the story to Mt. Aetna may be the result of combining the story of his source with another tradition or of inventing a new version.

In the chapter on Sardinia (*Mir.* 100), Carthaginian brutality seems to have been enhanced and perhaps another cruel law has been invented.

In the chapter on Elba (*Mir.* 93), not enough text is available to decide whether *Mir.* follows the same source as Diodorus. If it does, the tradition has been modified in *Mir.* as well. In this case, the *thaumasion* has been reduced to one mine and, at the same time, has been made more spectacular by having a different metal grow. However, here we have to be cautious because a dependence cannot be proven.[186]

For the amber island (*Mir.* 81), the differences rather suggest that our author did not follow the same source as Diodorus. This is important because it shows that a common topic in Diodorus' book 5 and *Mir.* is no sufficient reason to assign a source to that chapter.[187]

Let us now, before this background, try to establish the transmission of the material from Diodorus' source to *Mir.* Things are complicated by the fact that Irene Pajón Leyra has made it very probable that in *Mir.* 78–121, a work *On Islands* has been used by our author.[188] She excludes 85–98, where a work on the wanderings of Heracles seems to have been the basis, and 106–110, where the connecting feature seems to be that they deal with the presence of the Trojan heroes in the West. She also, and rightly, notes, that in the passages based on the work *On Islands*, some chapters from other works and on topics different from islands have been inserted and that the Heracles section also contains some entries related to islands (88; 93). As Heracles' journey was overland, she rightly considers the possibility that these entries originally stemmed from the work *On Islands* and that our compiler inserted them in the Heracles section where they fitted geographically.

If we now combine her interpretation with the results of the study above, we see that the chapters that have parallels in Diodorus appear in the part of *Mir.* claimed by her for the work *On Islands* (82, 84, 100) but also in the Heracles section (88, perhaps also 93) and her assumed island book contains material attested for Timaeus (82, 102) but also information for which this source has been excluded (79, 81). In the Heracles section, the chapter on the Gymnesian Islands (88) contains material from Timaeus and shares a common source with Diodorus' chapter on the same islands, and the chapter on Aethalia (93) may or may not go back to the same source as Diodorus's description of it. These happen to be the only two chapters in this section of *Mir.* that do not fit Heracles' journey on the mainland and may originally stem from the book *On Islands*. The certain Timaean chapter on Cumae (102), dealing with a city on the mainland, does not force us to assume that the work *On Islands* was not the only source through which Timaean material

---

186 If *Mir.* 93 is a reworking of an earlier tradition, this would have an impact on the discussion above about the possible use of Timaeus (via Posidonius) in Strabo's chapter on Populonia. Then the differences between *Mir.* and Strabo can no longer be regarded as an argument to exclude such a transmission.

187 For example, Flashar (1972) 119 does with *Mir.* 101: the Liparian Islands were treated by Timaeus (*re vera* Diod. Sic. 5.10), thus the source of *Mir.* 101 is "wahrscheinlich" Timaeus.

188 Pajón Leyra (in this volume), 10–31.

entered *Mir.* and that this chapter was inserted here by the compiler from another work. It belongs together with the following chapter 103 on the Sirinusae islands whose location is even described with reference to Cumae. It is perfectly possible that in a work *On Islands* information was included, by the way of addition, about a location on the mainland close to the coast and the islands it treated. The idea of a work *On Islands*, in which Timaeus was one source besides others, as the main source of the first part of the historical section of *Mir.* is thus compatible with our interpretation, if we assume that the author of *Mir.* used this work not only in the "island section" (78–84, 99–105, 111–121) but also for additions in the "Heracles section" (85–99).

So the situation is the following: Diodorus is based on Timaeus or a work *On Islands* in which Timaeus was used as one source among others. *Mir.* uses a work *On Islands*, in which Timaeus figures among the sources as well, and it is clear that the accounts in *Mir.* are reworkings of the stories in Diodorus' source. The question is, therefore, whether the succession is

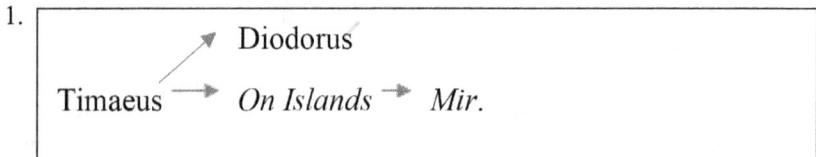

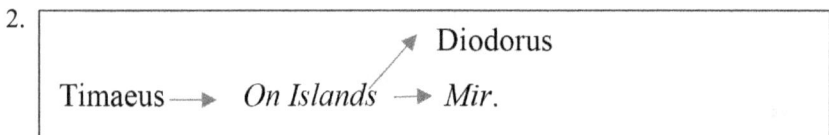

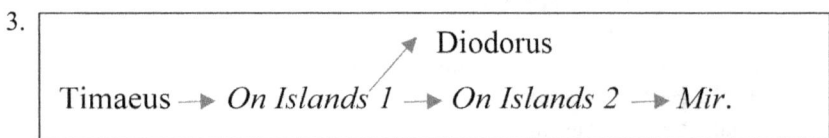

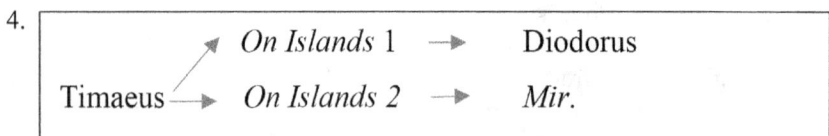

1.

```
                        ↗ Diodorus
    Timaeus ──→ On Islands ──→ Mir.
```

2.

```
                              ↗ Diodorus
    Timaeus ──→  On Islands ──→ Mir.
```

3.

```
                              ↗ Diodorus
    Timaeus ──→ On Islands 1 ──→ On Islands 2 ──→ Mir.
```

4.

```
              ↗ On Islands 1  ──→    Diodorus
    Timaeus ──→ On Islands 2  ──→    Mir.
```

To answer this question, we need to establish who is responsible for the reworking of the information in *Mir.* At first sight, one might think that it was our paradoxographer because the strong tendency to make the stories more spectacular and to synthesize the information would fit such an author. We need not assume that he was a mindless compiler without a literary agenda, so he may have been the one who made these modifications. And indeed, there are some instances in the other parts of the collection where we may detect his interventions. But the problem is that in our case, the modifications go far beyond those we notice elsewhere. Especially, if we compare the material from Theopompus in *Mir.*, we see that some minor adaptions may have been made but nothing that would come close to the reworkings in our section. But there is no reason why our author should have used the material from Theopompus differently than that in our section. It thus seems more likely that the modifications were already made in the work that was excerpted by the author of *Mir.*, thus the book *On Islands* that Irene Pajón Leyra has identified as the source. If this is correct, it excludes scenario 2 above. In addition, it also makes scenario 3 improbable for chronological reasons: *Mir.* 100 provides the *terminus ante quem* of 238 for *Mir.*'s source. The period in which Timaeus wrote and published his large work is difficult to establish. Parts of it perhaps date to the end of the 4th century (the introductory books that interest us here?) and he may have worked on it until late in his life († 360s or 350s).[189] Thus, the time window between Timaeus and *Mir.*'s source is at best ca. 300–240; assuming a late publication of Timaeus' early books, it is even smaller. If the source of Diodorus was not Timaeus, it was, at any rate, the work of an author of an *On Islands* who used Timaeus in his book. But between him and *Mir.*, we have to assume another intermediary source, the one that is responsible for the reworking (i.e., Irene Pajón Leyra's author *On Islands*). It does not seem realistic to me to squeeze two *On Islands* between the publication of Timaeus' first books (between 300 and the 250s) and 238 (the *terminus ante quem* for *Mir.*'s source).

Thus, the only scenarios that remain as possible are 1 and 4. Although scenario 4 cannot be excluded, it does not seem very likely and perhaps unnecessarily complicates things. Its main advantage would be that it easily explains the discrepancies between named fragments of Timaeus and Diodorus. But as we have seen above, they are very small and may be due to simple misunderstandings or sloppy summarizing. On the other side, it would force us to assume that two authors of *On Islands* used a rather high amount of the same information from Timaeus. Furthermore, as the traditions in *Mir.* are obviously reworkings of those in Diodorus, we would have to assume that in the course of the transmission from Timaeus via *On Islands* to Diodorus, not much was changed or added, which is not very likely with such an additional intermediary source. And finally, it is much more plausible that the parallel with Cicero's description of Enna is due to the common use of Timaeus than that of a work On *Islands*.

Thus, the most likely scenario is that Diodorus used Timaeus and that Timaeus was used and reworked in an unknown work *On Islands* before 238, which was then excerpted by the author of *Mir.* (scenario 1). If we could, however, show that our paradoxographer reworked his source material thoroughly elsewhere, it would be possible to conclude that Diodorus used a work *On Islands* which was in part drawing on Timaeus and that this work was then excerpted and reworked by the author of *Mir.* Based on my knowledge of *Mir.*, this was, however, not the case. Looking at the strong coherence of the first islands section in Diodorus and

---

189 Cf. Schorn (2018) 17 with n. 84.

his general way of summarizing his sources, the conclusion imposes itself that we should abandon Jacoby's idea of the description of the western islands in Timaeus and we should conclude that already Timaeus' *Sikelika* contained such a thematic section on islands which was then adopted by Diodorus and used, together with other sources, by the unknown author of *On Islands* that *Mir.*'s compiler excerpted.

Is it possible to identify this work *On Islands*, written between 300/250 and 238? Authors who would fit chronologically are Philostephanus of Cyrene (*On Islands*), Callimachus the Younger (*Nesias*), and Semos of Delos (*Nesias*),[190] while Xenagoras (*On Islands*) is probably a bit too late.[191] As Callimachus wrote in verse, he can be excluded. Among Philostephanus' fragments without a book title, one (*FGrHist* 1751 F 30) deals with the Eridanos and another (F 34) seems to state that he spoke of the "Trinakrian land of the Siceli." He thus also treated Sicily. In addition, the lake in F 34 also appears in *Mir.* 112 and F 22 deals with Daedalus' flight to the West, as does *Mir.* 81. There are thus common interest but, unfortunately, there is no significant overlap that would point to him as *Mir.*'s direct source. Too little is known about this work and there were probably other works dealing with islands that have totally disappeared from the tradition or are known but regarded as undatable which may have been used by *Mir.*

5. Concluding remarks

What have we learned about Timaeus in *Mir.* at the end of this overlong chapter? The first result is that he was not used directly but through a work *On Islands*. We thus need not look for a series of chapters coming from his *Sikelika*. Even if Irene Pajón Leyra's idea that a work *On Islands* was the main source of the ethnographical section of *Mir.* should prove wrong, it is to be assumed that there was an intermediary source which included material from Timaeus, but also other traditions. The second point is that we may use *Mir.* as evidence for Timaeus only with extreme caution, because the traditions that have reached us through the work *On Islands* have been thoroughly reworked by its author. Only in a few cases may we be able to improve our knowledge of Timaeus by supplementing it with information in *Mir.* The same adaption may be assumed for material this intermediary source took from other sources. So in addition, if a chapter in the ethnographical section going back the work *On Islands* shows overlap with, for example, Lycus, its value for reconstructing this author is no less limited than for Timaeus. Third, regarding the assessment of works *On Islands* as historical sources, we may state that at least one author of such a work used the traditions at his disposal in a very nonchalant way, creating new traditions by freely rearranging existing material and adding elements of his own.[192]

---

190 Philostephanus, *FGrHist* 1751 F 4–6; Callimachus iunior, *FGrHist* 1755 T 1; Semus of Delos, *FGrHist* 1756 F 1.

191 Xenagoras, *FGrHist* 1757 F 1.

192 I would like to thank Ciro Giacomelli, Oliver Hellmann, Robert Mayhew, Irene Pajón Leyra, and Arnaud Zucker for valuable suggestions and inspiring discussions and Robert Mayhew for correcting my English.

Works cited

Acconcia, V. and Giuffré, E.M. 2009. "Lo scavo della Spiaggia di Baratti: campagne 2007–2008." In Ghizzani Marcìa, F. and Megale, C. eds. *Materiali per Populonia*. Vol. 8. Pisa: Edizioni ETS. 127–160.

Aly, W. 1957. *Strabonis Geographica. Strabons Geographika in 17 Büchern. Vol. 4. Strabon von Amaseia. Untersuchungen über Text, Aufbau und Quellen der Geographika*. Bonn: Rudolf Habelt.

Ambaglio, D. 2008. "Introduzione alla *Biblioteca storica* die Diodoro." In Ambaglio, D., Landucci, F. and Bravi, L. eds. *Diodoro Siculo. Biblioteca storica. Commento storico. Introduzione generale*. Milan: Vita e Pensiero. 3–102.

Ameling, W. 1993. *Karthago. Studien zu Militär, Staat und Gesellschaft*. Munich: C.H. Beck.

Amiotti, G. 1982. "Lico di Reggio e *l'Alessandra* di Licofrone." *Athenaeum* 60: 452–460.

Amiotti, G. 1988. "Le Isole Fortunate: mito, utopia, realtà geografica." In Sordi, M. ed. *Geografia e storiografia nel mondo antico*. Milan: Vita e pensiero. 166–177.

Anello, P. 2014. "Tradizioni etnografiche e storiografiche sulla Sardegna". *Hormos. Ricerche di storia antica* 6: 1–20.

Antonelli, L. 1990. *I Greci oltre Gibilterra. Rappresentazioni mitiche dell'estremo occidente e navigazioni commerciali nello spazio atlantico fra VIII e IV secolo a.C.* Rome: L'Erma di Bretschneider.

Aranguren, B. *et al.* 2004. "Attività metallurgica negli insediamenti costieri dell'Etruria centrale fra VI e V secolo a.C. Nuovi dati di scavo." In Lehoërff, A. ed. *L'artisanat métallurgique dans les sociétés anciennes en Méditerranée occidentale. Techniques, lieux et formes de production*. Rome: École Française de Rome. 323–339.

Arnott, W.G. 1996. *Alexis: The Fragments. A Commentary*. Cambridge: University Press.

Beck, H. and Walter, U. 2004. *Die frühen römischen Historiker. Vol. 2. Von Coelius Antipater bis Pomponius Atticus*. Darmstadt: Wissenschaftliche Buchgesellschaft.

Bernardini, P. 2004. "Gli eroi e le fonti." In Zucca, R. ed. *Λόγος περὶ τῆς Σάρδους. Le fonti classiche e la Sardegna. Atti del Convegno di Studi – Lanusei 29 dicembre 1998*. Rome: Carocci editore. 39–62.

Bernardini, P. and Zucca, R. eds. 2005. *Il Mediterraneo di Herakles. Studi e ricerche*. Rome: Carocci editore.

Bethe, E. 1903. "Diomedes 1." *Paulys Realencyclopaedie der classischen Altertumswissenschaft* V(1): 815–826.

Bianchetti, S. 1990. *πλωτὰ καὶ πορευτά. Sulle tracce di una periegesi anonima*. Florence: Università degli Studi di Firenze, Dipartimento di Storia.

Bianchetti, S. 1996. "Plinio e la descrizione dell'Oceano settentrionale in Pitea di Marsiglia." *Orbis Terrarum* 2: 73–84.

Bianchetti, S. 1998. *Pitea di Massilia. L'oceano. Introduzione, testo, traduzione e commento*. Pisa; Rome: Istituti Editoriali e Poligrafici Internazionali.

Bloch, L. 1890–1894. "Kora." In Roscher, W. ed. *Ausführliches Lexicon der griechischen und römischen Mythologie*. II.1. Leipzig: Teubner. 1284–1379.

Bömer, F. 1957–1958. *P. Ovidius Naso. Die Fasten. Herausgegeben, übersetzt und kommentiert*. 2 vols. Heidelberg: Carl Winter Universitätsverlag.

Bömer, F. 1976. *P. Ovidius Naso. Metamorphosen. Kommentar. Buch IV-V*. Heidelberg: Carl Winter Universitätsverlag.

Boncquet, J. 1987. *Diodorus Siculus (II, 1–34) over Mesopotamië. Een historisch kommentaar*. Brussels: Paleis der Academiën.

Borca, F. 2000. *Terra mari cincta. Insularità e cultura romana*. Rome: Carocci editore.

Braccesi, L. ²1977. *Grecità adriatica. Un capitolo della colonizzazione greca in occidente.* Bologna: Pàtron.

Bräuninger, F. 1937. "Persephone." *Paulys Realencyclopaedie der classischen Altertums-wissenschaft* XIX(1): 944–972.

Breglia Pulci Doria, L. 1981. "La Sardegna arcaica tra tradizioni euboiche ed attiche." In *Nouvelle contribution à l'étude de la société et de la colonisation eubéennes*. Naples: Centre Jean Bérard. 61–95 (I quote from the open access version in which no page numbers are given but the paragraphs have been numbered: <https://books.openedition.org/pcjb/213?lang=de>).

Breglia Pulci Doria, L. 2005. "La Sardegna arcaica e la presenza greca: nuove riflessioni sulla tradizione letteraria." In Bernardini, P. and Zucca, R. eds. *Il Mediterraneo di Herakles. Studi e ricerche*. Rome: Carocci editore. 61–86.

Brown, T.S. 1952. "Timaeus, and Diodorus' Eleventh Book." *American Journal of Philology* 73: 337–355.

Brown, T.S. 1958. *Timaeus of Tauromenium*. Berkeley; Los Angeles: University of California Press.

Burton, A. 1972. *Diodorus Siculus. Book 1. A Commentary*. Leiden: E.J. Brill.

Camilli, A. 2016. "La lavorazione del ferro a Populonia. Considerazioni topografiche e cronologiche." *Res Antiquae* 13: 37–58.

Camilli, A. 2019. "Populonia tra necropoli e scorie. Appunti topografici sulla conca di Baratti." *Rassegna di Archeologia* 26: 87–132.

Casevitz, M. and Jacquemin, A. 2015. *Diodore de Sicile. Bibliothèque historique. Vol. V. Livre V. Livre des îles. Texte établi et traduit par. M.C., présenté et commenté par A.J.* Paris: Les Belles Lettres.

Castiglioni, M.P. 2008. "The Cult of Diomedes in the Adriatic: Complementary Contributions from Literary Sources and Archaeology." In Carvalho, J. ed. *Bridging the Gap: Sources, Methodology and Approaches to Religion in History*. Pisa: Pisa University Press. 9–27.

Ceccarelli, P. 1989. "I Nesiotika." *Annali della Scuola Normale Superiore di Pisa III* 19: 903–935.

Chamoux, F., Bertrac, P. and Vernière, Y. eds. 2002. *Diodore de Sicile. Bibliothèque historique. Vol. I. Introduction générale par F.Ch. et P.B. Livre I. Texte établi par P.B., traduit par Y.V.* Paris: Les Belles Lettres.

Champion, C. 2010. "Timaios (566)." In Worthington, I. ed. *Jacoby Online. Brill's New Jacoby, Part III*. Leiden: Brill.

Charles-Picard, G. and Charles-Picard, C. 1958. *La vie quotidienne à Carthage au temps d'Hannibal. IIIᵉ siècle avant Jésus-Christ*. Paris: Hachette.

Chiai, G.F. 2004. "Sul valore storico della tradizione dei Daidaleia in Sardegna (a proposito dei rapporti tra la Sardegna e i Greci in età arcaica)." In Zucca, R. ed. *Λόγος περὶ τῆς Σάρδους. Le fonti classiche e la Sardegna. Atti del Convegno di Studi – Lanusei 29 dicembre 1998*. Rome: Carocci editore. 112–127.

Coppola, A. 1995. *Archaiologhía e propaganda*. Rome: L'Erma di Bretschneider.

Cordiano, G. and Zorat, M. 2014. *Diodoro Siculo. Biblioteca storica. Volume secondo. Libri IV-VIII. Testo greco a fronte*. Milan: BUR.

Cornell, T.J. ed. 2013. *The Fragments of the Roman Historians*. 3 vols. Oxford: University Press.

Corretti, A. 2004. "Per un riesame delle fonti greche e latine sull'isola d'Elba." In Gualandi, M.L. and Mascione, C. eds. *Materiali per Populonia*. Vol. 3. Florence: Edizioni all'insegna del giglio. 269–289.

Corretti, A. 2009. "Siderurgia in ambito elbano e populoniese: un contributo dalle fonti letterarie." In Cambi, F., Cavari, F. and Mascione, C. eds. *Materiali da costruzione e produzione del ferro. Studi sull'economia populoniese fra periodo etrusco e romanizzazione.* Bari: Edipuglia. 133–139.

Corretti, A., Chiarantini, L., Benvenuti, M. and Cambi, F. 2014. "The Aithle Project: Men, Earth and Sea in the Tuscan Archipolago (Italy) in Antiquity. Perspectives, Aims and First Results." In Cech, B. and Rehren, T. eds. *Early Iron in Europe.* Montagnac: éditions Monique mergoil. 181–195.

Corssen, P. 1913. "Ist die Alexandra dem Tragiker Lykophron abzusprechen?" *Rheinisches Museum für Philologie n.F.* 68: 321–335.

Curchin, L. 2007. "Boiotians in the Baleares: The Origin of a Greek Myth." *Ordia Prima* 6: 83–89.

Curchin, L. 2010. "A Babel of Tongues. The Ancient Names of the Balearic Islands and their Meaning." *Philologus* 154: 155–160.

Didu, I. 2002. *I Greci e la Sardegna: il mito e la storia.* Cagliari: Scuola Sarda Editrice.

Didu, I. 2005. "Iolao nipote di Eracle e Sardo figlio di Maceride in Sardegna: assimilazione, mutuazione, distinzione." In Bernardini, P. and Zucca, R. eds. *Il Mediterraneo di Herakles. Studi e ricerche.* Rome: Carocci editore. 53–60.

Dini, A., Corretti, A., Innocenti, F., Rocchi, S. and Scott Westerman, D. 2007. "'Sooty Sweat Stains or Tourmaline Spots.' The Argonauts on the Island of Elba (Toscany) and the Spread of Greek Trading in the Mediterranean Sea." *Myth and Geology* 273: 227–243.

Dudziński, A. 2016. "Diodorus' Use of Timaeus." *Ancient History Bulletin* 30: 43–76.

Enmann, A. 1880. *Untersuchungen über die Quellen des Pompeius Trogus für die griechische und sicilische Geschichte.* Dorpat: Schnakenberg's litho- und typographische Anstalt.

Flashar, H. 1972. *Aristoteles. Mirabilia, übersetzt von H. F. De audibilibus, übersetzt von U. Klein.* Berlin: Akademie-Verlag; [2]1981 with corrigenda to *Audib.*, but not to *Mir.*

Fusillo, M., Hurst, A. and Paduano, G. 1991. *Licofrone. Alessandra.* Milan: Guerini e associati.

Galvagno, E. 2004. "La Sardegna vista dalla Sicilia: Diodoro Siculo." In Zucca, R. ed. *Λόγος περὶ τῆς Σάρδους. Le fonti classiche e la Sardegna. Atti del Convegno di Studi – Lanusei 29 dicembre 1998.* Rome: Carocci editore. 27–38.

Geffcken, J. 1892. *Timaios' Geographie des Westens.* Berlin: Weidmannsche Buchhandlung.

Genovese, G. 2009. *Nostoi, tradizioni eroiche e modelli mitici nel meridione d'Italia.* Rome: L'Erma di Bretschneider.

Giacomelli, C. 2021. *Ps.-Aristotele, De mirabilibus auscultationibus. Indagini sulla storia della tradizione e ricezione del testo.* Berlin; Boston: De Gruyter.

Giacomelli, C. 2023. *Pseudo-Aristotele, De mirabilibus auscultationibus. Edizione critica, traduzione e commento filologico a cura di C.G.* Roma: Bardi edizioni.

Giannini, A. 1965. *Paradoxographorum Graecorum reliquiae recognovit, brevi adnotatione critica instruxit, latine reddidit.* Milan: Istitto Editoriale Italiano.

Gigante Lanzara, V. 2000. *Licofrone. Alessandra. Introduzione, traduzione e note. Testo greco a fronte.* Milan: BUR Rizzoli.

Grilli, A. 1983. "La documentazione sulla provenienza dell'ambra in Plinio." *Acme* 36: 5–17.

Günther, P. 1889. *De ea, quae inter Timaeum et Lycophronem intercedit, ratione.* Doctoral thesis. Leipzig: Max Hoffmann.

Hau, L.I. 2009. "The Burden of Good Fortune in Diodoros of Sicily: A Case for Originality?" *Historia* 58: 171–197.

Hau, L.I., Meeus, A. and Sheridan, B. 2018. "Introduction." In Hau, L.I., Meeus, A. and Sheridan, B. eds. *Diodoros of Sicily. Historiographical Theory and Practice in the Bibliotheke.* Leuven; Paris; Bristol, CT: Peeters. 1–12.

Heinze, R. 1919. *Ovids elegische Erzählung*. Leipzig: B.G. Teubner.

Hennig, R. ²1944. *Terrae incognitae. Eine Zusammenstellung und kritische Bewertung der wichtigsten vorcolumbischen Entdeckungsreisen an Hand der darüber vorliegenden Originalberichte*. Leiden: Brill.

Herter, H. 1941. "Ovids Persephone-Erzählungen und ihre hellenistischen Quellen." *Rheinisches Museum für Philologie n.F.* 90: 236–268.

Hett, W.S. 1936. *Aristotle. Minor Works*. London: William Heineman; Cambridge, MA: Harvard University Press.

Holm, A. 1870. *Geschichte Siciliens im Alterthum*. Vol. 1. Leipzig: Wilhelm Engelmann.

Honigmann, E. 1931. "Strabon (3) von Amaseia." *Paulys Realencyclopaedie der classischen Altertumswissenschaft* IV(A.1): 76–155.

Hornblower, S. 2015. *Lykophron. Alexandra. Greek Text, Translation, Commentary, and Introduction*. Oxford: University Press.

Huß, W. 1985. *Geschichte der Karthager*. Munich: C.H. Beck.

Hülsen, C. 1903. "Dikaiarcheia." *Paulys Realencyclopaedie der classischen Altertumswissenschaft* V(1): 546.

Hunter, R. ed. 2015. *Apollonius of Rhodes. Argonautica. Book IV*. Cambridge: University Press.

Janni, P. 2016. "The Sea of the Greeks and Romans." In Bianchetti, S., Cataudella, M.R. and Gehrke, H.-J. eds. *Brill's Companion to Ancient Geography. The Inhabited World in Greek and Roman Tradition*. Leiden; Boston: Brill. 21–42.

Jones, H.L. 1923. *The Geography of Strabo. With an English Translation*. Vol. II. London: William Heinemann; Cambridge, MA: Harvard University Press.

Knaack, G. 1902–1909. "Phaethon 2." In Roscher, W. ed. *Ausführliches Lexikon der griechischen und römischen Mythologie*. Vol. III.2. Leipzig: B.G. Teubner. 2177–2200.

Lacroix, L. 1969. "Les Béotiens, ancêtres des Baléares." *Latomus* 28: 393–403.

La Penna, A. 1968. *Sallustio e la rivoluzione romana*. Milan: Feltrinelli.

Laqueur, R. 1936. "Timaios (3) von Tauromenion." *Paulys Realencyclopaedie der classischen Altertumswissenschaft* VI(A.1): 1076–1203.

Livrea, E. 1973. *Apollonii Rhodii* Argonauticon *liber quartus. Introduzione, testo critico, traduzione e commento*. Florence: La Nuova Italia.

Lorimer, W.L. 1925. *Some Notes on the Text of Pseudo-Aristotle "De Mundo."* London: Humphrey Milford.

Magnani, S. 2012. *Il viaggio di Pitea sull'oceano*. Bologna: Pàtron.

Malten, L. 1910. "Ein Alexandrinisches Gedicht vom Raube der Kore." *Hermes* 45: 506–553.

Manfredi, V. ²1996. *Le Isole Fortunate. Topografia di un mito*. Rome: L'Erma di Bretschneider.

Manni, E. 1990. "Licofrone, Callimaco e Timeo." In Σικελικὰ καὶ Ἰταλικά. *Scritti minori di storia antica della Sicilia e dell'Italia meridionale*. Vol. 2. Rome: Giorgio Bretschneider Editore. 533–544 (first published in *Kokalos* 7, 1961, 3–14).

Marasco, G. 2004. "Timeo, la Sicilia e la scoperta delle Baleari." *Sileno* 33: 163–174.

Mariotta, G. and Magnelli, A. 2012. *Diodoro Siculo. Biblioteca storica. Libro IV. Commento storico*. Milan: Vita e Pensiero.

Martinez Pinna, J. 1996. "Helanico y el motivo del incendio de los barcos: un hecho troyano." *Giornale italiano di filologia* 48: 21–53.

Mastino, A. 2004. "I miti classici e l'isola felice." In Zucca, R. ed. Λόγος περὶ τῆς Σάρδους. *Le fonti classiche e la Sardegna. Atti del Convegno di Studi – Lanusei 29 dicembre 1998*. Rome: Carocci editore. 11–26.

Meeus, A. 2022. *The History of the Diadochoi in Book XIX of Diodoros' 'Bibliotheke'. A Historical and Historiographical Commentary.* Berlin; Boston: De Gruyter.

Meister, K. 1967. *Die sizilische Geschichte bei Diodor von den Anfängen bis zum Tod des Agathokles. Quellenuntersuchungen zu Buch IV-XXI.* Doctoral thesis, Munich: Blasaditsch.

Meister, K. 1990. *Die griechische Geschichtsschreibung: von den Anfängen bis zum Ende des Hellenismus.* Stuttgart; Berlin; Cologne: Verlag W. Kohlhammer.

Meloni, P. 1942–1944. "Gli Iolei ed il mito di Iolao." *Studi sardi* 6: 43–66.

Mette, H.J. 1952. *Pytheas von Massilia, collegit. H.J.M.* Berlin: De Gruyter.

Meyer, D. 2014. "3.11. Lykophron." In Zimmermann, B. and Rengakos, A. eds. *Handbuch der griechischen Literatur der Antike. Vol. 2. Die Literatur der klassischen und hellenistischen Zeit.* Munich: C.H. Beck. 90–100.

Morgan, M.G. 1969. "The Roman Conquest of the Balearic Isles." *California Studies in Classical Antiquity* 2: 217–231.

Morr, J. 1926. *Die Quellen von Strabons drittem Buch.* Leipzig: Dietrich'sche Verlagsbuchhandlung.

Müllenhoff, K. ¹1871. *Deutsche Altertumskunde.* Vol. 1. Berlin: Weidmannsche Buchhandlung. ²1890 ("neuer vermehrter Abdruck besorgt durch M. Roediger").

Müller, C. and Müller, T. 1841. *Fragmenta historicorum graecorum.* Vol. 1. Paris: Didot.

Nawratil, K. 1959. "Βαθύκολπος." *Wiener Studien* 72: 165–168.

Neri, F. 2002. "Dedalo, i *Daidaleia* e Aristeo: considerazioni sulla presenza mitica di Dedalo in Sardegna." *Annali dell'Istituto Italiano per gli Studi Storici* 19: 21–46.

Nicosia, F. 1981. "La Sardegna nel mondo classico." In *Ichnussa. La Sardegna dalle origini all'età classica.* Milan: Garzanti, Scheiwiller. 419–476.

Niese, B. 1893. Review of Geffcken 1892. In *Göttingische Gelehrte Anzeigen.* 353–360.

Öhler, H. 1913. *Paradoxographi Florentini anonymi opusculum de aquis mirabilibus.* Doctoral thesis. Tübingen: Heckenhauer.

Oldfather, C.H. 1933. *Diodorus of Sicily. With an English Translation. Vol. 1. Books I and II, 1–34.* London: William Heinemann; Cambridge, MA: Harvard University Press.

Ottone, G. 2002. "Lico di Reggio e la storiografia sulla Libia." In Vattuone, R. ed. *Storici greci d'occidente.* Bologna: Società editrice il Mulino. 411–437.

Pais, E. 1881. *La Sardegna prima del dominio romano. Studi storici ed archeologici.* Rome: Coi tipi del Salviucci.

Palm, J. 1955. *Über Sprache und Stil des Diodoros von Sizilien. Ein Beitrag zur Beleuchtung der hellenistischen Prosa.* Lund: CWK Gleerup.

Panzram, S. 2013. "Kleine Geschichte der Balearen." *Klio* 95: 5–39.

Parmeggiani, G. 2011. *Eforo di Cuma. Studi di storiografia greca.* Bologna: Pàtron.

Pearson, L. 1987. *The Greek Historians of the West. Timaeus and His Predecessors.* Atlanta: The Americal Philological Association.

Pearson, L. 1991. "The Character of Timaeus' History, as it is Revealed by Diodorus." In Galvagno, E. and Molè Ventura, C. eds. *Mito, storia, tradizione: Diodoro Siculo e la storiografia classica. Atti del convegno internazionale, Catania-Agira, 7–8 dicembre 1984.* Catania: Edizioni del Prisma. 17–29.

Peter, H. 1906. *Historicorum Romanorum reliquiae.* Vol. 1. Leipzig: B.G. Teubner (repr. with bibliographical addenda Stuttgart: Teubner 1967).

Pfeiffer, R. 1949. *Callimachus. Vol. 1. Fragmenta, edidit R.P.* Oxford: Clarendon Press.

Pietrina, A. 2014. "Tradizioni etnografiche e storiografiche sulla Sardegna (Diod. IV 29–30; 82; V 15)." *Hormos n.s.* 6: 1–20.

Preller, L. and Robert, C. 1926. *Griechische Mythologie. Zweiter Band. Die Heroen (Griechische Heldensage). Drittes Buch. Die großen Heldenepen. 2. Abteilung. Zweite Hälfte. Der Troische Kreis. Die Nosten.* Berlin: Weidmann.

Radt, S. 2007. *Strabonis Geographika. Vol. 6. Buch V-VIII: Kommentar.* Göttingen: Vandenhoek & Ruprecht.

Rathmann, M. 2016. *Diodor und seine "Bibliotheke". Weltgeschichte aus der Provinz.* Berlin; Boston: De Gruyter.

Rebuffat, R. 1976. "Arva beata petamus arva divites et insulas." In *L'Italie et la Rome républicaine. Mélanges offerts à Jacques Heurgon.* Vol. 2. Rome: École Française de Rome. 877–902.

Rohde, E. ³1914. *Der griechische Roman und seine Vorläufer,* ed. by Schmid. W. Leipzig: B.G. Teubner.

Roller, D.W. 2006. *Through the Pillars of Herakles. Greco-Roman Exploration of the Atlantic.* New York; London: Routledge.

Romano, D. 1980. "Cicerone e il ratto di Proserpina." *Ciceroniana n.s.* 4: 191–201.

Rose, V. 1854. *De Aristotelis librorum ordine et auctoritate commentatio.* Berlin: Georg Reimer.

Rose, V. 1863. *Aristoteles pseudepigraphus.* Leipzig: Teubner.

Rossbach, O. 1912. *Castrogiovanni. Das alte Henna in Sizilien nebst einer Untersuchung über griechische und italische Todes- und Frühlingsgötter.* Leipzig: B.G. Teubner.

Rozokoki, A. 2019. *The Negative Presentation of the Greeks in Lycophron's Alexandra and the Dating of the Poem.* Athens: Coralli (non vidi; see the review by Hurst, A. Rivista di filologia e di istruzione classica 149, 2021, 260–263).

Sacks, K.S. 1990. *Diodorus Siculus and the First Century.* Princeton: Princeton University Press.

Sandbach, F.H. 1965. *Plutarch's Moralia in Fifteen Volumes. Vol. 11. 845 E-874 C, 911 C-919 F.* With an English translation by Pearson, L. and Sandbach, F.H. London: William Heinemann; Cambridge, MA: Harvard University Press.

Sanna, S. 2004. "La figura di Aristeo in Sardegna." In Zucca, R. ed. *Λόγος περὶ τῆς Σάρδους. Le fonti classiche e la Sardegna. Atti del Convegno di Studi – Lanusei 29 dicembre 1998.* Rome: Carocci editore. 99–111.

Scardino, C. 2014. "V. Historiographie." In Zimmermann, B. and Rengakos, A. eds. *Handbuch der griechischen Literatur der Antike. Vol. 2. Die Literatur der klassischen und hellenistischen Zeit.* Munich: C.H. Beck. 617–677.

Schade, G. 1999. *Lykophrons 'Odyssee' Alexandra 648–819, übersetzt und kommentiert.* Berlin; New York: De Gruyter.

Schorn, S. 2018. *Studien zur hellenistischen Biographie und Historiographie.* Berlin; New York: De Gruyter.

Schrader, H. 1868. "Über die Quellen der pseudoaristotelischen Schrift ΠΕΡΙ ΘΑΥΜΑΣΙΩΝ ΑΚΟΥΣΜΑΤΩΝ." *Jahrbücher für classische Philologie* 14: 217–232.

Schulten, A. ²1974. *Iberische Landeskunde. Geographie des antiken Spanien.* Baden-Baden: Koerner.

Schütrumpf, E. *et al.* 2008. *Heraclides of Pontus. Texts and 7 Translation.* New Brunswick; London: Transaction Publishers.

Schwabe, L. 1870. "Zum vierten buch der Verrinen (*de signis*)." *Philologus* 30: 311–346.

Schwartz, E. 1903. "Diodoros (38) von Agyrion." *Paulys Realencyclopaedie der classischen Altertumswissenschaft* V(1): 663–704.

Seibert, J. 1983. *Das Zeitalter der Diadochen.* Darmstadt: Wissenschaftliche Buchgesellschaft.

Smith, D.G. 2013. "Lykos (Boutheras) of Rhegium (570)." In Worthington, I. ed. *Jacoby Online. Brill's New Jacoby*, Part III. Leiden; Boston: Brill.

Sørensen, S. 2022. "1680. *Paradoxographus Florentinus*." In Schorn, S. ed. *Felix Jacoby. Die Fragmente der Griechischen Historiker Continued. Part IV. Biography and Antiquarian Literature. E. Paradoxography and Antiquities. Fasc. 2. Paradoxographers of Imperial Times and Undated Authors [Nos. 1667–1693]*. Leiden; Boston: Brill. 633–785.

Štrmelj, D. 2015. "East Adriatic in Pseudo-Aristotle's *De Mirabilibus Auscultationibus*." *Electryone* 3: 51–74.

Stylianou, P.J. 1998. *A Historical Commentary on Diodorus Siculus Book 15*. Oxford: Clarendon Press.

Sulimani, I. 2011. *Diodorus' Mythistory and the Pagan Mission. Historiography and Culture-Heroes in the First Pentad of the Bibliotheke*. Leiden; Boston: Brill.

Theiler, W. 1982. *Poseidonios. Die Fragmente*. 2 vols. Berlin; New York: De Gruyter.

Tovar, A. 1989. *Iberische Landeskunde. Vol. 2. Las tribus y las ciudades de la antigua Hispania. Tomo 3. Tarraconensis*. Baden-Baden: Verlag Valentin Koerner.

Trapanis, C.A. 1960. "A New Collection of Epigrams from Chios." *Hermes* 88: 69–74.

Türk, G. 1938. "Phaethon." *Paulys Realencyclopaedie der classischen Altertumswissenschaft* XIX(2): 1508–1515.

van Wees, H. 2005. "Clothes, Class and Gender in Homer." In Cairns, D. ed. *Body Language in the Greek and Roman Worlds*. Swansea: The Classical Press of Wales. 1–36.

Vanotti, G. 2007. *Aristotele. Racconti meravigliosi. Testo greco a fronte. Introduzione, traduzione, note e apparti*. Milan: Bompiani.

Vattuone, R. 1991. *Sapienza d'occidente. Il pensiero storico di Timeo de Tauromenio*. Bologna: Pàtron.

Vimercati, E. 2004. *Posidonio. Testimonianze e frammenti. Testo latino a fronte. Introduzione, traduzione, commentario e apparati*. Milan: Bompiani.

Vogel, F. 1890. *Diodori Bibliotheca historica*. Vol. 2. Leipzig: B.G. Teubner.

von Holzinger, C. 1895. *Lykophron's Alexandra. Griechisch und deutsch mit erklärenden Anmerkungen*. Leipzig: B.G. Teubner.

von Wilamowitz-Moellendorff, U. 1883. *De Lycophronis Alexandra commentatiuncula*. Index lectionum Greifswald 1883.

Wehrli, F. ²1969. *Die Schule des Aristoteles. Texte und Kommentar. Vol. 7. Herakleides Pontikos*. Basel; Stuttgart: Schwabe.

Westermann, A. 1839. *ΠΑΡΑΔΟΞΟΓΡΑΦΟΙ. Scriptores rerum mirabilium Graeci*. Braunschweig: Georg Westerman; London: Black and Armstrong.

Wirth, G., Veh, O. and Nothers, T. 1993. *Diodoros. Griechische Weltgeschichte. Buch I-X. Zweiter Teil. Übersetzt von G.W. (Buch I-III) und O.V. (Buch IV-X). Eingeleitet und kommentiert von Th.N.* Stuttgart: Anton Hiersemann.

Ziegler, K. 1912. "Henna." *Pauly's Realencyclopaedie der classischen Altertumswissenschaft* VIII(1): 284–287.

Zinzow, W. 1906. *De Timaei Tauromenitani apud Ovidium vestigios*. Doctoral thesis. Greifswald: Ex officina Iul. Abel.

Zucca, R. 1998. *Insulae Baliares. Le isole Baleari sotto il dominio romano*. Rome: Carocci editore.

Zucca, R. ed. 2004. *Λόγος περὶ τῆς Σάρδους. Le fonti classiche e la Sardegna. Atti del Convegno di Studi – Lanusei 29 dicembre 1998*. Rome: Carocci editore.

3   Pseudo-Aristotle, *De mirabilibus auscultationibus* 122–138 and Theopompus' *Philippica*

*Pietro Zaccaria**

1. Introduction

The 4th-century BC historian Theopompus of Chios is never mentioned as a source by the compiler of *De mirabilibus auscultationibus* (henceforth *Mir.*). Nonetheless, it has often been supposed that a large section of this paradoxographical collection ultimately depends on Theopompus' *Philippica*, a universal history of 58 books concerning the period 359–336 BC and focusing on the figure of the Macedonian king Philip II. A few hundred fragments of the work are extant (*FGrHist/BNJ* 115).[1]

Needless to say, the hypothesis that Theopompus was used by the compiler of *Mir.* rests mainly on parallels and similarities between *Mir.* and various passages that are explicitly attributed to the historian by other sources. However, the specific reasons for attributing individual chapters of *Mir.* to Theopompus vary. The aim of this chapter is to ascertain which sections of *Mir.* can be reasonably attributed to Theopompus on textual grounds and to what extent *Mir.* can be regarded as a source for reconstructing his work.

In view of its aim and scope, the present analysis will not focus on those passages of *Mir.* that, despite sharing some topics of interest or views with the fragments of Theopompus, are not connected to them by close verbal parallels. This is the case for some isolated chapters regarding the geography and ethnography of the Adriatic area, namely, *Mir.* 80–81 and 104–105,[2] which have at times been

* Citations from *Mir.*, references to manuscript readings, and chapter numbering are based on Giacomelli (2023). Translations of *Mir.* are from Hett (1936), modified, though I found Mayhew's draft translation helpful. I am sincerely grateful to Oliver Hellmann, Robert Mayhew, Stefan Schorn, and Arnaud Zucker for inviting me to the International Conference "The Aristotelian *Mirabilia*" at the Université Côte d'Azur (Nice, 21–23 April 2022). I would also like to thank Robert Mayhew, Stefan Schorn, and Ciro Giacomelli for their valuable remarks on a previous version of this chapter, as well as the Research Foundation – Flanders (FWO: file number 1207720N) for its generous support.

1 By universal history, I mean a historical work with a broad thematic scope that was not limited to Greek history: see recently Ottone (2018) 518–519 with n. 29, who speaks of a "prospettiva 'ecumenica'." Some new Theopompan fragments have been identified by scholars after Jacoby's edition: see the bibliographical references collected in Biagetti (2017) 139–140 n. 6. For a critical assessment of Jacoby's edition of Theopompus, see Ottone (2004).

2 In the following, I will not mention passages that have been attributed to Theopompus on purely speculative grounds, such as *Mir.* 118, which has been tentatively connected to Theopompus by Vanotti (1977) 167–168; (2007) 204, followed by Occhipinti (2011) 291–292.

DOI: 10.4324/9781003437819-4

connected to Theopompus because the compiler of *Mir*. and the historian seem to share a geographical and ethnographic interest in that region as well as some specific views.[3] In what follows, I briefly rehearse the content of these chapters and the reasons why some scholars have connected them to Theopompus.

- *Mir*. 80 concerns the fertility of the Umbrians' earth, cattle, and women. Theopompus also mentioned the fertility of the Umbrian land (Ath. 12.526f-527a = Theopomp., *FGrHist/BNJ* 115 F 132).[4]
- *Mir*. 81 contains detailed geographical, ethnographic, and mythical information concerning the so-called Electrides Islands and the area of the Eridanus River (the myth of Phaethon and the Heliades and that of Daedalus and Icarus). Various elements in this description have parallels in other ancient sources, including pseudo-Scymnus' description of the Adriatic Sea (369–397). Pseudo-Scymnus used Theopompus (*FGrHist/BNJ* 115 F 130) as a source, but only the description of the position (*thesis*) of the Adriatic Sea (371–374)[5] is explicitly attributed to the historian, while the ethnographic and mythical reports that follow are introduced by verbs such as ἱστοροῦσι (375), φασιν (389, 392 [396]), and λέγουσι (394 [398]). Therefore, it is uncertain to what extent pseudo-Scymnus' description of the Adriatic Sea actually derives from Theopompus.[6]
- *Mir*. 104 mentions first a certain mountain called Δέλφιον (Mt. Haemus?), located μεταξὺ τῆς Μεντορικῆς καὶ τῆς Ἰστριανῆς ("between Mentorica and Istriana"), from the top of which the Mentores living near the Adriatic Sea could observe ships sailing into the Pontus. This description recalls a fragment of Theopompus according to which both the Adriatic and the Pontus are visible from

---

3 The Theopompan provenance of these passages has been recently, and somewhat uncritically, suggested by Štrmelj (2015). Schrader (1868) 226–227 ascribed *Mir*. 104–105 to Theopompus. Müllenhoff (1870) 429–434 assigned *Mir*. 80–81 and 104–105 to Lycus, whom he supposed had elaborated on and corrected Theopompus' account. Müllenhoff's views have been critically reviewed by Enmann (1880) 193–203; von Gutschmid (1893) 123–138. Geffcken (1892) 91–94, 129–133, followed by Ziegler (1949) 1152, attributed *Mir*. 80–81 and 104–105 to Timaeus, whom he supposed had argued against Theopompus. Jacoby (1955a, 600, 1955b, 334 n.) 352 rejected Geffcken's attribution of *Mir*. 81 to Timaeus and suggested Theopompus and maybe another author (perhaps Lycus) as possible sources. (Attributing *Mir*. 81 to Timaeus is also regarded as unlikely by Schorn [in this volume] 68–71.) Jacoby (1955b) 341 n. 443 also rejected Geffcken's attribution of *Mir*. 105 to Timaeus and considered this chapter "bestenfalls ein konglomerat aus verschiedenen quellen." Fraser (1972) II 1079 n. 383, 1080 n. 385 attributed *Mir*. 80–81 to Timaeus and 104–105 to Theopompus. According to Flashar (1972) 45, 105–108, it is difficult to identify the source of *Mir*. 80–81: these chapters may derive from either Timaeus or Lycus (together with the entire section 78–81), but Theopompus may still be a point of reference (especially for *Mir*. 80); Flashar (1972) 45, 121–124 ascribed *Mir*. 104–105 to an unidentifiable author (not Timaeus) who was improving on Theopompus' account. Occhipinti (2011) 292 attributed *Mir*. 104–105 to Theopompus.
4 On *Mir*. 80, cf. Jacoby (1930) 378; Flashar (1972) 107; Vattuone (2000) 24–25, 28–30; Vanotti (2007) 166–167; Morison (2014) on F 132.
5 The Electrides Islands are mentioned here along with the Apsyrtides and the Liburnides.
6 On *Mir*. 81, see Jacoby (1930) 378; Braccesi (1968); Musso (1976); Braccesi (1977) 30–55; Flashar (1972) 108; Briquel (1984) 60–62; Bianchetti (1990) 77–108; Vattuone (2000) 22–27; Braccesi (2001) 51–54, 118–119; Giacomelli (2023) 260–261. For a commentary on pseudo-Scymnus' verses, see Marcotte (2000) 194–200; Korenjak (2003) 83–85.

a certain mountain (Strab. 7.5.9.316–317C = Theopomp., *FGrHist/BNJ* 115 F 129).

- The second part of *Mir.* 104 speaks of a common market located halfway between the Adriatic and the Pontus, in which Lesbian, Chian, and Thasian things were sold by merchants coming up from the Pontus, while Corcyraean amphorae were sold by merchants from the Adriatic. In the same fragment of Theopompus preserved by Strabo mentioned above, we read that the two seas are connected by an underground channel, as can be deduced from the fact that both Chian and Thasian ceramics have been found in the Naron River (Strab. 7.5.9.316–317C = Theopomp., *FGrHist/BNJ* 115 F 129).
- Finally, *Mir.* 105.1, which discusses the flow of the Ister in connection with the route followed by Jason and the Argonauts, claims that the river "divides, and one part flows into the Pontus, and the other into the Adriatic." Theopompus was one of the ancient authors who reported that "one of the mouths of the Ister empties into the Adriatic" (Strab. 7.5.9.316–317C = Theopomp., *FGrHist/BNJ* 115 F 129).[7]

The possible relationship between these chapters of *Mir.* and Theopompus surely deserves attention and should be taken into account when interpreting the information contained in both texts.[8] However, since we lack substantial verbal parallels between them, any hypothesis postulating direct textual dependence on Theopompus by the compiler of *Mir.* with regard to these sections must remain speculative.[9]

The present analysis will rather focus on a group of contiguous chapters of *Mir.* in which a surprisingly high number of substantial verbal parallels with Theopompus' fragments can be detected, namely, *Mir.* 122–138. As is to be expected, not all of these chapters exhibit parallels with Theopompus' fragments: in some cases the parallels concern only part of the information, while in other cases there are no parallels at all. Nonetheless, (most of) this section is traditionally attributed to Theopompus on the assumption that the text of *Mir.* can be divided into large continuous blocks, each deriving from a different source. On this interpretation, Theopompus would thus be the source for a series of reports concerning the eastern Mediterranean, as opposed to those regarding the western Mediterranean, which are traditionally attributed to the historians Lycus and Timaeus.[10]

---

7 On *Mir.* 104–105, see Osann (1848); Jacoby (1930) 378; Flashar (1972) 121–124; Braccesi (1977) 111–119; Bianchetti (1990) 130–134, 144–153; Vattuone (2000) 19–22; Braccesi (2001) 117–120; Vanotti (2007) 184–187; Morison (2014) on F 129; Cordano (2014) 17–23; (2015) 57–59; Habaj (2018); Giacomelli (2023) 281–285.

8 For Theopompus, see especially Bianchetti (1990) 100–108, 130–154; Shrimpton (1991) 94–101; Vattuone (2000); Braccesi (2001) 115–121. On the role played by the Adriatic area in the fragments of the Greek historians, see Ambaglio (2002).

9 Cf. the prudent approach of Vattuone (2000) 21–22. *Mir.* 104, for example, clearly resembles the account of Theopompus, but at the same time significantly differs from it: see, with different arguments, Müllenhoff (1870) 433; Geffcken (1892) 91; Braccesi (1977) 111–119; Bianchetti (1990) 130–132; Habaj (2018). In such cases, *pace* Cordano (2015) 57–59, we cannot consider the version reported by *Mir.* as coming from Theopompus, even if the historian might have been the point of reference for the compiler of *Mir.* (or for the compiler's source).

10 On Timaeus and *Mir.*, see Schorn (in this volume).

This theory goes back to Rose, who attributed the entire block under discussion to Theopompus.[11] Schrader soon raised doubts about the soundness of this method and argued that some chapters within this block may derive from other sources: according to Schrader, only *Mir.* 123, 125, 128–131, and 133–135 surely derive from Theopompus, while 126 and 132, although probably Theopompan, may ultimately derive from Aristotle, and 127 is from both Theopompus and Lycus.[12] Later scholars, however, reacted against Schrader's "atomistic" analysis of *Mir.* and went back to Rose's hypothesis, supposing that contiguous chapters that are somehow connected to each other derive as a rule from one and the same source.

Müllenhoff argued that since *Mir.* 125, 127–131, and 133–136 surely derive from Theopompus, we should also attribute to him 122–138.[13] Guenther, followed by Susemihl, approved of this hypothesis, but ascribed 127 to Lycus, instead of to Theopompus.[14] With a better knowledge of Theopompus' fragments, Geffcken then hypothesized that since 123, 125, 127–131, and 133–137 are Theopompan, "die Wahrscheinlichkeit, dass auch das mit 115 [i.e., 123] eng verbundene 116 [i.e., 124], ferner 118 [i.e., 126] . . . und 124 [i.e., 132] aus derselben Quelle stammen, ist unabweisbar."[15] Later scholars of *Mir.* mostly limited themselves to repeating, with minor modifications, the results reached by the 19th-century *Quellen-forschung* and usually attributed 123–137 to Theopompus, sometimes with the addition of 122 and 138 and with the exception of 126.[16]

Only Flashar, in his influential annotated German translation of *Mir.* (1972), devoted a new study to the sources of *Mir.* Flashar was aware that Rose's approach to the sources of *Mir.* (the so-called *Reihenargument*) is too mechanical and is contradicted by some cases in which a block of chapters supposedly deriving from a given source is interrupted by one or more chapters of a different origin. In the footsteps of Schrader, Flashar claimed that only a detailed case-by-case analysis could reveal the sources that had been used in the individual chapters. Still, Flashar conceded that chapters that are closely connected to each other may be prudently attributed to one and the same source, even when that provenance cannot be shown on internal grounds.[17] With regard to Theopompus, Flashar thus concluded that *Mir.* 123–138 derive from the historian, with the exception of 126, which derives from Aristotle or Theophrastus.[18]

---

11 Rose (1863) 255, 280. Cf. also Rose (1854) 54–56.
12 Schrader (1868) 226–230.
13 Müllenhoff (1870) 428. Von Gutschmid (1893) 132 approved of Müllenhoff's approach.
14 Guenther (1889) 24–27; Susemihl (1891) 478 n. 94b.
15 Geffcken (1892) 89.
16 Ziegler (1949) 1152: *Mir.* 123–137 derive from Theopompus. Giannini (1964) 135 n. 219: *Mir.* 122–138 are almost entirely from Theopompus. Giannini (1965) 281: *Mir.* 123–137 are almost entirely Theopompan; however, 126–127 may derive from Theophrastus, 133 from either Theopompus or Theophrastus, and 138 from either Theopompus or Timaeus. Fraser (1972) II 1079 n. 383, 1080 n. 385: *Mir.* 123–137 (or 135) are from Theopompus. Theopompus is presented as a source of *Mir.* (without further specification) by Moraux (1951) 260; Sassi (1993) 458; Schepens and Delcroix (1996) 380 n. 18; Pajón Leyra (2011) 274.
17 Flashar (1972) 39–40.
18 Flashar (1972) 41, 47–48. His results are repeated by Vanotti (2007) 43; Gómez Espelosín (2008) 235 n. 119, 236 n. 120.

Flashar's commentary surely represents an important milestone in the study of the sources used by the compiler of *Mir*. However, his approach led to the (to my mind) unjustified attribution of some chapters to Theopompus, while in other cases the attribution to Theopompus may be put forward on the basis of stronger arguments. In what follows, the individual chapters of *Mir*. belonging to the block 122–138 will be systematically analyzed from an historiographical point of view, in order to assess their relationship to Theopompus' extant fragments as well as to other parallel traditions, which in some cases also seem to go back to the historian. Because of its rather large scope, the present analysis cannot exhaustively treat all the historical and philological issues arising from the passages under discussion, and it will only occasionally deal with the *realia* mentioned in the texts or with the role that these reports may have played in the original account of Theopompus. Yet the prudent, textual approach adopted here will provide, as I hope, a safer textual basis for further research on both *Mir*. and Theopompus.

2. Historiographical analysis of *Mir*. 122–138

2.1 Mir. 122

1. Ἐν τῇ Πηδασίᾳ τῆς Καρίας θυσία τῷ Διὶ συντελεῖται, ἐν ᾗ πέμπουσιν αἶγά τινα, περὶ ἣν θαυμαστόν τί φασι γίγνεσθαι· βαδίζουσα γὰρ ἐκ Πηδάσων σταδίους ἑβδομήκοντα δι᾽ ὄχλου πολλοῦ τοῦ θεωροῦντος οὔτε διαταράττεται κατὰ τὴν πορείαν οὔτ᾽ ἐκτρέπεται τῆς ὁδοῦ, δεδεμένη δὲ σχοινίῳ προπορεύεται τοῦ τὴν ἱερωσύνην ἔχοντος. 2. θαυμαστὸν δ᾽ ἐστὶ καὶ τὸ δύο κόρακας εἶναι διὰ τέλους περὶ τὸ τοῦ Διὸς ἱερόν, ἄλλον δὲ μηδένα προσιέναι πρὸς τὸν τόπον, καὶ τὸν ἕτερον αὐτῶν ἔχειν τὸ πρόσθεν τοῦ τραχήλου λευκόν.

1. In Pedasia in Caria a sacrifice is offered to Zeus, in which they take a she-goat in procession, concerning which a marvel is related. For, when walking seventy stades from the Pedasi through a large crowd of watchers, it is not disturbed on its journey, nor does it turn out of the road, but tied with a rope it walks in front of the man who is conducting the sacrifice. 2. There is also a wonderful thing, in that there are two crows always about the temple of Zeus, and that no other approaches the spot, and that one of them has a white patch in the front of its neck.

A story similar to the first marvel related in this chapter is contained in Apollonius, *Historiae mirabiles* (*FGrHist* 1672) 13, where we read that

in Halicarnassus, when a sacrifice to Zeus Askraios is celebrated, a herd of goats is brought and set before the temple; when the prayers are completed, one goat goes forward, led by no one, and approaches the altar, and the priest takes it and sacrifices it with good omens.[19]

---

19 Apollon. *Mir.* (*FGrHist* 1672) 13: ἐν Ἁλικαρνασσῷ θυσίας τινὸς τῷ Διὶ τῷ Ἀσκραίῳ συντελουμένης ἀγέλην αἰγῶν ἄγεσθαι πρὸ τοῦ ἱεροῦ καὶ ἵστασθαι· τῶν δὲ κατευχῶν συντελεσθεισῶν, προβαίνειν μίαν αἶγα ὑπὸ μηδενὸς ἀγομένην καὶ προσέρχεσθαι τῷ βωμῷ· τὸν δὲ ἱερέα λαβόμενον αὐτῆς καλλιερεῖν. Trans. Spittler, with minor modifications.

According to Apollonius, this story was originally found in ἐν τῷ Κατὰ τόπους μυθικῷ, but the name of its author is unfortunately lost in a lacuna and cannot be recovered with any certainty.[20]

The two versions of the story, despite some general similarities (they occur in the same area[21] and in both cases a goat willingly participates in the sacrifice), clearly represent two independent traditions and must therefore depend on different sources.[22] Rose and other 19th-century scholars speculatively attributed the version of *Mir.* to Theopompus because its oriental setting distinguishes it from the preceding chapters of *Mir.*, which concern the Phoenicians and are traditionally assigned to Timaeus, and because it precedes a chapter which can be attributed to Theopompus on internal grounds (i.e., 123).[23] However, this story has no parallels in the extant fragments of Theopompus and cannot therefore be shown to derive from him. More cautiously, Flashar left the problem open ("Quelle unbekannt").[24]

The second marvel of this chapter cannot be attributed to a specific source either. Stories about pairs of crows are widespread in the ancient sources, but no source speaks of two crows around a temple of Zeus in Caria. That Theopompus spoke of crows living in Thessalian Crannon (see below on *Mir.* 134) cannot be a reason to attribute to him the marvel concerning Pedasia.

2.2 Mir. 123

Λέγεται περὶ τὴν τῶν Σιντῶν καὶ Μαιδῶν [Μαιδῶν Sylburg: μεδῶν ω] χώραν καλουμένην τῆς Θρᾴκης ποταμόν τινα εἶναι Πόντον προσονομαζόμενον, ἐν ᾧ καταφέρεσθαί τινας λίθους οἳ καίονται καὶ τοὐναντίον πάσχουσι τοῖς ἐκ τῶν ξύλων ἄνθραξι· ῥιπιζόμενοι γὰρ σβέννυνται ταχέως, ὕδατι δὲ ῥαινόμενοι ἀναλάμπουσι καὶ ἀνάπτουσι κάλλιον. παραπλησίαν δὲ ἀσφάλτῳ, ὅταν καίωνται, καὶ πονηρὰν οὕτω ὀσμὴν καὶ δριμεῖαν ἔχουσιν ὥστε μηδὲν τῶν ἑρπετῶν ὑπομένειν ἐν τῷ τόπῳ καιομένων αὐτῶν.

The story goes that in the district of Thrace called (the land) of the Sinti and Maedi there is a river called Pontus, which rolls down stones which burn and behave in the opposite way to charcoal made from wood; for when the flame is fanned these stones are quickly quenched, but when soaked in water they light up and kindle finely. When they burn they have a smell like pitch, just as unpleasant and acrid, so that no reptile can stay in the place while they are burning.

---

20 See especially Spittler (2022) 461.
21 Hdt. 1.175.1; Strab. 13.1.59.611C locate Pedasa in the territory of Halicarnassus: cf. Flashar (1972) 140.
22 Spittler (2022) 462. An inscription from a sanctuary of Zeus Panamaros in Caria (*I. Stratonikeia* 266) describes an ox that preceded a priest to the *bouleuterion* and disappeared after a sacrifice: see Flashar (1972) 140; Şahin (1981) 138–139 no. 266; Spittler (2022) 463 n. 270, with *status quaestionis* and previous literature. For a recent interpretation of the passages from *Mir.* and Apollonius against the background of Carian cults, see Rivault (2021) 311–321.
23 See Rose (1863) 255, 280; Müllenhoff (1870) 428; Guenther (1889) 24–27; Susemihl (1891) 478 n. 94b. Even according to the traditional sequence of the chapters, *Mir.* 122 still precedes a Theopompan chapter, namely, 138. Rose (1854) 55–56 also argued that Theopompus should be identified with the author of the work cited by Apollonius the paradoxographer.
24 Flashar (1972) 41, 140.

This chapter of *Mir.* is quoted by Stephanus of Byzantium, Σ 174 Billerbeck, *s.v.* Σιντία; a scholium on Nicander, *Theriaca* 45–47 (pp. 52.14–53.4 Crugnola)[25]; and Leo Choirosphactes, *De thermis* 151–156 = 160–165 Gallavotti. Aelian, *Natura animalium* 9.20, who reports the same information citing Aristotle as his source (F 270.27 Gigon), may also come from *Mir.* The chapter of *Mir.* also has a close parallel in pseudo-Antigonus, *Mir.* 136,[26] which ultimately depends on Theopompus (*FGrHist/BNJ* 115 F 268a) via Callimachus' paradoxographical collection (F 407 VIII Pfeiffer):

> περὶ δὲ τὴν τῶν Ἀγριέων [Ἀγριέων Salmasius: Ἀγρίων P] Θρᾳκῶν χώραν φησὶν ποταμὸν προσαγορευόμενον Πόντον καταφέρειν λίθους ἀνθρακώδεις· τούτους δὲ κάεσθαι μέν, πᾶν δὲ τοὐναντίον πάσχειν τοῖς ἐκ τῶν ξύλων ἀνθρακευομένοις· ὑπὸ μὲν γὰρ τῶν ῥιπίδων πνευματιζομένους σβέννυσθαι, τῷ δὲ ὕδατι ῥαινομένους βέλτιον κάεσθαι. τὴν δ᾽ ὀσμὴν αὐτῶν οὐδὲν ὑπομένειν ἑρπετόν.[27]

> He (Callimachus) says that in the vicinity of the Agriean Thracians there is a river called Pontus that discharges stones the color of coal; they burn, but act entirely opposite from coals made from wood. For when they are blown upon by a fan, they are extinguished, but they burn better when they are sprinkled with water. No animal can at all endure their smell. (trans. Morison, with modifications)

Attribution to Theopompus is confirmed by *Paradoxographus Palatinus* (*FGrHist* 1681) 19 (= Theopomp., *FGrHist/BNJ* 115 F 268a), which closely resembles pseudo-Antigonus' testimony:[28]

> Θεόπομπός φησιν ἐν τῇ τῶν Ἀγριέων [Ἀγριέων De Stefani: ἀγρισίων A: αγρι(ν)σίων B] Θρᾳκῶν χώρᾳ ποταμὸν εἶναι ὀνομαζόμενον Πόντον, ὃν καταφέρειν λίθους ἀνθρακώδεις· τούτους δὲ ἀναφθέντας ὑπὸ μὲν τῶν ῥιπιδίων ῥιπιζομένους <οὐ> καίεσθαι, ὑπὸ δὲ ὕδατος ῥαινομένους ἀναλάμπειν. οὐδὲν δὲ ἑρπετὸν τὴν ὀσμὴν αὐτῶν ὑπομένειν.[29]

> Theopompus says that in the territory of the Agriean Thracians there is a river called Pontus, which carries downstream stones that are the color of coal. When these have been lit, they do not burn when fanned by small bellows, but flare up when sprinkled with water. No animal can stand the smell of them. (trans. Sørensen, with modifications)

Other sources speak of this marvelous stone, which may be identified as a form of lignite or asphaltic bitumen.[30] Theophrastus, *De lapidibus* 2.12–13 (p. 60

---

25 Cf. Giacomelli (2021) 349–350.

26 On the uncertain date of pseudo-Antigonus, see most recently Dorandi (2017) 61–71; Giacomelli (2021) 18–20; Greene (2022) 654 with n. 36, with previous literature.

27 Ed. Musso (1985) 61; cf. also Eleftheriou (2018a) 194. Note that Theopompus is cited by name at the end of 137: ἱστορεῖν δὲ ταῦτα Θεόπομπου.

28 It is usually assumed that the *Paradoxographus Palatinus* depends on pseudo-Antigonus, but there is no certainty about this dependence.

29 Ed. Sørensen (2022b) 788.

30 See also de Martini (2021) 472–478, with a table on p. 476.

Eichholz), distinguishes between a stone found near Binae, which is carried down a river and has a poor smell, and another stone known as *spinos*, which burns fiercely when sprayed with water.[31] Pliny, *Naturalis historia* 33.94, and Alexander of Aphrodisias, *Problemata* 2.64, also mention a Thracian stone which burns when sprayed with water. Like Theopompus, Nicander, *Theriaca* 45–50, locates the Thracian stone in a Thracian river called Pontus.[32] The same report is found in Dioscorides, *Materia medica* 5.129.1, and pseudo-Dioscorides, *De lapidibus* 11, which also specify that the river Pontus flows in the region of the Sinti (κατὰ Σιντίαν).[33]

Comparison between *Mir.* 123, pseudo-Antigonus, *Mir.* 136, and *Paradoxographus Palatinus* 19 shows that the three passages are very similar in content and vocabulary:

Mir. *123*	Ps.-Antig. Mir. *136 = Callim. F 407 VIII Pfeiffer = Theopomp., FGrHist/BNJ 115 F 268a*	Par. Pal. *19 = Theopomp., FGrHist/BNJ 115 F 268a*
Λέγεται περὶ τὴν τῶν Σιντῶν καὶ Μαιδῶν χώραν καλουμένην τῆς Θρᾴκης ποταμόν τινα εἶναι Πόντον προσονομαζόμενον, ἐν ᾧ καταφέρεσθαί τινας λίθους οἳ καίονται καὶ τοὐναντίον πάσχουσι τοῖς ἐκ τῶν ξύλων ἄνθραξι· ῥιπιζόμενοι γὰρ σβέννυνται ταχέως, ὕδατι δὲ ῥαινόμενοι ἀναλάμπουσι καὶ ἀνάπτουσι κάλλιον. παραπλησίαν δὲ ἀσφάλτῳ, ὅταν καίωνται, καὶ πονηρὰν οὕτω ὀσμὴν καὶ δριμεῖαν ἔχουσιν ὥστε μηδὲν τῶν ἑρπετῶν ὑπομένειν ἐν τῷ τόπῳ καιομένων αὐτῶν	περὶ δὲ τὴν τῶν Ἀγριέων Θρακῶν χώραν φησὶν ποταμὸν προσαγορευόμενον Πόντον καταφέρειν λίθους ἀνθρακώδεις· τούτους δὲ κάεσθαι μέν, πᾶν δὲ τοὐναντίον πάσχειν τοῖς ἐκ τῶν ξύλων ἀνθρακευομένοις· ὑπὸ μὲν γὰρ τῶν ῥιπίδων πνευματιζομένους σβέννυσθαι, τῷ δὲ ὕδατι ῥαινομένους βέλτιον κάεσθαι. τὴν δ᾽ ὀσμὴν αὐτῶν οὐδὲν ὑπομένειν ἑρπετόν	Θεόπομπός φησιν ἐν τῇ τῶν Ἀγριέων Θρακῶν χώρᾳ ποταμὸν εἶναι ὀνομαζόμενον Πόντον, ὃν καταφέρειν λίθους ἀνθρακώδεις· τούτους δὲ ἀναφθέντας ὑπὸ μὲν τῶν ῥιπιδίων ῥιπιζομένους <οὐ> καίεσθαι, ὑπὸ δὲ ὕδατος ῥαινομένους ἀναλάμπειν. οὐδὲν δὲ ἑρπετὸν τὴν ὀσμὴν αὐτῶν ὑπομένειν

31 Binae is also mentioned in Thphr. *Lap.* 2.15 (p. 62 Eichholz). The *spinos* is also described in *Mir.* 33 and 41 (perhaps from Theophrastus). According to Regenbogen (1940) 1407–1408, followed by Eichholz (1965) 96, Theophrastus corrected Theopompus, whom Theophrastus supposed had confused the two stones. Flashar (1972) 129 considers it possible that Theophrastus was drawing on Theopompus, or vice versa. On this problem, see also Sharples (1998) 184–185 with n. 532. According to Caley and Richards (1956) 80–82, the two stones mentioned by Theophrastus may have been varieties of the same mineral substance. Cf. also Kostov (2007); Amigues (2018) 37–40.

32 See also Gal. *SMT* XII pp. 203–204 Kühn. Cf. Gow and Scholfield (1953) 171; Jacques (2002) 83–84 n. 8; Overduin (2015) 203–204.

33 See also Isid. *Etym.* 16.4.8.

Because of these striking similarities, it has been reasonably concluded that the compiler of *Mir.*, like pseudo-Antigonus and the *Paradoxographus Palatinus*, depends on Theopompus.[34] Nevertheless, this attribution has been rejected by Boshnakov because of two divergences between *Mir.* and pseudo-Antigonus (and the *Paradoxographus Palatinus*).[35] First, *Mir.* locates the river Pontus in that part of Thrace inhabited by the Sinti and the Maedi, while pseudo-Antigonus locates it in the region of the Agriean Thracians. Second, the comparison between the Thracian stone and pitch, reported by *Mir.* (παραπλησίαν δὲ ἀσφάλτῳ), is absent from pseudo-Antigonus' account. Boshnakov concluded that these differences should prevent us from regarding Theopompus as a common source and suggested considering the similarities between the accounts of *Mir.*, pseudo-Antigonus, and Theophrastus as the result of "ein intensiver mündlicher Informationaustausch."[36]

The arguments put forward by Boshnakov, however, are unconvincing. The second difference mentioned by the scholar cannot be used as an argument, since comparison with pitch may have been either added by the compiler of *Mir.* or omitted by Callimachus or pseudo-Antigonus. The only factual difference between the passages under examination concerns the location of the marvel either among the Sinti and the Maedi or among the Agrieis. However, even this (minor) difference cannot be used to draw far-reaching conclusions, since the Agrieis (or Agrianes), the Sinti, and the Maedi were all settled not far from each other in the Strymon valley (see, for example, Strab. 7 F 16a Radt).[37] Unfortunately, the river Pontus cannot be identified with certainty, since it is otherwise unknown,[38] though it is usually thought to be a tributary of the Strymon River.[39]

Pace Boshnakov, the close verbal parallels between *Mir.* and pseudo-Antigonus cannot be explained as due to an oral exchange of information; they rather imply textual dependence on a shared source. This source can only be Theopompus, whose original topographical indication might have been either selectively reported or slightly adapted by the citing authors.[40] That *Mir.* depends on pseudo-Antigonus[41]

---

34 See Rose (1863) 255; Schrader (1868) 227; Geffcken (1892) 89; Grenfell and Hunt (1909) F 257b; Jacoby (1930) 391; Regenbogen (1940) 1407–1408; Eichholz (1965) 96; Giannini (1965) 281; Flashar (1972) 129; Vanotti (2007) 197; Gauger and Gauger (2010) 232. Overduin (2015) 120 suggested that the compiler of *Mir.* may have used Nicander as a source; this hypothesis has been rightly rejected by Giacomelli (2023) 311.

35 Boshnakov (2003) 283–285, who also speculates about the possible role played by Eudoxus of Cnidus in the development of this tradition.

36 Boshnakov (2003) 284.

37 See Talbert (2000) 49 E-F.

38 Cf. Pape and Benseler (1911) 1235.

39 See Eichholz (1965) 97: modern Strumitza; Sharples (1998) 185 n. 533: modern Radovitz. Sørensen (2022b) 821 considers the possibility that there was a mistake in Theopompus' text and that "Pontus" refers to the Black Sea region.

40 We know that Theopompus mentioned the Agrieis/Agrianes and that he used the form Ἀγριεῖς: see Steph. Byz. α 47 Billerbeck, *s.v.* Ἀγρίαι = Theopomp., *FGrHist/BNJ* 115 F 268b. From a syntactical point of view, the phrase περὶ τὴν τῶν Σιντῶν καὶ Μαιδῶν χώραν καλουμένην τῆς Θράκης is not above suspicion. The mention of a specific toponym may be lost in a lacuna: see Giacomelli (2023) 311.

41 As tentatively suggested by Giannini (1965) 281; Eleftheriou (2018b) 264.

is improbable, not only because of the difference concerning the geographical indi-cation, but also because *Mir.* contains in general more information. Moreover, and more importantly, the section of *Mir.* under examination preserves much Theo-pompan material that is not paralleled in pseudo-Antigonus.

We do not know in which context Theopompus originally spoke of this stone, since neither pseudo-Antigonus nor the *Paradoxographus Palatinus* mention a specific book of the *Philippica*. According to Morison,[42] this fragment comes from a digression concerning Philip's II foundation of either Philippopolis (cf. *FGrHist/BNJ* 115 F 110, from book 13) or Kabyle/Kalybe (cf. *FGrHist/BNJ* 115 F 220, from book 47), where, according to Strabo (7.6.2.320C), the king "settled the most villainous people of his kingdom" (πόλις Καλύβη, Φιλίππου τοῦ Ἀμύντου τοὺς πονηροτάτους ἐνταῦθα ἱδρύσαντο; trans. Jones).[43]

Even though this attribution remains speculative, it seems probable that Theo-pompus was interested in this area because of Philip's II expansion and founda-tions. It is worth noting that the place mentioned by Theophrastus, *De Lapidibus* 2.12–13 (p. 60 Eichholz), namely, Binae (*Binai*, perhaps to be identified with the Thracian city of Benna referred to by Stephanus of Byzantium, B 67 Biller-beck, *s.v.* Βέννα[44]) is also mentioned in *Etym. Magn. Genuinum* 118 Lasserre and Livadaras, *s.v.* Βίνη: πόλις. μέμνηται δὲ ταύτης Ἡρόθεος [Ἡρόθεος Α: Ηρ()δ() Β: damn. Lasserre and Livadaras], καί φησιν ὠνομάσθαι ὑπὸ Φιλίππου οἰκισθεῖσαν ἀπὸ τῶν ἐν αὐτῇ συνοικισθέντων μοιχῶν. οὕτως Ὧρος. ("Binae: a city. Herotheus mentions it saying that it was so called because it was founded by Philip who gath-ered adulterers there. So Orus.").[45]

Binae is thus said to have been founded by Philip II and to have been named after the verb βινέω ("to screw") because the king collected adulterers there. The identity of the ultimate source for this report remains mysterious. The transmitted name Ἡρόθεος, which Lasserre and Livadaras regarded as corrupt, was tentatively emended to Δωρόθεος by Ritschl.[46] The latter had Dorotheus the lexicographer in mind, but a more plausible candidate might be the rather obscure namesake who is mentioned by Athenaeus 7.276f as the author of *Histories of Alexander*, in at least six books. According to the only preserved fragment of this work, both Philip II and Alexander were lovers of song (*FGrHist/BNJ* 145 F 1).[47] We may also consider the (venturesome) hypothesis that Theopompus was the source behind this report, since the historian sarcastically renamed Philippopolis as "Poneropolis" (*FGrHist/BNJ* 115 F 110 = *Suda* Δ 1423, *s.v.* Δούλων πόλις), suggesting that Philip II had founded that city by collecting there "those discredited for wickedness, such as informers, perjurers, and lawyers and about two thousand other wicked people"

---

42 Morison (2014) on F 268a.
43 Morison (2014) on F 268a.
44 Cf. also Procop. *Aed.* 4.4 (Βίνεος).
45 See also *Etym. Magn. auctum* 143 Lasserre and Livadaras, *s.v.* Βίνη; *Symeonis Etym.* 101–102 Lasserre and Livadaras, *s.v.* Βίνη; Tzetz. *Versus iambici* 1.161 Leone (πόλει δὲ μοιχῶν κλῆσις ἦν βινηρία). Cf. Oberhummer (1897) 475; Zambon (2000) 72 n. 4.
46 Ritschl (1866) 662; Lasserre and Livadaras (1992) 438.
47 Dorotheus is also mentioned by Plin. *NH* 1.12; 13 = *FGrHist/BNJ* 145 T 1.

(trans. Morison).[48] However, emendation is perhaps unnecessary in the end, since Ἡρόθεος, which is attested as a proper name in the Hellenistic period, may have been an otherwise unknown writer.[49]

2.3 Mir. 124

Εἶναι δέ φασι καὶ τόπον τινὰ παρ' αὐτοῖς οὐ λίαν μικρόν, ἀλλ' ὡς ἂν εἴκοσί που σταδίων, ὃς φέρει κριθάς, αἷς οἱ μὲν ἄνθρωποι χρῶνται, οἱ δ' ἵπποι καὶ βόες οὐκ ἐθέλουσιν αὐτὰς ἐσθίειν, οὐδ' ἄλλο οὐδέν· ἀλλ' οὐδὲ τῶν οἰῶν οὐδὲ τῶν κυνῶν οὐδεμία τολμᾷ γεύσασθαι τῆς κόπρου τῶν ἀνθρώπων, οἵτινες ἂν ἐκ τῶν κριθῶν τούτων μᾶζαν φαγόντες ἢ ἄρτον ἀφοδεύσωσι, †τῷ θνήσκειν†.

They also say that there is a district among them, not too small, but somewhere about twenty stades, which bears the barley which men use, but horses and cattle will not eat it, nor will any other animal; nor will any pigs[50] nor dogs venture to touch the excrement of men who void after eating meal or bread made from this barley, †because death follows†.

The only parallel for this story is provided by Theophrastus, *De odoribus* 2.4:

ἐπεὶ τοῖς γε ἄλλοις καὶ τὰ παντελῶς ἄοδμα φαινόμενα δίδωσί τινα ὀσμήν, ὥσπερ αἱ κριθαὶ τοῖς ὑποζυγίοις αἱ ἐκ τῆς Κεδροπόλιος, ἃς οὐκ ἐσθίουσιν διὰ τὴν κακωδίαν.[51]

Thus things which appear to us to have no odour give forth an odour of which other animals are conscious: for instance beasts of burden can smell the barley of Kedropolis, and refuse to eat it because of its evil odour.

(trans. Hort)

Even though both Theophrastus and *Mir.* clearly refer to the same phenomenon, their descriptions are rather different with regard to content and vocabulary. In particular, the second part of this chapter of *Mir.* is missing in the Theophrastean account. It seems therefore improbable that *Mir.* depends on Theophrastus.

---

48 Theopomp., *FGrHist/BNJ* 115 F 110 = *Suda* Δ 1423, *s.v.* Δούλων πόλις: ἔστι δέ τις καὶ περὶ Θρᾴκην Πονηρόπολις, ἣν Φίλιππόν φασι συνοικίσαι, τοὺς ἐπὶ πονηρίᾳ διαβαλλομένους αὐτόθι συναγαγόντα, συκοφάντας, ψευδομάρτυρας καὶ τοὺς συνηγόρους καὶ τοὺς ἄλλους πονηροὺς ὡς δισχιλίους, ὡς Θεόπομπος ἐν ιγ´ τῶν Φιλιππικῶν φησι. See also Plin. *HN* 4.41. On these foundations, see Hammond and Griffith (1979) 558–559; Zambon (2000).

49 See *LGPN* V1–73576 = *IG* XII (9) 1189.10 (I f. Ἡρόθεος II); V1–73577 = *IG* XII (9) 1189.10 (II s. Ἡρόθεος I); V2–31031 = *IG* II² 7015 = *PA* 6533; V5a-30419 = *IEph* 3429, 33 ([Η]ρόθεος). In his unpublished notes for *FGrHist* IV-VI, Jacoby accepted the existence of a writer named Herotheus (I owe this information to Stefan Schorn).

50 Note that Hett accepts the form ὑῶν ("pigs"), while Giacomelli prints οἰῶν ("sheep"): on this textual issue, see Giacomelli (2023) 313.

51 Ed. Eigler and Wöhrle (1993) 22, with commentary on p. 61.

Since this chapter of *Mir.* concerns the same area as the previous chapter (cf. παρ' αὐτοῖς, i.e., the Sinti and Maedi mentioned in *Mir.* 123),[52] which derives from Theopompus, it has been assumed that the present chapter also derives from the historian.[53] Given the lack of parallels in the extant fragments of Theopompus, however, this attribution remains uncertain. Interestingly enough, the toponym indicated by Theophrastus, that is, Kedropolis, recurs in a passage from Aristotle's *Historia animalium* 8(9) that seems to be the ultimate source of *Mir.* 126 (see Section 2.5).

2.4 Mir. 125

Ἐν δὲ Σκοτούσαις τῆς Θετταλίας φασὶν εἶναι κρηνίδιόν τι μικρόν, ἐξ οὗ ῥεῖ τοιοῦτον ὕδωρ ὃ τὰ μὲν ἕλκη καὶ θλάσματα ταχέως ὑγιεινὰ ποιεῖ καὶ τῶν ἀνθρώπων καὶ τῶν ὑποζυγίων, ἐὰν δέ τις ξύλον μὴ παντάπασι συντρίψας ἀλλὰ σχίσας ἐμβάλῃ, συμφύεται καὶ πάλιν εἰς τὸ αὐτὸ καθίσταται.

In Scotussae in Thessaly they say that there is a little spring, from which a kind of water flows, which quickly heals wounds and bruises both of men and beasts, but if one puts a log of wood into it without completely crushing it, but only breaking it in half, it grows again and returns to its original state.

This chapter has parallels in two fragments of Theopompus preserved by pseudo-Antigonus (via Callimachus) and Pliny:[54]

Ps.-Antig. *Mir.* 142 = Callim. F 407 XIV Pfeiffer = Theopomp., *FGrHist/ BNJ* 115 F 271a: ἐν Σκοτούσσῃ δ' εἶναι κρήνην ἰδίαν [κρήνην ἰδίαν P: κρηνίδιον Bentley] οὐ μόνον ἀνθρώπων ἕλκη, ἀλλὰ καὶ βοσκημάτων ὑγιάζειν δυναμένην. κἂν ξύλον δὲ σχίσας ἢ θραύσας ἐμβάλῃς, συμφύειν.[55]

In Scotussa there is a peculiar spring that is capable of healing the wounds not only of humans, but also of livestock. And if you throw a piece of wood into it, which is split or shattered, it grows together. (my translation)

Plin. *HN* 31.17 = Theopomp., *FGrHist/BNJ* 115 F 271b: *Theopompus in Scotusaeis lacum esse dicit qui volneribus medeatur.*[56]

Theopompus says that among the people of Scotussa is a lake that heals wounds.

(trans. Jones)

---

52 See Giacomelli (2023) 312.
53 See Giannini (1965) 283; Flashar (1972) 129; Vanotti (2007) 197.
54 This chapter is also paraphrased in Leon. Mag. *Therm.* 87–90 = 92–95 Gallavotti.
55 Ed. Musso (1985) 63; cf. also Eleftheriou (2018a) 198.
56 Ed. Mayhoff (1897) 7–8.

Mir. 125 is also paralleled in *Paradoxographus Florentinus* 9, which is said to depend on the paradoxographer Isigonus of Nicaea (*FGrHist* 1659 F 5):

περὶ Σκοτοῦσσαν τῆς Θεσσαλίας κρηνίδιόν ἐστι μικρόν, ὃ τὰ ἕλκη πάντα θεραπεύει καὶ τῶν ἀλόγων ζῴων· εἰς ὃ ἐάν τις ξύλον μὴ λίαν συντρίψας, ἀλλὰ σχίσας ἐμβάλη, ἀποκαθίσταται· οὕτως κολλῶδες ἔχει τὸ ὕδωρ, ὥς φησιν Ἰσίγονος.[57]

Near Scotussa in Thessaly there is a small little spring that heals all wounds, even those of irrational animals. If someone casts a branch into it, one that has not been entirely shattered but split, it is restored. So glutinous is its water, as Isigonus says. (trans. Greene, with modifications)

The remarkable similarities in both content and vocabulary highlighted in the following table (in bold) suggest that these passages ultimately derive from the same source, which can be reasonably identified with Theopompus, who is named as a source by both pseudo-Antigonus and Pliny:[58]

Mir. *125*	Par. Flor. *9*	Ps.-Antig. Mir. *142*	Plin. HN *31,17*
Ἐν δὲ **Σκοτούσαις τῆς Θετταλίας** φασὶν **εἶναι κρηνίδιόν** τι **μικρόν**, ἐξ οὗ ῥεῖ τοιοῦτον ὕδωρ ὃ **τὰ** μὲν **ἕλκη** καὶ θλάσματα ταχέως **ὑγιεινὰ** ποιεῖ καὶ **τῶν ἀνθρώπων** καὶ τῶν ὑποζυγίων, **ἐὰν** δέ **τις ξύλον μὴ** παντάπασι **συντρίψας ἀλλὰ σχίσας ἐμβάλη**, **συμφύεται** καὶ πάλιν εἰς τὸ αὐτὸ **καθίσταται.**	**περὶ Σκοτοῦσσαν τῆς Θεσσαλίας κρηνίδιόν ἐστι μικρόν, ὃ τὰ ἕλκη** πάντα θεραπεύει καὶ τῶν ἀλόγων ζῴων· εἰς ὃ **ἐάν τις ξύλον μὴ** λίαν **συντρίψας, ἀλλὰ σχίσας ἐμβάλη,** ἀποκαθίσταται· οὕτως κολλῶδες ἔχει τὸ ὕδωρ, ὥς φησιν Ἰσίγονος.	**ἐν Σκοτούσση** δ' εἶναι κρήνην ἰδίαν [κρηνίδιον?] οὐ μόνον **ἀνθρώπων ἕλκη**, ἀλλὰ καὶ βοσκημάτων **ὑγιάζειν** δυναμένην. **κἂν ξύλον δὲ σχίσας** ἢ θραύσας **ἐμβάλης, συμφύειν.**	*Theopompus in* **Scotusaeis** *lacum esse dicit qui volneribus medeatur.*

57 Ed. Greene (2022) 632.
58 See Schrader (1868) 226 n. 11; Müllenhoff (1870) 428; Grenfell and Hunt (1909) F 260c; Öhler (1913) 71; Jacoby (1930) 391; Eleftheriou (2018b) 275; Greene (2022) 685–686. The shared source is not Theophrastus, as apparently suggested by Flashar (1972) 130, though I suspect that this mention of "Theophrastus" by Flashar is just a typographical error for "Theopompus." On p. 41, Flashar claims that *Mir.* 119–129 are from Theopompus. Gauger and Gauger (2010) 232–233 repeat Flashar's erroneous attribution to Theophrastus, adding that the information ultimately derives from Theopompus.

Still, the mutual relationships between these passages are difficult to recover because of the uncertainty surrounding their relative chronology and the relationship between *Mir.* and the *Paradoxographus Florentinus*.[59] As noted above, the direct source of the *Paradoxographus Florentinus*, which has been dated to the first or 2nd century AD, but which may be later,[60] is said to be Isigonus of Nicaea's *Unbelievable Things* (*FGrHist* 1659 F 5), which is usually dated to the 1st centuries BC or AD.[61] The version of the *Paradoxographus Florentinus* is slightly shorter than *Mir.*'s, but the two passages clearly provide the same information. The additional report concerning the glutinous nature of this marvelous water can be an addition by either the compiler of the *Paradoxographus Florentinus* or, more probably, by Isigonus.[62] We may then conclude that either Isigonus depends on *Mir.* (which would imply a Hellenistic date for *Mir.*) or both texts depend on a common source (as was supposed by Öhler).[63] Alternatively, Giacomelli has recently considered the possibility that the attribution of this chapter of the *Paradoxographus Florentinus* to Isigonus might be the result of a misunderstanding on the part of a supposed intermediate author who combined material from *Mir.* with material from Isigonus.[64] *Non solvitur aenigma.* In any case, it is impossible that Isigonus (or the *Paradoxographus Florentinus*' intermediate source) and the compiler of *Mir.* depend on the shorter version of pseudo-Antigonus, which was reasonably assigned by Öhler to another branch of the tradition.[65]

It is possible that *Mir.* and the *Paradoxographus Florentinus*, which both specify that the piece of wood could only be restored to its original state if it had not been completely shattered (μὴ παντάπασι συντρίψας ἀλλὰ σχίσας [*Mir.*]; μὴ λίαν συντρίψας, ἀλλὰ σχίσας [*Par. Flor.*]), are closer to Theopompus' original text, while pseudo-Antigonus' σχίσας ἢ θραύσας seems to be a shortening of the original account on the part of Callimachus or pseudo-Antigonus himself.

Pseudo-Antigonus and Pliny do not specify in which book of the *Philippica* Theopompus related this story. While Jacoby did not assign it to a specific book, Wichers, Müller, and Morison argued that it was part of Theopompus' description of Thessalian geography in book 9.[66] However, some doubts may arise about identifying with the Thessalian city the Scotussae (or Scotussa) mentioned by Theopompus.

59 See recently Giacomelli (2021) 350–355.
60 See recently Giacomelli (2021) 350–351 n. 141; Greene (2022) 645, with *status quaestionis* and previous literature.
61 See recently Giacomelli (2021) 353 n. 145; Greene (2022) 644–645 with n. 16; Sørensen (2020) Introduction (arguing for a date in the first century BC), with previous bibliography.
62 See Greene (2022) 686.
63 See Öhler (1913) 71.
64 Giacomelli (2021) 355.
65 Öhler (1913) 71. Öhler also suggested that Pliny used Varro as his immediate source, and that this branch of the tradition, just like pseudo-Antigonus, goes back to Callimachus; but see Greene (2022) 686 n. 189.
66 Wichers (1829) 75–76; Müller (1841) 291–292; Morison (2014) on F 271. Modern scholars have usually explained the reports about this marvelous spring by appealing to Thessaly's connection to Asclepius: see Stählin (1927) 615; Flashar (1972) 130; Greene (2022) 689. Also noteworthy is *IG* IX 2 397, an unfinished decree assigned to either Scotussa (so *IG*) or Pherae (so *SEG* XLII 533),

Since a report concerning Thessaly seems to interrupt as it were a series of chapters (*Mir.* 123–130) related to Thrace (in particular, the Strymon valley and the area north of the Chalcidice)[67] and since the two ancient authorities who cite Theopompus by name, namely, pseudo-Antigonus and Pliny, do not mention Thessaly, one may wonder whether Theopompus may have meant not the Thessalian Scotussa, but a lesser-known homonymous city which the ancient sources locate in Thrace, in the lower valley of the Strymon, not far from Herakleia Sintike (thus in an area which would perfectly fit the Thracian setting of the surrounding chapters).[68] The ancient Thracian settlement of Scotussa cannot be identified with certainty, but a location in the vicinity of Lake Tachinos, and more specifically at Sidirokastron, has been suggested.[69] Interestingly enough, this place is known for its healing hot springs.[70] If we suppose that Theopompus originally spoke of the Thracian Scotussa, we may then consider the reference to Thessaly to be a later addition by the compiler of *Mir.* (or by a common source used by *Mir.* and the *Paradoxographus Florentinus*), who may have identified the rather obscure place described by Theopompus with the better-known Thessalian city. A possible objection to this interpretation might be that if the compiler of *Mir.* followed a geographical pattern (cf. Section 3), it would then be difficult to explain why he changed the original Thracian location (which would fit his own geographical sequence), substituting for it a reference to Thessaly (which interrupts the geographical sequence). The same objection, however, remains valid even if we suppose that Theopompus spoke of the Thessalian city: for even in that case there is no clear reason why the compiler would have decided to insert the story concerning the Thessalian city among the chapters concerning Thrace, thereby interrupting his own geographical pattern. So, even if it remains possible that the compiler of *Mir.*, for whatever reason, decided to insert a report concerning Thessaly into a section about Thrace, we may still cautiously consider the possibility that Theopompus originally spoke of the Thracian Scotussa and that the compiler of *Mir.* (or a common source used by *Mir.* and the *Paradoxographus Florentinus*), when excerpting information from Theopompus' description of Thrace (directly or through an intermediate source), added a

---

which mentions a cult of Asclepius: cf. also Mili (2015) 144 n. 262. The healing waters in Tempe are mentioned by Ael. *VH* 3.1 (*FGrHist/BNJ* 115 F 80).

67 With the exception of Mir. 127, about the Heneti, which may have been inserted into the 'Thracian series' because that chapter elaborates on the same topic as 126, that is, the relationship between humans and birds: see below Sections 2.5 and 2.6. Note that in the collection of pseudo-Antigonus as well, the report concerning Scotussa follows another fragment of Theopompus concerning a place in Thrace (ἐν †κιγχρωψωσιν† τοῖς Θρᾳξὶν), which has a parallel in *Mir.* 129.

68 See Strab. 7 F 16a Radt; Plin. *HN* 4.35; 4.42; Ptol. *Geog.* 3.13.31. The *Tabula Peutingeriana* locates the city to the southwest of Herakleia Sintike. See TM Geo 34021 = Talbert (2000) 50 D2. Cf. Oberhummer (1927) 617; Papazoglou (1988) 381–382; Loukopoulou (2004) 857.

69 See Oberhummer (1927) 617; Loukopoulou (2004) 857. For the identification with Sidirokastron, see especially Papazoglou (1988) 381–382.

70 See www.spa.gr. Stählin (1927) 615 noticed that no healing spring is known to be found in the vicinity of the Thessalian Scotussa: the spring has been identified with that of Brousonia, which, however, does not seem to have healing properties.

reference to Thessaly in order to correct the historian's account or simply because he confused the Thracian city with the Thessalian one.

2.5 Mir. *126*

1. Περὶ δὲ τὴν Θρᾴκην τὴν ὑπὲρ Ἀμφίπολιν φασὶ γενέσθαι <τι> τερατῶδες καὶ ἄπιστον τοῖς μὴ τεθεαμένοις. ἐξιόντες γὰρ οἱ παῖδες ἐκ τῶν κωμῶν καὶ τῶν ἐγγὺς χωρίων ἐπὶ θήραν τῶν ὀρνιθαρίων συνθηρεύειν παραλαμβάνουσι τοὺς ἱέρακας, καὶ τοῦτο ποιοῦσιν οὕτως· ἐπειδὰν προέλθωσιν εἰς τόπον ἐπιτήδειον, καλοῦσι τοὺς ἱέρακας ὀνομαστὶ κεκραγότες· οἱ δ᾽ ὅταν ἀκούσωσι τῶν παίδων τὴν φωνήν, παραγινόμενοι κατασοβοῦσι τοὺς ὄρνιθας· οἱ δὲ δεδιότες ἐκείνους καταφεύγουσιν εἰς τοὺς θάμνους, ὅπου αὐτοὺς οἱ παῖδες ξύλοις τύπτοντες λαμβάνουσιν. 2. Ὁ δὲ πάντων ἄν τις μάλιστα θαυμάσειεν· οἱ μὲν γὰρ ἱέρακες ὅταν αὐτοί τινα λάβωσι τῶν ὀρνίθων, καταβάλλουσι τοῖς θηρεύουσιν, οἱ δὲ παῖδες ἁπάντων τῶν ἁλόντων μέρος τι τοῖς ἱέραξιν ἀποδόντες ἀπέρχονται.

1. In Thrace above Amphipolis they say that there is a remarkable occurrence, which is incredible to those who have not seen it. For boys, coming out of the villages and places round to hunt small birds, take hawks with them, and behave as follows: when they have come to a suitable spot, they call the hawks addressing them by name; when they hear the boys' voices, they swoop down on the birds. The birds fly in terror into the bushes, where the boys catch them by knocking them down with sticks. 2. But there is one most remarkable feature in this; when the hawks themselves catch any of the birds, they throw them down to the hunters, and the boys after giving a portion of all that is caught to the hawks go home.

This chapter is unparalleled in the extant fragments of Theopompus. It does have parallels, however, in Aristotle, *Historia animalium* 8(9).36.620a33–620b8; pseudo-Antigonus, *Mir.* 27–28; Pliny, *Naturalis historia* 10.23; Philo of Alexandria, *De animalibus* 37 (only preserved in Armenian), and Aelian, *De natura animalium* 2.42:

Arist. *HA* 8(9).36.620a33–620b8: ἐν δὲ Θρᾴκῃ τῇ καλουμένῃ ποτὲ Κεδρειπόλιος ἐν τῷ ἕλει θηρεύουσιν οἱ ἄνθρωποι τὰ ὀρνίθια κοινῇ μετὰ τῶν ἱεράκων· οἱ μὲν γὰρ ἔχοντες ξύλα σοβοῦσι τὸν κάλαμον καὶ τὴν ὕλην ἵνα πέτωνται τὰ ὀρνίθια, οἱ δ᾽ ἱέρακες ἄνωθεν ὑπερφαινόμενοι καταδιώκουσιν· ταῦτα δὲ φοβούμενα κάτω πέτονται πάλιν πρὸς τὴν γῆν· οἱ δ᾽ ἄνθρωποι τύπτοντες τοῖς ξύλοις λαμβάνουσι, καὶ τῆς θήρας μεταδιδόασιν αὐτοῖς· ῥίπτουσι γὰρ τῶν ὀρνίθων, οἱ δὲ ὑπολαμβάνουσιν. καὶ περὶ τὴν Μαιῶτιν δὲ λίμνην τοὺς λύκους φασὶ συνήθεις εἶναι τοῖς ποιουμένοις τὴν θήραν τῶν ἰχθύων· ὅταν δὲ μὴ μεταδιδῶσι, διαφθείρειν αὐτῶν τὰ δίκτυα ξηραινόμενα ἐν τῇ γῇ.[71]

In the part of Thrace once named *Kedreipolios*, men hunt the small birds in the marsh in partnership with the hawks. The men hold sticks and stir the reeds and brushwood to make the small birds fly, while the hawks from

---

71 Ed. Balme (2002) 428–429.

above appear overhead and chase them down. In fear they fly down again to the ground; the men strike them with the sticks and take them, and give the hawks a share in the prey: they throw them some of the birds and the hawks catch them. And around Maeotis Lake too they say the wolves are habituated to the men who bring in the catch of fish, and when they do not give them a share they destroy their nets as they are drying on the ground. (trans. Balme, with modifications)

Ps.-Antig. *Mir.* 27–28: Φησὶν περὶ Κωνώπιον τῆς Μαιώτιδος λίμνης τοὺς λύκους παρὰ τῶν ἁλιέων λαμβάνοντας τροφὴν φυλάττειν τὴν θήραν· ἂν δ᾽ ὑπολάβωσίν τι ἀδικεῖσθαι, λυμαίνεσθαι τὰ λίνα καὶ τοὺς ἰχθύας αὐτῶν. Ἐν Θράκῃ δὲ τῇ κληθείσῃ ποτὲ †Κεδριπολι† τοὺς ἀνθρώπους καὶ τοὺς ἱέρακας κοινῇ θηρεύειν τὰ ὀρνιθάρια· τοὺς μὲν γὰρ σοβεῖν τοῖς ξύλοις, τοὺς δὲ ἱέρακας καταδιώκειν, τὰ δὲ φεύγοντα εἰς τοὺς ἀνθρώπους ἐμπίπτειν· διὸ καὶ μεταδιδόναι τοῖς ἱέραξιν αὐτοὺς τῶν ληφθέντων.[72]

He (sc. Aristotle) says that around Conopion at Maeotis Lake wolves guard the prey after having received food from the fishermen; but if they think that they have suffered injustice, they destroy their nets and fish. And in the part of Thrace once named †Kedripolit†, men and hawks together hunt the small birds. For the men chase them away with sticks, and then the hawks chase them down; the small birds, trying to escape from them, fall prey to the men. For this reason the men give the hawks a share in the prey. (my translation)

Plin. *HN* 10.23: *In Thraciae parte super Amphipolim homines et accipitres societate quadam aucupantur. hi ex silvis et harundinetis excitant aves, illi supervolantes deprimunt rursus; captas aucupes dividunt cum iis. traditum est missas in sublime ibi excipere eos et, cum sit tempus capturae, clangore ac volatus genere invitare ad occasionem. simile quiddam lupi ad Maeotim paludem faciunt. nam nisi partem a piscantibus suam accepere, expansa eorum retia lacerant.*[73]

In the district of Thrace inland from Amphipolis, men and hawks have a sort of partnership for fowling: the men put up the birds from woods and reed-beds, and the hawks flying overhead drive them down again; the fowlers share the bag with the hawks. It is reported that when the birds have been put up, the hawks intercept them in the air, and by their screaming and their way of flying the hawks invite the fowlers to seize the opportunity when it is time for a catch. Wolves at the Maeotic Marsh act somewhat in the same way, for unless they get a share from the fishermen they tear their nets when spread. (trans. Rackham, with modifications)

Philo *Anim.* 37 (English translation from the Armenian by Terian): One would think that the story of the Thracian falcons is most fascinating. When I heard it for the first time I did not believe it, not until many natives of the

---

72   Ed. Musso (1985) 32. Eleftheriou (2018a) 142 accepts Giannini's δ᾽ ἐν τῇ κληθείσῃ ποτὲ Κεδριπόλ<ε>ι.

73   Ed. Mayhoff (1909) 225–226.

province and some of the foreigners who visit us – those simple-minded and unpremeditating men – told whatever they know about them. They said that these work concertedly, aiding the fowlers by preying swiftly upon the catch. They live in the forests, as is proper, in thick, tangled woods where a variety of birds is often found, and cooperate with those who come to fowl, especially those who approach cautiously and who do not refuse to share their reward, as is only fair for comrades in arms. What an example they set of mutual aid! First of all, the fowlers shake the trees. Because the weak and frail birds cannot withstand the shaking, they fall from fear and try to fly away. But the falcons swoop down and knock them to the ground by pecking, preparing easy game for those who have come to fowl. When they are forcibly thrown from the branches by the shaking, they flutter fearfully as a result of the pounce and are thrown into a state of utter confusion and terror. Then they are easily caught, even by the fowlers' hands. Nevertheless the fowlers are glad to share the catch partly to compensate the falcons for their cooperation and partly to invite their aid for future expeditions.[74]

Ael. *NA* 2.42: ἀκούω δὲ ὅτι ἐν τῇ Θρᾴκῃ καὶ ἀνθρώποις εἰσὶ (*sc.* ἱέρακες) σύνθηροι ἐν ταῖς ἑλείοις ἄγραις. καὶ ὁ τρόπος, οἱ μὲν ἄνθρωποι τὰ δίκτυα ἁπλώσαντες ἡσυχάζουσιν, οἱ δὲ ἱέρακες ὑπερπετόμενοι καὶ φοβοῦσι τοὺς ὄρνις καὶ συνωθοῦσιν εἰς τὰς τῶν δικτύων περιβολάς. τῶν οὖν ᾑρημένων οἱ Θρᾷκες μέρος ἀποκρίνουσι καὶ ἐκείνοις, καὶ ἔχουσιν καὶ αὐτοὺς πιστούς· μὴ δράσαντες δὲ τοῦτο ἑαυτοὺς τῶν συμμάχων ἐστέρησαν.[75]

And I am told that in Thrace they (*sc.* the hawks) even join with men in the pursuit of marsh-fowl. And this is how they do it. The men spread their nets and keep still while the hawks fly over them and scare the fowl and drive them into the circle of nets. For this the Thracians allot a portion of their catch to the hawks and find them trusty; if they do not do so, they at once deprive themselves of helpers. (trans. Scholfield, with modifications)

Even though these passages differ on various points, it seems that the entire tradition ultimately depends on the anecdote reported in book 8(9) of Aristotle's *Historia animalium*.[76] Pseudo-Antigonus explicitly quotes Aristotle and provides the version closest to his, often using the same vocabulary and reporting Aristotle's story concerning wolves at Lake Maeotis immediately before our anecdote (thus reversing Aristotle's order).[77] The toponym attested in Aristotle's passage, Κεδρειπολιός (with minor variants in the manuscript tradition),[78] is paralleled in the corrupt reading Κεδριπολι that is reported by the *testis unicus* of pseudo-Antigonus and in the form Κεδροπόλιος used by Theophrastus, *De odoribus* 2.4.

74 Terian (1981) 84–85.
75 Ed. Garcia Valdes, Llera Fueyo and Rodriguez-Noriega Guillen (2009) 47.
76 Cf. Longo (1986–1987).
77 According to Eleftheriou (2018b) 80, "cette inversion peut indiquer l'existence d'une source intermédiaire, présentant les notices dans cet ordre, car Ps.-Antigonos est en général très fidèle à ses sources." On the anecdote concerning Lake Maeotis, see recently Schnieders (2019) 878–879.
78 See Balme (2002) 428; Schnieders (2019) 876; Giacomelli (2021) 301.

Theophrastus, however, mentions that place not in connection with the hunting habit described by Aristotle, but with regard to beasts of burden refusing to eat barley because of its evil smell (see Section 2.3).

The version of *Mir.*, which admittedly presents some differences with respect to Aristotle's (especially in its first part), still seems to presuppose it. *Mir.* and Pliny probably share an intermediate source, which in turn was based on Aristotle's *Historia animalium*. On the one hand, Pliny's version is generally closer to the Aristotelian passage (as noted above, it also reports the additional anecdote about wolves at Lake Maeotis) but shares with *Mir.* (unlike Aristotle) the topographical notation concerning the region of Thrace "above Amphipolis" (Περὶ δὲ τὴν Θρᾴκην τὴν ὑπὲρ Ἀμφίπολιν – *In Thraciae parte super Amphipolim*). On the other hand, *Mir.* still has some words of the original Aristotelian passage (e.g., ξύλοις τύπτοντες λαμβάνουσιν – τύπτοντες τοῖς ξύλοις λαμβάνουσι) that are unparalleled in Pliny's Latin text. We may thus regard those elements of *Mir.* that are in contrast with the versions of both Aristotle and Pliny (namely, the presence of boys, the boys' shouting, and the hawks seizing some of the birds and throwing them down to the hunters) as innovations or interpolations by the compiler of *Mir.*, which might reflect another kind of hunting[79] and were perhaps meant to make the report even more impressive and sensational.[80] This motivation is also suggested by the compiler's emphasis on the sensational character of the report as well as on its veracity (allegedly based on eyewitness testimony: <τι> τερατῶδες καὶ ἄπιστον τοῖς μὴ τεθεαμένοις . . . Ὃ δὲ πάντων ἄν τις μάλιστα θαυμάσειεν).[81] Aelian's version is the furthest removed from Aristotle's and merely locates the hunting habit in Thrace without further specification.[82] Philo, who also elaborates on Aristotle's story, simply speaks of "Thracian falcons."[83]

Kedropolis (or Kedreipolis/Kedripolis) must refer to a place in Thrace, but the precise location can no longer be recovered.[84] It seems improbable that this was the old name of Amphipolis, as was cautiously suggested by Flashar.[85] More persuasively, but still speculatively, Dittenberger argued that this toponym may not

---

79 As suggested by Longo (1986–1987).

80 As argued by Lindner (1973) 112–113; Schnieders (2019) 877. According to Eleftheriou (2018a) 232, by contrast, "la version du Ps.-Aristote est plus longue et offre d'informations supplémentaires, ce qui témoigne l'existence d'une version du texte hors la version d'Aristote, issue, probablement, d'une contamination des sources."

81 Cf. Longo (1986–1987) 43. The compiler of *Mir.* seems to suggest here a sort of paradoxical climax, which culminates in a report that is presented as even more marvelous than the preceding ones: cf. Pajón Leyra (2011) 32 with n. 11. Another example is provided by *Mir.* 135 (ὃ καὶ θαυμάσειεν ἄν τις μάλιστα): see below Section 2.14.

82 Aelian reports the episode concerning wolves at Lake Maeotis at *NA* 6.65.

83 On Philo's dependence on Aristotle, see Terian (1981) 56, 151. Longo (1986–1987) 44–45 does not exclude the possibility that Philo also heard this story from some eyewitnesses (as Philo claims at the beginning of the passage).

84 On this toponym, see also Giacomelli (in this volume) 242–243.

85 Flashar (1972) 130. *Contra* Giacomelli (2023) 314: "La realtà storica del toponimo Cedropoli/ Cedripoli è controversa: Flashar, sulla scorta di ipotesi precedenti, propone di considerarlo antico nome di Anfipoli, ma non vi è alcuna indicazione in questo senso nei testi antichi."

refer to the former name of a city, but to the area once ruled by (and therefore named after) the Thracian dynast Ketriporis in the 350s.[86] However, Dittenberger's attempt to identify this area with "die Gegend östlich vom Strymon" on the basis of *Mir.* (περὶ δὲ τὴν Θράκην τὴν ὑπὲρ Ἀμφίπολιν) is doubtful. Since *Mir.* and Pliny seem ultimately to depend on Aristotle, it is possible that the reference to the area "above Amphipolis," reported in their shared source, originated as an attempt to make sense of an obscure toponym by replacing it with a reference to a better-known city (Amphipolis) that was located in the same area.[87]

We can thus conclude that *Mir.* 126 probably ultimately depends on Aristotle's *Historia animalium* 8(9).[88] That Theopompus was the intermediate source used by *Mir.* and Pliny, as tentatively suggested by Giannini,[89] is unlikely, because that source seems to have been primarily interested in animal behavior, as is shown by the fact that Pliny also includes the anecdote concerning wolves at Lake Maeotis.

2.6 Mir. *127*

Θαυμαστὸν δέ τι καὶ παρὰ τοῖς Ἐνετοῖς φασὶ γίνεσθαι. ἐπὶ γὰρ τὴν χώραν αὐτῶν πολλάκις κολοιῶν ἀναριθμήτους μυριάδας ἐπιφέρεσθαι καὶ τὸν σῖτον αὐτῶν σπειράντων καταναλίσκειν· οἷς τοὺς Ἐνετοὺς πρὸ τοῦ ἐφίπτασθαι μέλλειν ἐπὶ τὰ μεθόρια τῆς γῆς προτιθέναι δῶρα, παντοδαπῶν καρπῶν καταβάλλοντας σπέρματα, ὧν ἐὰν μὲν γεύσωνται οἱ κολοιοί, οὐχ ὑπερβαίνουσιν ἐπὶ τὴν χώραν αὐτῶν, ἀλλ' οἴδασιν οἱ Ἐνετοὶ ὅτι ἔσονται ἐν εἰρήνῃ· ἐὰν δὲ μὴ γεύσωνται, ὡσεὶ πολεμίων ἔφοδον αὐτοῖς γινομένην οὕτω προσδοκῶσιν.

They relate a remarkable occurrence among the Heneti; for countless thousands of jackdaws come to their country and consume their grain, when they have sown it; before they are about to fly over there the Heneti put out gifts for the birds on their boundaries, putting down seeds of all kinds of fruits; if the jackdaws taste these, they

---

86 Dittenberger (1879), followed by Balme (1991) 309 n. a; Kullmann (2014) 94; Schnieders (2019) 876–877. Ketriporis is mentioned in *IG* II² 127 (from 356 BC) and in numismatic evidence: see Peter (1997) 143–146; Rhodes and Osborne (2003) 255–258, with previous literature. According to Schnieders (2019) 876, Aristotle and Theophrastus may have visited that place together. Cf. also Kullmann (2014) 94.

87 Cf. Longo (1986–1987) 43. According to Dittenberger (1879) 302; Schnieders (2019) 877, the compiler of *Mir.* took the topographical notation from a source other than Aristotle.

88 See Ziegler (1949) 1151; Giannini (1964) 135 n. 219; Flashar (1972) 130; Gómez Espelosín (2008) 236 n. 120. Schrader (1868) 222, 227 suggested that Theopompus used Aristotle as a source, while Müllenhoff (1870) 428 n. ** argued that Aristotle cannot be the source used by *Mir.* That Theophrastus played a role in the creation or in the transmission of this story, as supposed by Regenbogen (1940) 1406 and Flashar (1972) 130, is a hypothesis that depends on the relationship between Theophrastus' lost work Περὶ ζῴων φρονήσεως καὶ ἤθους (*On the Intelligence and Habits of Living Creatures*), Aristotle's *Historia animalium* 8(9), and *Mir.* 1–15: on this problem, and on the debated authorship of Aristotle's *Historia animalium* 8(9), see Flashar (1972) 42–44; Balme (1991) 1–13; Sharples (1995) 45–47; Pajón Leyra (2011) 116–17, 273–274; Dorandi (2017) 75; Schnieders (2019) 97–108, 200–214; Giacomelli (2021) 15–17; Mayhew (2021); (forthcoming). On *Mir.* 1–15, see also Hatzimichali (forthcoming).

89 Giannini (1965) 281–283.

do not pass over the border into their country, and the Heneti know that they will be in peace; but if they do not taste them, they expect as it were an invasion of the enemy.

This passage, which elaborates on the same topic as the previous one, namely, the relationship between humans and birds, has two parallels in Theopompus' fragments. The first is preserved by pseudo-Antigonus, *Mir.* 173, who cites Theopompus (*FGrHist/BNJ* 115 F 274a) via Callimachus (F 407 XLIV Pfeiffer). Pseudo-Antigonus' *testis unicus*, however, breaks off in the course of the narration, so that its conclusion is irremediably lost to us:

Τοὺς δὲ περὶ τὸν Ἀδρίαν ἐνοικοῦντας Ἐνετοὺς Θεόπομπον φάσκειν κατὰ τὸν σπόρου καιρὸν τοῖς κολοιοῖς ἀποστέλλειν δῶρα, ταῦτα δ᾽ εἶναι ψαιστὰ καὶ μάζας. Προθέντας δὲ τοὺς ταῦτα κομίζοντας ἀποχωρεῖν, τῶν δὲ ὀρνέων τὸ μὲν πλῆθος ἐπὶ τοῖς ὁρίοις μένειν τῆς χώρας συνηθροισμένον, δύο δ᾽ ἢ τρεῖς προσπτάντας καὶ καταμαθόντας ἀφίπτασθαι πάλιν καθάπερεί τινας πρέσβεις ἢ κατασκόπους. Ἐὰν μὲν οὖν τὸ πλῆ. . .[90]

(Callimachus says) that Theopompus claims that during the planting season the Heneti living on the Adriatic Sea send out gifts to the jackdaws; these are barley cakes and breads. When those who bring these things have offered them and depart, the multitude of birds remain massed on the borders of the country, but two or three fly forward and once they have inspected (the gifts), they fly back again, just like some ambassadors or scouts. If then . . . (trans. Morison, with modifications)

The second parallel is provided by Aelian, *Natura animalium* 17.16, which relates the same story, citing Theopompus (*FGrHist/BNJ* 115 F 274b) and Lycus (*FGrHist/BNJ* 570 F 4) as sources:

Θεόπομπος λέγει τοὺς περὶ τὸν Ἀδρίαν οἰκοῦντας Ἐνετούς, ὅταν περὶ τὸν ἄροτον τρίτον καὶ σπόρου ἡ ὥρα ᾖ, τοῖς κολοιοῖς ἀποστέλλειν δῶρα· εἴη δ᾽ ἂν τὰ δῶρα ψαιστὰ ἄττα καὶ μεμαγμέναι μάζαι καλῶς τε καὶ εὖ. βεβούλευται δὲ ἄρα ἡ τῶνδε τῶν δώρων πρόθεσις μειλίγματα τοῖς κολοιοῖς εἶναι καὶ σπονδῶν ὁμολογίαι, ὡς ἐκείνους τὸν καρπὸν τὸν Δημήτρειον μὴ ἀνορύττειν καταβληθέντα εἰς τὴν γῆν μηδὲ παρεκλέγειν. Λύκος δὲ ἄρα καὶ ταῦτα μὲν ὁμολογεῖ, καὶ ἐκεῖνα δὲ ἐπὶ τούτοις προστίθησι, . . . [lac. stat. Jacobs, Hercher: <προτιθέναι> Jacobs] καὶ φοινικοῦς ἱμάντας τὴν χρόαν, καὶ τοὺς μὲν προθέντας ταῦτα εἶτα ἀναχωρεῖν. καὶ τὰ μὲν τῶν κολοιῶν νέφη τῶν ὅρων ἔξω καταμένειν, δύο δὲ ἄρα ἢ τρεῖς προηρημένους κατὰ τοὺς πρέσβεις τοὺς ἐκ τῶν πόλεων πέμπεσθαι κατασκεψομένους τῶν ξενίων τὸ πλῆθος· οἵπερ οὖν ἐπανίασι θεασάμενοι, καὶ καλοῦσιν αὐτούς, ἢ πεφύκασιν οἱ μὲν καλεῖν, οἱ δὲ ὑπακούειν. ἔρχονται μὲν κατὰ νέφη, ἐὰν

---

90 Ed. Musso (1985) 71; cf. also Eleftheriou (2018a) 212.

δὲ γεύσωνται τῶν προειρημένων, ἴσασιν οἱ Ἐνετοὶ ὅτι ἄρα αὐτοῖς πρὸς
τοὺς ὄρνιθας τοὺς προειρημένους ἔνσπονδά ἐστιν, ἐὰν δὲ ὑπερίδωσι καὶ
ἀτιμάσαντες ὡς εὐτελῆ μὴ γεύσωνται, πεπιστεύκασιν οἱ ἐπιχώριοι ὅτι τῆς
ἐκείνων ὑπεροψίας ἐστὶν αὐτοῖς λιμὸς τὸ τίμημα. ἄγευστοι γὰρ ὄντες οἱ
προειρημένοι καὶ ἀδέκαστοί γε ὡς εἰπεῖν ἐπιπέτονταί τε ταῖς ἀρούραις καὶ τό
γε πλεῖστον τῶν κατεσπαρμένων συλῶσι πικρότατά γε ἐκεῖνοι, σὺν τῷ θυμῷ
καὶ ἀνορύττοντες καὶ ἀνιχνεύοντες.[91]

Theopompus says that at the season of the third ploughing and sowing the Heneti who live on the shores of the Adriatic dispatch presents to
the jackdaws, and these presents would be some barley cakes and breads
kneaded both beautifully and well. The exposition of these presents is
meant to placate the jackdaws and to declare a truce, so that they shall
refrain from digging up and collecting the fruit of Demeter sown in the
soil. And *Lycus* confirms this adding further the following details . . . and
scarlet thongs, and after setting them out they withdraw. And the clouds of
jackdaws remain outside the boundaries, while two or three birds, selected
like ambassadors from cities, are sent to take a good look and see how
many presents there are. After their inspection they return and summon
the birds, giving the call which is natural for them to utter and for the
others to respond to. And the birds come in clouds, and if they eat the
aforesaid presents, the Heneti know that there is a truce between them and
the aforesaid birds. If however they ignore and scorn them as skimpy and
refuse to eat them, the inhabitants are confident that a famine will be the
price they have to pay for this rejection. For if the aforesaid birds do not
eat and are, so to say, unbribed, they swoop upon the ploughlands and pil-
lage in the most distressing way the greater part of what has been sown,
digging up and tracking out the seeds in their anger. (trans. Scholfield,
with modifications)

The account of *Mir.* seems to be a shorter version of those by pseudo-Antigonus
and Aelian, which closely resemble each other. Pseudo-Antigonus mentions Theopompus as a source, while Aelian, after citing Theopompus at the very beginning
of the passage, mentions Lycus as an additional source. However, the scope of
Lycus' fragment is difficult to establish because of a probable lacuna in Aelian's
text shortly after the mention of Lycus.

It is worth noting that the similarities between the accounts of pseudo-
Antigonus and Aelian concern not only the first part of the story, which both
sources explicitly attribute to Theopompus, but also the second part, which in
Aelian's text follows the mention of Lycus. If we consider Lycus to be the source
of the entire section which follows his mention by Aelian, then we should consider the possibility that Callimachus, as suggested by Schrader, created a new

91 Ed. García Valdés, Llera Fueyo and Rodríguez-Noriega Guillén (2009) 412–413.

version of the story by conflating those of Theopompus and Lycus, but ascribed it only to Theopompus.[92]

However, the probable lacuna and the uncertainty surrounding the scope of Lycus' fragment in Aelian's text both suggest caution. As argued by Jacoby, followed by Flashar, it is possible that Lycus was cited by Aelian only for additional information concerning the use of scarlet thongs by the Heneti[93] (a detail that was apparently omitted by Theopompus) and that Theopompus was Aelian's main source for the entire story.[94]

Since the compiler of *Mir.* does not mention scarlet thongs, Jacoby concluded that he must have used Theopompus as a source. However, the summary character of *Mir.* makes such arguments from silence uncertain at best. All in all, the textual uncertainty surrounding Aelian's text does not allow us to reach firm conclusions concerning the respective scope of Theopompus' and Lycus' fragments, while the fact that Theopompus' and Lycus' versions were apparently similar to each other makes it all the more difficult to distinguish between them (apart from the detail concerning the use of scarlet thongs).[95] It is thus difficult to completely exclude the possibility that *Mir.* depends here on Lycus, even though a Theopompan origin probably remains the most likely possibility, since Theopompus is often used as a source in the section of *Mir.* under discussion (although probably *not* in the previous chapter).[96] According to Jacoby and other editors of Theopompus' fragments,[97] this story was originally part of Theopompus' description of the Adriatic area in book 21 (*FGrHist/BNJ* 115 F 128–136), which is possible, but cannot be verified.

2.7 Mir. *128*

Ἐν δὲ τῇ Χαλκιδικῇ τῇ ἐπὶ Θρᾴκης πλησίον Ὀλύνθου φασὶν εἶναι Κανθαρόλεθρον ὀνομαζόμενον τόπον, μικρῷ μείζονα τὸ μέγεθος ἅλω, εἰς ὃν τῶν μὲν ἄλλων ζῴων

92 Schrader (1868) 227–228. *Contra* Müllenhoff (1870) 428–429 n. **.

93 In *FGrHist* 570 F 4, Jacoby accepts Jacobs' supplement <προτιθέναι>: Λύκος δ' ἄρα καὶ ταῦτα μὲν ὁμολογεῖ, καὶ ἐκεῖνα δὲ ἐπὶ τούτοις προστίθησι, <προτιθέναι> φοινικοῦς ἱμάντας τὴν χρόαν. In *FGrHist* 115 F 274b, the supplement is simply recorded in the *apparatus* without further commentary.

94 Jacoby (1955a) 600; Flashar (1972) 131. See also Gauger and Gauger (2010) 233.

95 Cf. Prosdocimi (1963–1964) 208; Santi Amantini (2005) 53–54; Smith (2013) on F 4. De Sensi Sestito (2013) 98, 107 does not discuss the scope of Lycus' fragment, but argues that Lycus elaborated on Theopompus' account.

96 Theopompus as a source: Müllenhoff (1870) 428; Grenfell and Hunt (1909) F 128b; Prosdocimi (1963–1964) 206 n. 15, 207; Giannini (1965) 283–285; Flashar (1972) 130–131; Gauger and Gauger (2010) 233; Eleftheriou (2018b) 321. Lycus as a source: Guenther (1889) 26, followed by Susemihl (1891) 478 n. 94b; Giannini (1964) 135 n. 219. Schrader (1868) 227–228 suggested that this chapter of *Mir.* was based on both Theopompus and Lycus, while Voltan (1985) 8 argued that the compiler of *Mir.* used either Theopompus or Lycus as a source.

97 See Müller (1841) 302–303; Grenfell and Hunt (1909) F 128b; Jacoby (1929) 594; Morison (2014) on F 274a. According to Voltan (1985) 8 n. 13, this story may derive from the account of *thaumasia* in book 8 of the *Philippica* (see Section 3).

ὅταν τι ἀφίκηται, πάλιν ἀπέρχεται, τῶν δὲ κανθάρων τῶν ἐλθόντων οὐδείς, ἀλλὰ κύκλῳ περιιόντες τὸ χωρίον λιμῷ τελευτῶσιν.

In Thracian Chalcidice near Olynthus they say that there is a place called Cantharole-thros [*sc.* "Death to Beetles"], a little larger in size than a threshing-floor; when any other animal enters into it, it departs again, but none of the beetles do so, but wheeling round and round the place die of hunger.

A parallel description of this marvel located in Thrace near Olynthus is found in a fragment of Theopompus (*FGrHist/BNJ* 115 F 266) preserved by pseudo-Antigonus, *Mir.* 14:

Θεόπομπος δέ φησιν κατὰ τοὺς ἐν Θράκῃ Χαλκιδεῖς εἶναί τινα τόπον τοιοῦτον, εἰς ὃν ὅ τι μὲν ἂν τῶν ἄλλων ζῴων εἰσέλθῃ, πάλιν ἀπαθὲς ἀπέρχεται, τῶν δὲ κανθάρων οὐδεὶς διαφεύγει, κύκλῳ δὲ στρεφόμενοι τελευτῶσιν αὐτοῦ· διὸ δὴ καὶ τὸ χωρίον ὀνομάζεσθαι Κανθαρώλεθρον.[98]

Theopompus says that a certain place among the Chalcidians in Thrace is such that whichever of the other animals enters into it, it comes out unaf-fected, but no beetle escapes, and turning round in a circle they die there; for this very reason the place is named Cantharolethros [*sc.* "Death to Beetles"]. (trans. Morison, with modifications)

Comparison between *Mir.* and pseudo-Antigonus suggests that Theopompus should be regarded as the source behind both passages, as is revealed by the striking similarities in both content and vocabulary (in bold in the following table).[99]

Mir. *128*	Ps.-Antig. Mir. *14* = FGrHist/BNJ *115 F 266*
Ἐν δὲ τῇ **Χαλκιδικῇ** τῇ ἐπὶ **Θράκης** *πλησίον Ὀλύνθου* φασὶν εἶναι **Κανθαρόλεθρον** ὀνομαζόμενον **τόπον**, *μικρῷ μείζονα τὸ μέγεθος ἅλω*, εἰς ὃν τῶν μὲν ἄλλων ζῴων ὅταν τι ἀφίκηται, **πάλιν ἀπέρχεται, τῶν δὲ κανθάρων** τῶν ἐλθόντων **οὐδείς, ἀλλὰ κύκλῳ** περιιόντες τὸ χωρίον *λιμῷ* **τελευτῶσιν**	Θεόπομπος δέ φησιν κατὰ τοὺς ἐν **Θράκῃ Χαλκιδεῖς** εἶναί τινα **τόπον** τοιοῦτον, εἰς ὃν ὅ τι μὲν ἂν τῶν ἄλλων ζῴων εἰσέλθῃ, **πάλιν ἀπαθὲς ἀπέρχεται, τῶν δὲ κανθάρων οὐδεὶς** διαφεύγει, **κύκλῳ** δὲ στρεφόμενοι **τελευτῶσιν** αὐτοῦ· διὸ δὴ καὶ τὸ χωρίον **ὀνομάζεσθαι Κανθαρώλεθρον**

98 Ed. Musso (1985) 25; cf. also Eleftheriou (2018a) 130. With Giacomelli (2023) 317, I accept the transmitted form Κανθαρόλεθρον, which is confirmed by *Mir.*, instead of the conjecture Κανθαρώλεθρον.

99 See Schrader (1868) 226 n. 11; Müllenhoff (1870) 428; Grenfell and Hunt (1909) F 255; Jacoby (1930) 391; Giannini (1965) 285; Flashar (1972) 131; Boshnakov (2003) 92–94. It is improbable that *Mir.* depends here on pseudo-Antigonus, since *Mir.* contains several additions: see Boshnakov (2003) 93 and below.

Other sources that preserve this story are Strabo, 7 F 15b Radt[100]; Pliny, *Naturalis historia* 11.99[101]; and Plutarch, *De tranquillitate animi* 15.473e.[102] Since they relate it in similar terms, it is possible that the entire tradition ultimately goes back to Theopompus, probably through intermediate sources.[103]

Three details provided by *Mir.* ("in italics" in the table above) are not mentioned by pseudo-Antigonus, namely, that the place in question was located near Olynthus; that it was a little larger in size than a threshing-floor; and that beetles entering it die of hunger. Comparison with the other parallel sources may help us evaluate the nature of these divergences and decide whether they are additions on the part of the compiler of *Mir.* or omissions by pseudo-Antigonus.

Since all other sources explicitly mention Olynthus (cf. especially Strabo's πλησίον Ὀλύνθου, which is identical to *Mir.*'s wording), it seems probable that the topographical indication found in *Mir.* was already found in Theopompus. The comparison between *Kantharolethros* and a threshing floor is more difficult to evaluate, since it has no clear parallel in the other sources. However, Pliny notes that the place in question was *parvus*, while Strabo speaks of a κοῖλον ("hollow place"). It thus seems probable that Theopompus spoke of a small place, but we cannot say with certainty whether the comparison with a threshing floor comes from the historian. Finally, that beetles entering *Kantharolethros* die of hunger is not reported by any other source and may thus be an explanation added by the compiler of *Mir.*[104]

Theopompus was obviously interested in the area of Olynthus because of Philip's II well-known war against that city, which he eventually completely destroyed in 348 BC. As plausibly suggested by Jacoby and others, this story, which pseudo-Antigonus does not assign to a specific book of the *Philippica*, may have been part of a digression concerning that region in books 20–25, which dealt with Philip's II military actions against Olynthus and the other Chalcidian cities.[105]

---

100   Strab. 7 F 15b Radt: Ὅτι πλησίον Ὀλύνθου χωρίον ἐστὶν κοῖλον καλούμενον Κανθαρώλεθρον ἐκ τοῦ συμβεβηκότος· τὸ γὰρ ζῷον ὁ κάνθαρος πέριξ τῆς χώρας γινόμενος ἡνίκα ψαύσῃ τοῦ χωρίου ἐκείνου, διαφθείρεται; ed. Radt (2003) 364.

101   Plin. *HN* 11.99: *in Threcia iuxta Olynthum locus est parvus, in quo unum hoc animal exanimatur, ob hoc Cantharolethrus appellatus*; ed. Mayhoff (1909) 314.

102   Plut. *De tranq. anim.* 15.473e: μᾶλλον δ᾽ ὥσπερ ἐν Ὀλύνθῳ τοὺς κανθάρους λέγουσιν εἴς τι χωρίον ἐμβαλόντας ὃ καλεῖται "Κανθαρώλεθρον," ἐκβῆναι μὴ δυναμένους ἀλλ᾽ ἐκεῖ στρεφομένους καὶ κυκλοῦντας ἐναποθνήσκειν, οὕτως εἰς τὴν τῶν κακῶν μνήμην ὑπορρυέντες ἀνενεγκεῖν μὴ θέλωσι μηδ᾽ ἀναπνεῦσαι; ed. Paton *et al.* (1972) 210.

103   See Grenfell and Hunt (1909) F 255; Boshnakov (2003) 264; Morison (2014) on F 266. That Strabo ultimately depends on Theopompus is accepted by Shrimpton (1991) 285 n. 3; Boshnakov (2003) 92–94, 264–265, 335 (possibly via Demetrius of Scepsis); Radt (2007) 357. Amiotti (1994) 205–208 also suggested Demetrius of Scepsis as Strabo's direct source. According to Flashar (1972) 131 and Boshnakov (2003) 93, pseudo-Antigonus cited Theopompus via Callimachus; however, this citation of Theopompus does not belong to the section of pseudo-Antigonus' collection that depends on Callimachus. Boshnakov (2003) 94 argued that Theophrastus may have been *Mir.*'s intermediate source for Theopompus.

104   See Amiotti (1994) 206.

105   See Jacoby (1929) 565, 593; Pédech (1989) 102–103 n. 93; Shrimpton (1991) 61; Boshnakov (2003) 92–94, 264; Morison (2014) on F 266. Alternatively, Amiotti (1994) 205 argued that this story may have been part of Theopompus' account of *thaumasia* in book 8 of the *Philippica* (see Section 3).

2.8 Mir. *129*

Ἐν δὲ †Κύκλωψι† τοῖς Θρᾳξὶ κρηνίδιόν ἐστιν ὕδωρ ἔχον ὃ τῇ μὲν ὄψει καθαρὸν καὶ διαφανὲς καὶ τοῖς ἄλλοις ὅμοιον, ὅταν δὲ πίῃ τι ζῷον ἐξ αὐτοῦ, παραχρῆμα διαφθείρεται.

Among the †Cyclopes† in Thrace there is a small spring with water which is clear and transparent to look at, and just like other water, but, when any animal drinks of it, it immediately dies.

The same report, with some minor variants, is consistently attributed to Theopompus by various sources:

Ps.-Antig. *Mir.* 141 = Callim. F 407 XIII Pfeiffer = Theopomp., *FGrHist/ BNJ* 115 F 270a: Θεόπομπον δέ φησιν γράφειν τῆς μὲν ἐν †κιγχρωψωσιν† τοῖς Θρᾳξὶν τὸν ἀπογευσάμενον τελευτᾶν εὐθύς.[106]
(Callimachus) says that Theopompus writes that when a person tastes (the water of the spring) among the Thracian †Kinchropes†, he immediately dies. (trans. Morison, with modifications)
Plin. *HN* 31.27 = Theopomp., *FGrHist/BNJ* 115 F 270d: *necare aquas Theopompus et in Thracia apud Cychros dicit.*[107]
Theopompus says that there are also deadly waters among the Cychri in Thrace. (my translation)
Par. Flor. (*FGrHist* 1680) 15 = Theopomp., *FGrHist/BNJ* 115 F 270c: Θεόπομπος ἱστορεῖ κρήνην ἐν Χρωψὶ τῆς Θρᾴκης, ἐξ ἧς τοὺς λουσαμένους παραχρῆμα μεταλλάσσειν.[108]
Theopompus records that there is a spring among the Chropsi in Thrace; those who have bathed in it immediately perish. (trans. Greene, with modifications)
Par. Vat. (*FGrHist* 1679) 38 = Theopomp., *FGrHist/BNJ* 115 F 270b: Θεόπομπος κρήνην ἐν Θρᾴκῃ λέγει εἶναι, ἐξ ἧς οἱ λουσάμενοι μεταλλάττουσι τὸν βίον.[109]
Theopompus says that there is a spring in Thrace in which those who have washed themselves take leave of life. (my translation)

Vitruvius, *De architectura* 8.3.15 also mentions this water and its effects, yet without explicitly citing Theopompus:

Et Chrobsi Thracia lacus, ex quo non solum qui biberint moriuntur, sed etiam qui laverint.[110]

---

106 Ed. Musso (1985) 62–63; cf. also Eleftheriou (2018a) 198.
107 Ed. Mayhoff (1897) 10.
108 Ed. Greene (2022) 634.
109 Ed. Sørensen (2022a) 584.
110 Ed. Callebat (1973) 18.

At Chrobs in Thrace there is a lake which brings death not only to those who drink of it, but also to those who bathe in it. (trans. Granger, with modifications)

Since four extant testimonies (pseudo-Antigonus, Pliny, the *Paradoxographus Florentinus*, and the *Paradoxographus Vaticanus*) explicitly mention Theopompus as a source and since all passages agree in the key points (the spring is located in Thrace, and it is fatal), the historian may be reasonably considered to be the ultimate source for the entire tradition concerning this spring, including *Mir*.[111] For despite some minor differences between *Mir*. and the other passages, its wording betrays a shared source (cf. Ἐν δὲ †Κύκλωψι† τοῖς Θραξὶ . . . παραχρῆμα διαφθείρεται *Mir*. ~ ἐν †κιγχρωψωσιν† τοῖς Θραξὶν . . . τελευτᾶν εὐθύς ps.-Antig. ~ ἐν Χρωψὶ τῆς Θράκης . . . παραχρῆμα μεταλλάσσειν: *Par. Flor.*). However, several problems, which cannot be solved with certainty, arise from these passages.[112]

First, even though all sources agree in locating the spring in Thrace, the original name of the Thracian tribe can no longer be recovered, since all the forms reported in the extant texts seem to be corrupt and cannot be identified with any known Thracian tribe (κύκλωψι *Mir*.; κιγχρωψωσιν pseudo-Antigonus; *Cychros* Pliny; Χρωψί *Paradoxographus Florentinus*; *Chrobsi* Vitruvius).[113]

Second, while according to *Mir*. and pseudo-Antigonus (who depends on Theopompus via Callimachus), the Thracian spring is lethal to those who *drink* from it,[114] the *Paradoxographus Florentinus* and the *Paradoxographus Vaticanus* claim that it is fatal to those who *bathe* in it.[115] Vitruvius mentions both possibilities, while Pliny just records the water's lethal character. Jacoby supposed that the variant γεύσασθαι-λούσασθαι may go back to an ancient corruption,[116] while Öhler argued that Theopompus, like Vitruvius, wrote that the spring was lethal both to those who drank from it and to those who bathed in it.[117]

---

111 See Schrader (1868) 226 n. 11; Müllenhoff (1870) 428; Grenfell and Hunt (1909) F 259; Öhler (1913) 80–81; Jacoby (1930) 391; Giannini (1965) 285; Flashar (1972) 132; Gauger and Gauger (2010) 232; Eleftheriou (2018b) 273; Greene (2022) 697–698.

112 See especially Öhler (1913) 80–82; Flashar (1972) 132; Stern (2008) 459; Sørensen (2022a) 617–618; Greene (2022) 697–700.

113 See Giacomelli (2023) 317.

114 See also Oribas. *Coll. med.* 5.3.29 Raeder: αὐτίκα ὕδωρ ἐν Λεοντίνοις ἔστιν, οὗ ἤν τις πίῃ, ἀποθνήσκει· τοῖον δ᾽ ἄλλο ἐν Φενεῷ τῆς Ἀρκαδίας, ὃ καλοῦσιν ὕδωρ Στυγός· τοῖον δ᾽ ἄλλο ἐν Θράκῃ.

115 The versions provided by the *Paradoxographus Florentinus* and the *Paradoxographus Vaticanus* are close to each other and may depend on a common source; the hypothesis by Öhler (1913) 82 that the *Paradoxographus Vaticanus* depends here on the *Paradoxographus Florentinus* remains uncertain: see Greene (2022) 699 n. 263. For the uncertain date of the *Paradoxographus Florentinus*, see n. 60 above. The date of the *Paradoxographus Vaticanus* is likewise uncertain: Sørensen (2022a) 591 prudently suggests a date in the second or third century AD.

116 Flashar (1972) 132 is skeptical about this hypothesis.

117 See Öhler (1913) 81 n. 1. Cf. also *Par. Flor.* (*FGrHist* 1680) 21 = Isigonus, *FGrHist* 1659 F 10: ἐν Συκαμίναις πόλει λίμνη ἐστίν, ἧς τῷ ὕδατι οἱ λουσάμενοι ἢ πιόντες ἀπ᾽ αὐτοῦ μαδῶσι τὰς τρίχας, τῶν δὲ ἀλόγων ζῴων αἱ ὁπλαὶ ἀποπίπτουσιν, ὡς ἱστορεῖ Ἰσίγονος.

Third, *Mir.* is the only source that records the water's normal appearance and that specifies that it was deadly for *animals*. On the one hand, since both details could be additions by the compiler of *Mir.*, one should be cautious in concluding that *Mir.* preserves here a fuller version of the Theopompan fragment than the rest of the tradition. On the other hand, it is still possible that Theopompus spoke of a spring whose water was deadly for men *and* animals and described the water's appearance, and that these details were omitted in the other extant sources.

Öhler speculatively assumed the existence of three main branches of the tradition that independently go back to Theopompus via various intermediate sources. The first would be represented by *Mir.*; the second by Pliny, Vitruvius, the *Paradoxographus Vaticanus*, and the *Paradoxographus Florentinus*; and the third by pseudo-Antigonus.[118] Even though this reconstruction remains to a certain extent hypothetical,[119] one may agree with Öhler that *Mir.* does not seem to depend on any of the other preserved sources.

None of the authors citing Theopompus as a source for this spring mention a book number. With due caution, Morison only says that the fragment was probably part of a digression in the *Philippica*.[120]

2.9 Mir. *130*

1. Φασὶ δὲ καὶ ἐν τῇ Κρηστωνίᾳ παρὰ τὴν Βισαλτῶν χώραν τοὺς ἁλισκομένους λαγὼς δύο ἥπατα ἔχειν, καὶ τόπον τινὰ εἶναι ὅσον πλεθριαῖον, εἰς ὃν ὅ τι ἂν εἰσέλθῃ ζῷον ἀποθνήσκει. 2. Ἔστι δὲ καὶ ἄλλο αὐτόθι ἱερὸν Διονύσου μέγα καὶ καλόν, ἐν ᾧ τῆς ἑορτῆς καὶ τῆς θυσίας οὔσης λέγεται, ὅταν μὲν ὁ θεὸς εὐετηρίαν μέλλῃ ποιεῖν, ἐπιφαίνεσθαι μέγα σέλας πυρός, καὶ τοῦτο πάντας ὁρᾶν τοὺς περὶ τὸ τέμενος διατρίβοντας, ὅταν δὲ ἀκαρπίαν, μὴ φαίνεσθαι τοῦτο τὸ φῶς, ἀλλὰ σκότος ἐπέχειν τὸν τόπον ὥς καὶ τὰς ἄλλας νύκτας.

1. They say that in Crestonia near the country of the Bisaltae hares which are caught have two livers, and that there is a place there about a *plethron* in extent, into which if any animal enters it dies. 2. There is also there a fine large temple of Dionysus, in which when a sacrifice and feast takes place, should the god intend to give a good season, it is said that a huge flame of fire appears and that all who go to the sacred enclosure see this, but when (the god is going to give) a fruitless season, this light does not appear, but darkness covers the place, just as on other nights.

The first report of this chapter, about the two livers possessed by hares that live in Crestonia, near the country of the Bisaltae,[121] probably derives from Theopompus,[122]

---

118 See Öhler (1913) 81–82.

119 Cf. Greene (2022) 699.

120 See Morison (2014) on F 270a, followed by Greene (2022) 697.

121 See Flensted-Jensen (2004) 810: "Bisaltia . . . was the district west of the lower Strymon, up to Herakleia Sintika."

122 See Schrader (1868) 226 n. 11; Müllenhoff (1870) 428; Grenfell and Hunt (1909) F 122; Schranz (1912) 55 n. 4; Jacoby (1930) 378, 391; Kroll (1940) 13; Giannini (1965) 285; Flashar (1972) 132; Sharples (1995) 59 n. 190; Gauger and Gauger (2010) 212. On the role possibly played by

as is shown by numerous parallel passages which attribute the same claim, with nearly the same words, to the historian:

Ath. 9.401a-b = Theopomp., *FGrHist/BNJ* 115 F 126a: Θεόπομπος δὲ ἐν τῇ εἰκοστῇ τῶν Ἱστοριῶν περὶ τὴν Βισαλτίαν φησὶ λαγωοὺς γίγνεσθαι [γίγνεσθαι A: γίνεσθαι CE: γενέσθαι Olson] δύο ἥπατα ἔχοντας.[123]

Theopompus says in the twentieth book of the *Histories* that in the area of Bisaltia there are hares that have two livers. (trans. Morison, with modifications)

Steph. Byz. B 103 Billerbeck, *s.v.* Βισαλτία = Theopomp., *FGrHist/BNJ* 115 F 126b = Favorin. F 61 Mensching = F 86 Barigazzi = F 89 Amato: Βισαλτία· πόλις καὶ χώρα Μακεδονίας . . . περὶ [περὶ QPN: παρὰ R] ταύτην οἱ λαγοὶ σχεδὸν πάντες ἁλίσκονται δύο ἥπατα ἔχοντες, ὡς Θεόπομπος ἱστορεῖ καὶ Φαβωρῖνος.[124]

Bisaltia: a city and region in Macedonia . . . in this region nearly all the hares that are caught have two livers, as Theopompus records and Favorinus. (trans. Morison, with modifications)

Ael. *NA* 5.27 = Theopomp., *FGrHist/BNJ* 115 F 126a: τοὺς ἐν τοῖς Βισάλταις λαγὼς διπλᾶ ἥπατα ἔχειν Θεόπομπος λέγει.[125]

Theopompus says that among the Bisaltae hares have a double liver. (my translation)

Ael. *NA* 11.40 (cf. Thphr. F 356 FHS&G): Πέρδικες οἱ Παφλαγόνες δικάρδιοί εἰσιν, ὥσπερ οὖν Θεόφραστος λέγει. καὶ Θεόπομπος λέγει τοὺς ἐν Βισαλτίᾳ λαγὼς διπλᾶ ἥπατα ἔχειν ἕκαστον.[126]

Partridges in Paphlagonia have two hearts, as Theophrastus says. And Theopompus says that hares in Bisaltia each have a double liver. (trans. Scholfield, with modifications)

Gell. *NA* 16.15.1 = Theopomp., *FGrHist/BNJ* 115 F 126a (cf. Thphr. F 356 FHS&G): *Theophrastus philosophorum peritissimus omnes in Paphlagonia perdices bina corda habere dicit, Theopompus in Bisaltia lepores bina iecora.*[127]

Theophrastus, most expert of philosophers, says that in Paphlagonia all the partridges have two hearts; Theopompus, that in Bisaltia the hares have two livers. (trans. Rolfe, with modifications)

Favorinus in the transmission of Theopompus' fragment, see Barigazzi (1966) 238; Amato (2010) 339–340 n. 788 (with a cautious approach).
123 Ed. Olson (2020) 134.
124 Ed. Billerbeck (2006) 350. Cf. also ps.-Zonar. *Lex. s.v.* Βισαλτία pp. 388.29–389.2 Tittmann; *Etymologicum Symeonis* p. 526.6 Lasserre and Livadaras, *s.v.* Βισαλτία.
125 Ed. García Valdés, Llera Fueyo and Rodríguez-Noriega Guillén (2009) 113. Cf. Prandi (2005) 58, 181.
126 Ed. García Valdés, Llera Fueyo and Rodríguez-Noriega Guillén (2009) 279.
127 Ed. Holford-Strevens (2020) 569.

Aristotle speaks of the same phenomenon, locating it near Lake Bolbe in the so-called Sycine, just south of Crestonia and Bisaltia (the region mentioned by Theopompus), and explains that these hares seem to have two livers, due to the length of the connections between the two parts of the liver, but in actuality have only one.[128] Aristotle seems to relate an independent tradition that cannot have served as a source for *Mir.* In this case, we know that Theopompus spoke of this marvel in book 20 of the *Philippica*, probably as part of a digression concerning the region of Thrace north of the Chalcidice in the context of the events that led to the war against Olynthus and the Chalcidians in around 350 BC.

There are no parallel sources for either of the other two stories related in this chapter of *Mir.*: the first concerns a certain place, located in the same area, into which whatever animal enters dies; the second describes a marvel that took place in a temple of Dionysus which was located in the same region (αὐτόθι). No temple of Dionysus in this area, however, is known from the ancient sources.[129] Moreover, *Mir.* mysteriously speaks of "another" temple of Dionysus, while no other temple is mentioned in the previous lines (Ἔστι δὲ καὶ ἄλλο αὐτόθι ἱερὸν Διονύσου).[130]

Cordano argued that the first temple of Dionysus mentioned by *Mir.* should be identified with the Delphion (i.e., Mt. Haemus) mentioned at *Mir.* 104 (see Section 1).[131] Since Herodotus 7.111 mentions a *manteion* of Dionysus located on the highest mountains of Thrace (i.e., on the Haemus), where a prophetess prophesied in the same way as in Delphi (κατά περ ἐν Δελφοῖσι), Cordano concluded that someone may have called the Haemus "Delphion" because of the Delphic character of its oracle. This hypothesis, however, has been rejected by Giacomelli, who rightly observes that Herodotus' passage does not allow us to conclude that the oracle of Dionysus was Apollinean; Giacomelli rather suggests the possibility of a scribal error such as AIMON > ΔΕΛ<Φ>ΙΟΝ.[132] In any case, Cordano's hypothesis is contradicted by the fact that the Delphion mentioned at *Mir.* 104 is considered by the compiler to be a mountain, not a temple of Dionysus. It seems more probable that the source of *Mir.* originally spoke of two temples of Dionysus located in that region and that the compiler of *Mir.* did not adapt the excerpt to the new context.[133] That this source was Theopompus, as Jacoby and Flashar are inclined to suppose,[134] is certainly plausible, since all the marvels contained in this chapter are linked to the same region; but it cannot be definitely proven without parallel passages in the historian's named fragments.

128 Arist. *HA* 2.17.507a17–19; *PA* 3.7.669b34–36. Plin. *HN* 11.190 locates this phenomenon "circa Briletum et Tharnem et in Cherroneso ad Propontidem;" Kroll (1940) 13 attributes this version to Theophrastus, but see Sharples (1995) 52, 59. Ziegler (1949) 1151 considers Aristotle's *Historia animalium* to be the source used by *Mir.*

129 See Iliev (2013) 64–65.

130 Cf. Vanotti (2007) 199; Giacomelli (2023) 318.

131 Cordano (2015) 57–59.

132 Giacomelli (2023) 282.

133 Cf. Giacomelli (2023) 318.

134 See Jacoby (1930) 378; Flashar (1972) 132.

2.10 Mir. 131

1. Ἐν Ἤλιδι λέγουσιν εἶναί τι οἴκημα σταδίους ἀπέχον ὀκτὼ μάλιστα τῆς πόλεως, εἰς ὃ τιθέασι τοῖς Διονυσίοις λέβητας χαλκοῦς τρεῖς κενούς. τοῦτο δὲ ποιήσαντες παρακαλοῦσι τῶν Ἑλλήνων τῶν ἐπιδημούντων τὸν βουλόμενον ἐξετάσαι τὰ ἀγγεῖα καὶ τοῦ οἴκου κατασφραγίζεσθαι τὰς θύρας. καὶ ἐπειδὰν μέλλωσιν ἀνοίγειν, ἐπιδείξαντες τοῖς πολίταις καὶ τοῖς ξένοις τὰς σφραγῖδας, οὕτως ἀνοίγουσιν. οἱ δ' εἰσελθόντες εὑρίσκουσι τοὺς μὲν λέβητας οἴνου πλήρεις, τὸ δὲ ἔδαφος καὶ τοὺς τοίχους ὑγιεῖς, ὥστε μηδεμίαν εἶναι ὑποψίαν λαβεῖν ὡς τέχνῃ τινὶ κατασκευάζουσιν. 2. Εἶναι δέ φασι παρ' αὐτοῖς καὶ ἰκτίνους, οἳ παρὰ μὲν τῶν διὰ τῆς ἀγορᾶς τὰ κρέα φερόντων ἁρπάζουσι, τῶν δὲ ἱεροθύτων οὐχ ἅπτονται.

1. In Elis they say there is a building about eight stades from the city into which at the Dionysia they place three empty bronze cauldrons. When they have done this they call upon any of the visiting Greeks who wishes to examine the vessels, and seal up the doors of the house. When they are going to open it, they show the seals to citizens and strangers, and then open it. Those that go in find the cauldrons full of wine, but the floor and walls intact, so that there is no suspicion that they effect it by any artifice. 2. They also say that there are kites among them which seize pieces of meat from those who are carrying them through the market-place, but they do not touch those which are offerings to the gods.

The first part of this chapter of *Mir.*, which concerns marvels occurring in Elis, has close parallels in a fragment of Theopompus preserved by Athenaeus and in a passage from Pausanias' description of Elis:

Ath. 1.34a = Theopomp., *FGrHist/BNJ* 115 F 277: ὅτι Θεόπομπος ὁ Χῖος τὴν ἄμπελον ἱστορεῖ εὑρεθῆναι ἐν Ὀλυμπίᾳ παρὰ τὸν Ἀλφειόν· καὶ ὅτι τῆς Ἠλείας τόπος ἐστὶν ἀπέχων ὀκτὼ στάδια, ἐν ᾧ οἱ ἐγχώριοι κατακλείοντες τοῖς Διονυσίοις χαλκοῦς λέβητας τρεῖς κενοὺς παρόντων τῶν ἐπιδημούντων ἀποσφραγίζονται καὶ ὕστερον ἀνοίγοντες εὑρίσκουσιν οἴνου πεπληρωμένους.[135]

Theopompus of Chios records that the vine was discovered in Olympia along the Alpheius River. He also reports that there is a spot in Elis eight stades from there, where the inhabitants at their Dionysia close three empty bronze cauldrons in the presence of the visitors; seal them shut; and when they open them later, find them full of wine. (trans. Olson, with modifications)

Paus. 6.26.1–2: ἀπέχει μέν γε τῆς πόλεως ὅσον τε ὀκτὼ στάδια ἔνθα τὴν ἑορτὴν ἄγουσι Θυῖα ὀνομάζοντες· λέβητας δὲ ἀριθμὸν τρεῖς ἐς οἴκημα ἐσκομίσαντες οἱ ἱερεῖς κατατίθενται κενούς, παρόντων καὶ τῶν ἀστῶν καὶ ξένων, εἰ τύχοιεν ἐπιδημοῦντες· σφραγῖδας δὲ αὐτοί τε οἱ ἱερεῖς καὶ τῶν ἄλλων ὅσοις ἂν κατὰ γνώμην ᾖ ταῖς θύραις τοῦ οἰκήματος ἐπιβάλλουσιν. ἐς δὲ τὴν ἐπιοῦσαν τά τε σημεῖα ἐπιγνῶναι πάρεστί σφισι καὶ ἐσελθόντες ἐς τὸ οἴκημα εὑρίσκουσιν οἴνου πεπλησμένους τοὺς λέβητας.[136]

---

135 Ed. Kaibel (1887) 78.
136 Ed. Rocha-Pereira (1977) 142.

The place where they hold the festival they name the Thyia is about eight stades from the city. Three pots are brought into the building by the priests and set down empty in the presence of the citizens and of any strangers who may chance to be in the country. The doors of the building are sealed by the priests themselves and by any others who may be so inclined. On the morrow they are allowed to examine the seals, and on going into the building they find the pots filled with wine.

(trans. Jones)

Pausanias could not observe this marvel himself, since he was not in Elis at the time of the festival of the Thyia. He seems to have known this story from a literary source and to have discussed it with the most respectable of the Eleans.[137] The numerous verbal parallels between the three passages (in bold in the table below) suggest that both Pausanias and *Mir.* ultimately depend on Theopompus.[138]

Mir. *131*	Ath. *1.34a* = Theopomp., FGrHist/BNJ *115 F 277*	Paus. *6.26.1–2*
Ἐν Ἤλιδι λέγουσιν εἶναί τι οἴκημα σταδίους **ἀπέχον ὀκτὼ μάλιστα τῆς πόλεως, εἰς ὃ τιθέασι τοῖς Διονυσίοις λέβητας χαλκοῦς τρεῖς κενούς.** τοῦτο δὲ ποιήσαντες παρακαλοῦσι τῶν Ἑλλήνων **τῶν ἐπιδημούντων** τὸν βουλόμενον ἐξετάσαι τὰ ἀγγεῖα καὶ τοῦ οἴκου **κατασφραγίζεσθαι τὰς θύρας.** καὶ ἐπειδὰν μέλλωσιν ἀνοίγειν, ἐπιδείξαντες **τοῖς πολίταις καὶ τοῖς ξένοις τὰς σφραγῖδας,** οὕτως **ἀνοίγουσιν.** οἱ δ᾽ **εἰσελθόντες εὑρίσκουσι** τοὺς μὲν λέβητας οἴνου πλήρεις, τὸ δὲ ἔδαφος καὶ τοὺς τοίχους ὑγιεῖς, ὥστε μηδεμίαν εἶναι ὑποψίαν λαβεῖν ὡς τέχνῃ τινὶ κατασκευάζουσιν	ὅτι Θεόπομπος ὁ Χῖος τὴν ἄμπελον ἱστορεῖ εὑρεθῆναι ἐν Ὀλυμπίᾳ παρὰ τὸν Ἀλφειόν· καὶ ὅτι **τῆς Ἠλείας** τόπος ἐστὶν **ἀπέχων ὀκτὼ στάδια,** ἐν ᾧ οἱ ἐγχώριοι κατακλείοντες **τοῖς Διονυσίοις χαλκοῦς λέβητας τρεῖς κενοὺς παρόντων τῶν ἐπιδημούντων ἀποσφραγίζονται** καὶ **ὕστερον ἀνοίγοντες εὑρίσκουσιν οἴνου πεπληρωμένους**	**ἀπέχει** μέν γε **τῆς πόλεως** ὅσον τε **ὀκτὼ στάδια** ἔνθα τὴν ἑορτὴν ἄγουσι Θυῖα ὀνομάζοντες· **λέβητας** δὲ ἀριθμὸν **τρεῖς** ἐς **οἴκημα** ἐσκομίσαντες οἱ ἱερεῖς **κατατίθενται κενούς, παρόντων** καὶ τῶν **ἀστῶν καὶ ξένων,** εἰ τύχοιεν **ἐπιδημοῦντες· σφραγῖδας** δὲ αὐτοί τε οἱ ἱερεῖς καὶ τῶν ἄλλων ὅσοις ἂν κατὰ γνώμην ᾖ ταῖς **θύραις** τοῦ **οἰκήματος** ἐπιβάλλουσιν. ἐς δὲ τὴν ἐπιοῦσαν τά τε σημεῖα ἐπιγνῶναι πάρεστί σφισι καὶ **ἐσελθόντες** ἐς τὸ οἴκημα **εὑρίσκουσιν** οἴνου πεπλησμένους τοὺς **λέβητας**

---

137 As noted by Flashar (1972) 133; Maddoli *et al.* (1999) 399; Gauger and Gauger (2010) 233. Cf. Pritchett (1999) 254–255. Pausanias' passage is printed by Jacoby (1930) 392 in his commentary on *FGrHist* 115 F 277.

138 See Jacoby (1930) 391–392; Giannini (1965) 287; Flashar (1972) 133; Vanotti (2007) 199; Gómez Espelosín (2008) 238 n. 123; Gauger and Gauger (2010) 233.

The second part of *Mir.* 131, which concerns the pious behavior of kites at Olympia, also seems to derive from Theopompus,[139] as is shown by a fragment of Theopompus (*FGrHist/BNJ* 115 F 76 = *FGrHist* 1652 F 1) cited by Apollonius, *Historiae mirabiles* (*FGrHist* 1672) 10, which reports this marvel with nearly the same words as *Mir.*:

Θεόπομπος δὲ ἐν τοῖς Θαυμασίοις, ἐν τῷ ἀγῶνι τῶν Ὀλυμπίων πολλῶν ἐπιπολαζόντων ἰκτίνων ἐν τῇ πανηγύρει καὶ διασυριζόντων, τὰ διαφερόμενα κρέα τῶν ἱεροθύτων ἀθιγῆ μένειν.[140]

Theopompus, in the *Thaumasia*, [says] that at the Olympic games, although many kites hover over the festival and whistle, the meat of the sacrifices that is carried remains untouched. (trans. Spittler, with modifications)

The pious behavior of kites at Olympia is also mentioned by other authors. In his description of Elis, Pausanias 5.14.1 identifies as a θαῦμα the fact that kites do not seize the meat of the sacrifices from the altar at Olympia.[141] Since Pausanias, as noted above, tacitly cites Theopompus for another marvel occurring in the same place, we may wonder whether his story concerning kites also depends on the historian. However, in the present case, there are no clear verbal parallels to prove or even suggest textual dependence on Theopompus by Pausanias (which does not exclude, of course, the possibility that Pausanias knew Theopompus' version of the story and elaborated on it). The same holds true for the stories reported by Aelian, *Natura animalium* 2.47,[142] and Pliny, *Naturalis historia* 10.28,[143] which belong to the same tradition but do not show any clear connection with Theopompus.

As attested by Apollonius, Theopompus spoke of the kites at Olympia ἐν τοῖς Θαυμασίοις, which may refer either to a section in book 8 of the *Philippica* concerning *thaumasia* or to a later collection of excerpts regarding marvels related by Theopompus throughout his work (see Section 3). In any case, the way in which Apollonius refers to Theopompus at the beginning of his collection (Θεόπομπος ἐν ταῖς Ἱστορίαις ἐπιτρέχων τὰ κατὰ τόπους θαυμάσια)[144] shows that Apollonius was

---

139 See Müller (1841) 291; Schrader (1868) 226 n. 11; Müllenhoff (1870) 428; Grenfell and Hunt (1909) F 76b; Giannini (1965) 287; Flashar (1972) 133; Vanotti (2007) 199; Spittler (2022) 458–459.

140 Ed. Spittler (2022) 392.

141 Paus. 5.14.1: Ὁ δὲ ἐν Ὀλυμπίᾳ βωμὸς παρέχεται καὶ ἄλλο τοιόνδε ἐς θαῦμα· οἱ γὰρ ἰκτῖνες πεφυκότες ἁρπάζειν μάλιστα ὀρνίθων ἀδικοῦσιν οὐδὲν ἐν Ὀλυμπίᾳ τοὺς θύοντας· ἢν δὲ ἁρπάσῃ ποτὲ ἰκτῖνος ἤτοι σπλάγχνα ἢ τῶν κρεῶν, νενόμισται τῷ θύοντι οὐκ αἴσιον εἶναι τὸ σημεῖον; ed. Rocha-Pereira (1977).

142 Ael. *NA* 2.47: Ἰκτῖνος εἰς ἁρπαγὴν ἀφειδέστατος. οἶδε [δὲ, εἰ δέοι, del. Hercher] τῶν μὲν ἐξ ἀγορᾶς ἐμποληθέντων κρεαδίων ἐὰν γένωνται κρείττους, ἥρπασαν προσπεσόντες, τῶν δὲ ἐκ τῆς τοῦ Διὸς ἱερουργίας οὐκ ἂν προσάψαιντο; ed. García Valdés, Llera Fueyo and Rodríguez-Noriega Guillén (2009) 50.

143 Plin. *HN* 10.28: *Milvi ex eodem accipitrum genere magnitudine differunt. notatum in his, rapacissimam et famelicam semper alitem nihil esculenti rapere umquam e funerum ferculis nec Olympiae ex ara, ac ne ferentium quidem manibus nisi lugubri municipiorum inmolantium ostento;* ed. Mayhoff (1909) 227.

144 Apollon. *Mir.* (*FGrHist* 1672) 1 = Theopomp., *FGrHist/BNJ* 115 F 67b.

well aware that the information he related was originally part of the *Philippica*.[145] Both the description of the kites' pious behavior and the marvel concerning wine (which is not explicitly assigned to a specific work or book title by Athenaeus or his epitomizer) were probably part of a digression about *thaumasia* occurring in Elis.[146]

2.11 Mir. 132

Ἐν Κορωνείᾳ δὲ τῆς Βοιωτίας λέγεται τοὺς ἀσπάλακας {τὸ ζῷον} μὴ δύνασθαι ζῆν μηδ' ὀρύσσειν τὴν γῆν, τῆς λοιπῆς Βοιωτίας πολὺ πλῆθος ἐχούσης.

In Coronea in Boeotia it is said that the moles cannot live, nor dig in the earth, though the rest of Boeotia has many of them.

No parallels for this story can be found among Theopompus' fragments. The same claim is repeated with nearly the same words by pseudo-Antigonus and Stephanus of Byzantium:

Ps.-Antig. *Mir.* 10.1: Τῆς δὲ Βοιωτίας ἐχούσης πλήθει πολλοὺς ἀσπάλακας, ἐν τῇ Κορωνειακῇ μόνῃ οὐ γίνεσθαι τοῦτο τὸ ζῷον, ἀλλὰ κἂν εἰσαχθῇ τελευτᾶν.[147]
While in Boeotia there are plenty of moles, in the region of Coronea alone this animal cannot be found, and even if it is introduced, it dies. (my translation)
Steph. Byz. Κ 180 Billerbeck, *s.v.* Κωρόνεια: ἐν ταύτῃ οὐ φαίνεται ἀσπάλαξ, πάσης τῆς Βοιωτίας ἐχούσης τὸ ζῷον, κἄν τις ἔξωθεν ἐνέγκῃ εἰς τὴν Κορωναίων χώραν, οὐ δύναται ζῆν.[148]
In Coronea there are no moles, while the animal can be found throughout the entire Boeotia, and even if one brings a mole to the region of the Coroneans from outside, it cannot survive. (my translation)

Aristotle, *Historia animalium* 7(8).28.605b31–606a2; Pliny, *Naturalis historia* 8.226; and Aelian, *De natura animalium* 17.10, provide a similar, yet slightly different version of the same phenomenon, claiming that the region of Orchomenus in Boeotia is full of moles, while the animal cannot be found in the region of Lebadea, and that if one takes them there, they refuse to dig or they simply die. It seems that *Mir.* and pseudo-Antigonus share a source and that this source was not (or at least not directly) Aristotle.[149] According to Flashar and others,[150] the source was Theopompus, since he is the source of numerous reports in the surrounding chapters

---

145 See Ottone (2018) 530 n. 56.
146 Jacoby (1929) 546, 595; (1930) 367, 392; Morison (2014) on F 76, 277; Spittler (2022) 459.
147 Ed. Musso (1985) 23–24; cf. Eleftheriou (2018a) 126.
148 Ed. Billerbeck (2014) 102.
149 See Flashar (1972) 133; Eleftheriou (2018a) 224; (2018b) 32. Ziegler (1949) 1151 considered Aristotle's *Historia animalium* to be the source used by *Mir.* Schrader (1868) 222, 227 suggested that Theopompus used Aristotle as a source. *Contra* Müllenhoff (1870) 428 n. **. According to Sharples (1995) 52, Theophrastus was not the source used by *Mir.*
150 Müllenhoff (1870) 428 n. **; Giannini (1965) 287; Flashar (1972) 133.

of *Mir.* Nevertheless, in the absence of parallel passages explicitly assigned to Theopompus, the attribution cannot be proven in this case either.

2.12 Mir. 133

1. Ἐν Λούσοις [Beckmann: Λουσοῖς Sylburg: τοῖς Λούσοις Giannini: κολουσοῖς ω] δὲ τῆς Ἀρκαδίας κρήνην εἶναί τινά φασιν, ἐν ᾗ χερσαῖοι μύες γίνονται καὶ κολυμβῶσι, τὴν δίαιταν ἐν ἐκείνῃ ποιούμενοι. 2. Λέγεται δ' αὐτὸ τοῦτο καὶ ἐν Λαμψάκῳ εἶναι.

1. At Lusi in Arcadia they say there is a spring in which there are land mice; they dive and live in it. The same thing is said to occur at Lampsacus.

This chapter of *Mir.* is quoted almost verbatim in *Paradoxographus Florentinus* (*FGrHist* 1680) 10,[151] which assigns it to Aristotle. The same chapter has parallels in a fragment of Theopompus preserved by pseudo-Antigonus (via Callimachus) and in a fragment of Theophrastus cited by Pliny:

Ps.-Antig. *Mir.* 137 = Callim. F 407 IX Pfeiffer = Theopomp., *FGrHist/BNJ* 115 F 269: τὴν δ' ἐν Λούσοις κρήνην, καθάπερ παρὰ τοῖς Λαμψακηνοῖς, ἔχειν ἐν ἑαυτῇ μῦς ὁμοίους τοῖς κατοικιδίοις· ἱστορεῖν δὲ ταῦτα Θεόπομπον.[152]

(Callimachus says) that the spring in Lusi, as among the people of Lampsacus, has in it mice that are like house mice. Theopompus records these things. (trans. Morison, with modifications)

Plin. *HN* 31.14 = Thphr. F 218A FHS&G: *at in Lusis Arcadiae quodam fonte (Theophrastus . . . dicit) mures terrestres vivere et conversari.*[153]

But in a certain spring at Lusi in Arcadia, land-mice, (Theophrastus says), live and have their homes. (trans. FHS&G, with modifications)

The similarity between the passages of pseudo-Antigonus (stemming from Theopompus) and Pliny (borrowed from Theophrastus) raises the question of the source used by *Mir.*[154] Steinmetz suggested emending pseudo-Antigonus' Θεόπομπον to Θεόφραστον on the basis of Pliny's passage and regarding Theophrastus as the ultimate source for the entire tradition.[155] This hypothesis is unconvincing, however,

---

151 *Par. Flor.* (*FGrHist* 1680) 10: ἐν Λούσοις [Λουσίοις codd.] τῆς Ἀρκαδίας φησὶν Ἀριστοτέλης κρήνην τινὰ εἶναι, ἐν ᾗ μῦς χερσαίους γίνεσθαι, καὶ τούτους κολυμβᾶν ἐν ἐκείνῃ τὴν δίαιταν ποιουμένους· ed. Greene (2022) 634. Cf. also *Exc. NA* 2.371 (p. 109.10–13 Lambros). The intermediate source used by the *Paradoxographus Florentinus* was either Isigonus, as supposed by Greene (2022) 646, 654, 689, or the anonymous compiler of a collection based on *Mir.* and Isigonus, as argued by Giacomelli (2021) 350–355; cf. also Giacomelli (2023) 319–320. Öhler (1913) 72 argued that Isigonus and *Mir.* relied on the same source.

152 Ed. Musso (1985) 61; cf. also Eleftheriou (2018a) 196.

153 Ed. Mayhoff (1897) 6.

154 See especially Öhler (1913) 71–73; Flashar (1972) 134; Sharples (1998) 216–217; Greene (2022) 689.

155 Steinmetz (1964) 273 n. 3.

not only for its arbitrariness, but also because the preceding chapter in pseudo-Antigonus (136) likewise derives from Theopompus (see Section 2.2).[156] Giannini and Vanotti claimed that *Mir.* may depend on either Theopompus or Theophrastus.[157] Nonetheless, since both pseudo-Antigonus and *Mir.* mention a similar phenomenon occurring in Lampsacus, we can reasonably suppose that *Mir.* ultimately depends on Theopompus.[158] Together with *Mir.* 123, this chapter shows that Theopompus and Theophrastus could be interested in and speak about the same natural phenomena.[159]

We are not told in which book Theopompus spoke of the spring in Lusi. According to Morison, he did so in a digression in books 32–33 or 46 of the *Philippica*, which concerned Peloponnesian history,[160] while Jacoby cautiously did not assign this report to any specific book.

2.13 Mir. *134*

Ἐν δὲ Κραννῶνι τῆς Θετταλίας φασὶ δύο κόρακας εἶναι μόνους ἐν τῇ πόλει. οὗτοι ὅταν ἐκνεοττεύσωσιν, ἑαυτοὺς μέν, ὡς ἔοικεν, ἐκτοπίζουσιν, ἑτέρους δὲ τοσούτους τῶν ἐξ αὐτῶν γενομένων ἀπολείπουσιν.

At Crannon in Thessaly they say that there are only two crows in the city. After they have nested apparently they migrate, and leave behind just the same number of the young birds they hatch.

This passage finds parallels, in nearly the same words, in two Theopompan fragments; one is preserved by pseudo-Antigonus and the other by Stephanus of Byzantium, who jointly cites Theopompus and Callimachus.

Ps.-Antig. *Mir.* 15a = Theopomp., *FGrHist/BNJ* 115 F 267a: ἐν δὲ Κραννῶνι τῆς Θετταλίας δύο φασὶν μόνον εἶναι κόρακας· διὸ καὶ ἐπὶ τῶν προξενιῶν τῶν ἀναγραφομένων τὸ παράσημον τῆς πόλεως, καθάπερ ἐστὶν ἔθιμον πᾶσι προσπαρατιθέναι, ὑπογράφονται δύο κόρακες ἐφ᾽ ἁμαξίου χαλκοῦ διὰ τὸ μηδέποτε πλείους τούτων ὦφθαι. ἡ δὲ ἅμαξα προσπαράκειται διὰ τοιαύτην αἰτίαν – ξένον γὰρ ἴσως ἂν καὶ τοῦτο φανείη – ἔστιν αὐτοῖς ἀνακειμένη χαλκῆ, ἥν, ὅταν αὐχμὸς ᾖ, σείοντες ὕδωρ αἰτοῦνται τὸν θεόν, καί φασι γίνεσθαι. τούτου δέ τι ἰδιαίτερον (ἰδιώτερον P) ὁ Θεόπομπος λέγει· φησὶν γὰρ ἕως τούτου διατρίβειν αὐτοὺς ἐν τῇ Κραννῶνι, ἕως ἂν τοὺς νεοττοὺς

156 Cf. Flashar (1972) 134; Greene (2022) 689.
157 Giannini (1965) 287; Vanotti (2007) 200.
158 See Schrader (1868) 226 n. 11; Müllenhoff (1870) 428; Grenfell and Hunt (1909) F 258b; Öhler (1913) 71–72 (who argued that the entire tradition, with the possible exception of Callimachus/ pseudo-Antigonus, goes back to Theopompus via Theophrastus); Jacoby (1930) 391; Fraser (1972) II 1080 n. 385; Flashar (1972) 134; Gauger and Gauger (2010) 232; Eleftheriou (2018b) 266; Greene (2022) 689.
159 See Flashar (1972) 134.
160 Morison (2014) on F 269.

ἐκνεοττεύσωσιν, τοῦτο δὲ ποιήσαντας τοὺς μὲν νεοττοὺς καταλείπειν, αὐτοὺς δὲ ἀπιέναι.[161]

In Thessalian Crannon they say that there are only two crows: and for this reason on the proxeny decrees bearing the emblem of the city, as is customary for all to append, two crows are depicted on a bronze chariot, because more than these have never been seen. The chariot – for this would probably also appear strange – is added for some such reason: they have a bronze chariot that they shake, whenever there is drought, to ask the god for water, and they say it happens. And Theopompus says something even more peculiar than this: for he says that they (*sc.* the crows) stay in Crannon until they hatch their young chicks, but having done this, they leave the young chicks behind and themselves depart. (my translation)

Steph. Byz. K 207 Billerbeck, *s.v.* Κραννών = Theopomp., *FGrHist/ BNJ* 115 F 267b = Callim. F 408 Pfeiffer: ἐν ταύτῃ δύο κόρακας εἶναί φασι μόνους, ὡς Καλλίμαχος ἐν τοῖς Θαυμασίοις καὶ Θεόπομπος. ὅταν δὲ ἄλλους ἐκνεοσσεύσωσιν, ἴσους αὐτοὺς καταλιπόντες ἀπέρχονται.[162]

In this city there are only two crows, as Callimachus in the *Thaumasia* and Theopompus (say). Whenever they hatch other (chicks), they depart, leaving behind the same number as themselves. (trans. Morison, with modifications)

Comparison among these passages strongly suggests that *Mir.*, like pseudo-Antigonus and Stephanus, depends here on Theopompus.[163] However, *Mir.* and Stephanus omit the explanation concerning the city's emblem and the religious ritual,[164] which is reported only by pseudo-Antigonus. The citation from Theopompus may have reached Stephanus through Callimachus' paradoxographical collection, here called *Thaumasia* (ἐν τοῖς Θαυμασίοις), but we cannot of course exclude the possibility that Stephanus read Theopompus directly. It is tempting to suppose, with Flashar,[165] that the passage of pseudo-Antigonus also derives from Callimachus, but pseudo-Antigonus' citation of Theopompus does not come from the section of his collection based on the Alexandrian scholar.[166]

Giacomelli finds it striking that Stephanus cites Callimachus, instead of *Mir.*, for this story, since Stephanus did have a direct knowledge of *Mir.*[167] He further observes (1) that Stephanus nowhere else cites Callimachus' paradoxographical collection in his *Ethnica*; (2) that the title *Thaumasia* is never used elsewhere with

---

161 Ed. Musso (1985) 25–26, with Κραννῶνι instead of Κράννωνι: see Giacomelli (2023) 320. Cf. also Eleftheriou (2018a) 130.
162 Ed. Billerbeck (2014) 116.
163 See Schrader (1868) 226 n. 11; Müllenhoff (1870) 428; Grenfell and Hunt (1909) F 256b; Jacoby (1930) 391; Giannini (1965) 287; Flashar (1972) 134; Vanotti (2007) 200; Gauger and Gauger (2010) 232; Giacomelli (2021) 325–326. Ziegler (1949) 1151 considered Aristotle's *Historia animalium* to be the source used by *Mir.* (see below).
164 On which cf. Morison (2014) on F 267a; Mili (2015) 185–187.
165 Flashar (1972) 134.
166 As rightly noted by Eleftheriou (2018b) 46; Giacomelli (2021) 326–327.
167 Giacomelli (2021) 323–328.

reference to Callimachus' collection; and (3) that Stephanus cites elsewhere material from *Mir.* as coming from Aristotle's *Thaumasia.* Giacomelli thus prudently suggests that the name "Callimachus" may be corrupt and that Stephanus may have written "Aristotle" (i.e., the compiler of *Mir.*) instead of "Callimachus."[168]

So, either we follow Giacomelli's interesting suggestion and think that Stephanus directly read Theopompus and *Mir.* ("Aristotle"), or we accept the transmitted text and conclude that Stephanus cited Theopompus and Callimachus. If this Theopompan story was reported by Callimachus, the question arises as to why it is not contained in the section of pseudo-Antigonus' work that depends on Callimachus (129–173), which otherwise contains several fragments of Theopompus (see Section 3). Perhaps pseudo-Antigonus did not include it in his excerpt from Callimachus since he had already included it (directly or via another intermediate source) in a previous part of his work (15a).[169]

Even though various ancient sources speak of couples of crows,[170] only the aforementioned passages and Pliny, *Naturalis historia* 10.31, relate the marvel regarding the two crows in Crannon, which are said to leave behind their young instead of banishing them from the nest.[171] We may thus consider the possibility that Pliny's claim about the two crows in Crannon, which is not attributed to a specific source, may also ultimately go back to Theopompus.[172] Jacoby also suspected that reports regarding a marvelous spring in Crannon (Plin. *HN* 31.20; Ath. 2.42c) may likewise derive from Theopompus, but he was aware that such provenance is impossible to prove.[173]

Neither pseudo-Antigonus nor Stephanus say in which book Theopompus related this marvel. Wichers and Morison argued that the historian did so in his

---

168 Giacomelli (2021) 327–328: "La testimonianza di Stefano di Bisanzio sull'opera paradossografica di Callimaco deve essere soppesata con cautela. Stabilita la sicura ascendenza teopompea di quanto si legge in *Mir.* <134> (126), rimane infatti da spiegare la ragione per la quale Stefano di Bisanzio – che conosceva certo direttamente i *Mirabilia* ps.-aristotelici – nel citare la notizia sui corvi di Crannon abbia mancato di segnalare la testimonianza della compilazione attribuita ad Aristotele e si sia invece preoccupato di precisare che anche Callimaco, in un'opera citata da Stefano in questa sola occasione con un titolo in tutto identico a quello della raccolta ps.-aristotelica, avrebbe riportato la stessa notizia. Sorge inevitabilmente il sospetto che qui un guasto della tradizione manoscritta (una svista del compilatore o dell'epitomatore medievale) abbia obliterato il nome di Aristotele, in luogo del quale si legge ora quello di Callimaco: il sospetto è accresciuto dal fatto che quella di Stefano di Bisanzio è l'unica testimonianza nella quale il titolo dell'opera di Callimaco sia offerto nella forma θαυμάσια."

169 Schepens and Delcroix (1996) 395 n. 68 claimed that pseudo-Antigonus preferred to cite the original source (i.e., Theopompus), rather than the intermediate one (i.e., Callimachus).

170 *Mir.* 122; Arist. *HA* 8(9).31.618b9–12; Ael. *NA* 2.49 = Arist. F 270.3 Gigon; Ael. *NA* 7.18; ps.-Antig. *Mir.* 15c.

171 As crows otherwise do, according to Arist. *HA* 8(9).31.618b9–12; Ael. *NA* 2.49 = Arist. F 270.3 Gigon; Plin. *HN* 10.31.

172 Plin. *HN* 10.31: *itaque parvis in vicis non plus bina coniugia sunt, circa Crannonem quidem Thessaliae singula perpetuo. genitores suboli loco cedunt*; ed. Mayhoff (1909) 228. According to Giacomelli (2021) 325, Pliny's version does not derive from Theopompus, but is closer to that of Aristotle and Aelian (see previous note).

173 Jacoby (1930) 391.

discussion of Thessalian geography in book 9 of the *Philippica*, while Jacoby left the problem open.[174]

2.14 Mir. 135

1. Ἐν δὲ Ἀπολλωνίᾳ τῇ πλησίον κειμένῃ τῆς τῶν Ἀτιντάνων [ἀτιντάνων Β[γρ]: ἀτλαντίνων ψ: ἀτλαντικῶν GR: ἀτλαντ(ιν)ῶν P *ut vid.*: ταυλαντίων Ο[sl]] χώρας φασὶ γίνεσθαι ἄσφαλτον ὀρυκτὴν καὶ πίσσαν, τὸν αὐτὸν τρόπον ἐκ τῆς γῆς ἀναπηδῶσαν τοῖς ὕδασιν, οὐδὲν διαφέρουσαν τῆς Μακεδονικῆς, μελαντέραν δὲ καὶ παχυτέραν πεφυκέναι ἐκείνης. 2. Οὐ πόρρω δὲ τούτου τοῦ χωρίου πῦρ ἐστι καιόμενον πάντα τὸν χρόνον, ὡς φασὶν οἱ κατοικοῦντες περὶ τὴν χώραν ἐκείνην. ὁ δὲ καιόμενος τόπος ἐστὶν οὐ πολύς, ὡς ἔοικεν, ἀλλ᾽ ὅσος μάλιστα πεντακλίνου τὸ μέγεθος. ὄζει δὲ θείου καὶ στυπτηρίας. 3. Καὶ πέφυκε περὶ αὐτὸν πόα τε βαθεῖα, ὃ καὶ θαυμάσειεν ἄν τις μάλιστα, καὶ δένδρα μεγάλα, οὐκ ἀπέχοντα τοῦ πυρὸς πήχεις τέσσαρας. 4. Καίεται δὲ συνεχῶς περὶ Λυκίαν καὶ Μεγάλην πόλιν τὴν ἐν Πελοποννήσῳ.

1. At Apollonia, which lies near to the country of the Atintanii, they say that bitumen and pitch is buried, and springs up out of the earth in the same way as water, in no way different from that in Macedonia, except that it is blacker and thicker. 2. Not far from this spot is a fire which burns perpetually, as those who live in the district testify. The burning place is apparently not large, about the size of five couches. It smells of sulphur and vitriol. 3. And around it grows thick grass, which is a most surprising fact, and there are huge trees not four cubits from the fire. 4. There is also continuous burning in Lycia and near Megalopolis in the Peloponnese.

This chapter of *Mir.* has a parallel in Aelian, *Varia historia* 13.16:

Ἀπολλωνιᾶται πόλιν οἰκοῦσι γείτονα Ἐπιδάμνου ἐν τῷ Ἰονίῳ κόλπῳ. καὶ ἐν τοῖς πλησίον αὐτῆς χωρίοις ἄσφαλτός [ἄσφαλτός Gesner: -άλτου x: -αλέστερ. V] ἐστιν ὀρυκτὴ καὶ πίττα [πίττα ex *Mir.* Kor.: πιμπλᾷ codd.] τὸν αὐτὸν ἐκ τῆς γῆς ἀνατέλλουσα τρόπον, ὃν καὶ αἱ πλεῖσται πηγαὶ τῶν ὑδάτων. οὐ πόρρω δὲ καὶ τὸ ἀθάνατον δείκνυται πῦρ. ὁ δὲ καιόμενος τόπος ἐστὶν [τόπος ἐστὶν ex *Mir.* Her.: ἐστι λόφος codd.] ὀλίγος καὶ οὐκ εἰς μέγα διήκει καὶ ἔχει περίβολον οὐ πολύν, ὄζει δὲ θείου καὶ στυπτηρίας. καὶ περὶ αὐτόν ἐστι δένδρα εὐθαλῆ καὶ πόα [πόα ex *Mir.* Kühn: πολλὰ codd.] χλωρά· καὶ τὸ πῦρ πλησίον ἐνακμάζον οὐδὲν λυπεῖ οὔτε τὴν τῶν φυτῶν βλάστην οὔτε τὴν τεθηλυῖαν <πόαν> [suppl. Her.]. καίεται δὲ τὸ πῦρ καὶ νύκτα καὶ μεθ᾽ ἡμέραν, καὶ διέλιπεν οὐδέποτε, ὡς Ἀπολλωνιᾶται λέγουσι, πρὶν τοῦ πολέμου τοῦ πρὸς Ἰλλυριοὺς συμβάντος αὐτοῖς.[175]

The city of Apollonia is a neighbor of Epidamnus on the Ionian gulf. In its territory, asphalt is dug up, and pitch comes out of the ground in the same way as most springs of water. Not far away the everlasting flame is pointed out; the burnt area is small, neither extensive nor with a large perimeter, and there is a

---

174 Wichers (1829) 75–76 F 85, 165–166; Jacoby (1930) 391; Morison (2014) on F 267.

175 Ed. Dilts (1974) 160–161. Note that Aelian's transmitted text has been emended by editors at various points on the basis of *Mir.*

smell of sulfur and alum. Around it, trees flourish and the grass is green; the fire nearby does no damage to the growing plants or the luxuriant grass. The fire burns night and day, and the men of Apollonia say it never went out before the war they fought against the Illyrians. (trans. Wilson, with modifications)

The following table shows the significant similarities (in bold) as well as the few differences ("in italics") between the two passages:[176]

Mir. *135*	Ael. VH *13.16*
1. Ἐν δὲ **Ἀπολλωνίᾳ** τῇ **πλησίον** *κειμένῃ τῆς τῶν Ἀτιντάνων χώρας* φασὶ γίνεσθαι **ἄσφαλτον ὀρυκτὴν καὶ πίσσαν, τὸν αὐτὸν τρόπον ἐκ τῆς γῆς** ἀναπηδῶσαν **τοῖς ὕδασιν**, *οὐδὲν διαφέρουσαν τῆς Μακεδονικῆς, μελαντέραν δὲ καὶ παχυτέραν πεφυκέναι ἐκείνης.* 2. **Οὐ πόρρω δὲ** τούτου τοῦ χωρίου **πῦρ ἐστι καιόμενον** πάντα τὸν χρόνον, **ὡς φασὶν οἱ κατοικοῦντες** περὶ τὴν χώραν ἐκείνην. **ὁ δὲ καιόμενος τόπος ἐστὶν οὐ** πολύς, ὡς ἔοικεν, ἀλλ’ ὅσος μάλιστα πεντακλίνου τὸ μέγεθος. **ὄζει δὲ θείου καὶ στυπτηρίας.** 3. **Καὶ πέφυκε περὶ** αὐτὸν **πόα** τε βαθεῖα, ὃ καὶ θαυμάσειεν ἄν τις μάλιστα, **καὶ δένδρα** μεγάλα, **οὐκ** ἀπέχοντα τοῦ πυρὸς πήχεις τέσσαρας. 4. **Καίεται δὲ** συνεχῶς *περὶ Λυκίαν καὶ Μεγάλην πόλιν τὴν ἐν Πελοποννήσῳ*	**Ἀπολλωνιᾶται** *πόλιν οἰκοῦσι γείτονα Ἐπιδάμνου ἐν τῷ Ἰονίῳ κόλπῳ.* **καὶ ἐν τοῖς πλησίον** αὐτῆς χωρίοις **ἄσφαλτός ἐστιν ὀρυκτὴ καὶ πίττα τὸν αὐτὸν ἐκ τῆς γῆς** ἀνατέλλουσα **τρόπον**, ὃν *καὶ αἱ πλεῖσται πηγαὶ τῶν ὑδάτων.* **οὐ πόρρω δὲ** καὶ *τὸ ἀθάνατον δείκνυται* **πῦρ.** **ὁ δὲ καιόμενος τόπος ἐστὶν** ὀλίγος καὶ **οὐκ** εἰς μέγα διήκει καὶ ἔχει περίβολον οὐ πολύν, **ὄζει δὲ θείου** καὶ **στυπτηρίας.** καὶ **περὶ** αὐτόν ἐστι **δένδρα** εὐθαλῆ καὶ **πόα** χλωρά· καὶ τὸ **πῦρ** πλησίον ἐνακμάζον οὐδὲν λυπεῖ οὔτε τὴν τῶν φυτῶν βλάστην οὔτε τὴν τεθηλυῖαν <**πόαν**>. **καίεται δὲ** τὸ **πῦρ** καὶ *νύκτα καὶ μεθ’ ἡμέραν, καὶ διέλιπεν οὐδέποτε, ὡς Ἀπολλωνιᾶται λέγουσι, πρὶν τοῦ πολέμου τοῦ πρὸς Ἰλλυριοὺς συμβάντος αὐτοῖς*

The numerous verbal parallels and the few differences in content clearly show that the two passages cannot depend on each other, but rather derive from the same source. On the one hand, *Mir.* claims that Apollonia is located not far from the region of the Atintanii[177]; that pitch in Apollonia is "in no way different from that in Macedonia, except that it is blacker and thicker;" and that other fires that burn continuously can be found near Lycia and Megalopolis. On the other hand, Aelian says that Apollonia is located close to Epidamnus on the Ionian gulf and that the fire never went out before a war between Apollonia and the Illyrians. Moreover, *Mir.* offers precise figures regarding dimensions or distances ("about the size of five couches," "not four cubits from the fire"), while Aelian provides

---

176 The close similarity between the two passages was noticed by Beckmann (1786) 282. Cf. also de Martini (2021) 468–469.

177 On the textual, historical, and geographical problems regarding the name of the tribe mentioned by *Mir.*, see Giacomelli (2023) 320–321, with *status quaestionis* and previous literature. Cf. also de Martini (2021) 468 n. 24. In any case, there can be no doubt as to the identification of Apollonia with Apollonia on the Aous, a well-known *polis* located in Illyria not far from Epidamnus (cf. Diosc. *Mat. med.* 1.73.1): see TM Geo 3603 = Talbert (2000) 49 B3.

fewer details (τόπος ἐστὶν ὀλίγος καὶ οὐκ εἰς μέγα διήκει καὶ ἔχει περίβολον οὐ πολύν, πλησίον).

Interestingly, *Mir.*'s claim that pitch in Apollonia is "in no way different from that in Macedonia" has a parallel in a fragment of Theopompus (*FGrHist/BNJ* 115 F 320) preserved by Pliny, *Naturalis historia* 16.59:

> *Theopompus scripsit in Apolloniatarum agro picem fossilem, non deteriorem Macedonica, inveniri.*[178]
>
> Theopompus wrote that in the territory of the Apolloniates a mineral pitch is found that is not inferior to that of Macedonia. (trans. Rackham)

In another fragment of Theopompus (*FGrHist/BNJ* 115 F 316) preserved by Pliny, *Naturalis historia* 2.237 (not mentioned by Flashar in his commentary), we read that the second place described by *Mir.* and Aelian, which was located not far from the asphalt well in the territory of Apollonia, was known as a Nymphaeum:[179]

> *similiter in Megalopolitanorum agro. nam si intermisit ille iucundus fron-*
> *demque densi supra se nemoris non adurens et iuxta gelidum fontem semper*
> *ardens Nymphaei crater, dira Apolloniatis suis portendit, ut Theopompus*
> *tradidit; augetur imbribus egeritque bitumen temperandum fonte illo ingus-*
> *tabili, et alias omni bitumine dilutius.*[180]
>
> Likewise in the territory of Megalopolis: for if that agreeable Bowl of Nymphaeus, which does not scorch the foliage of the thick wood above it and though near a cold stream is always glowing hot, ceases to flow, it por-tends horrors to its neighbours in the town of Apollonia, as Theopompus has recorded. It is augmented by rain, and sends forth asphalt to mingle with that unappetizing stream, which even without this is more liquid than ordinary asphalt. (trans. Rackham)

Comparing these passages of Pliny with those of *Mir.* and Aelian not only reveals that Theopompus described the place where a fire was burning all the time, sur-rounded by grass and trees (the Nymphaeum), such comparison also suggests that the "dire consequences for the people of Apollonia" mentioned by Pliny allude to

---

178 Ed. Mayhoff (1892) 16.

179 For the Nymphaeum, see also Plin. *HN* 2.240; 3.145 (*Apollonia . . . cuius in finibus celebre Nym-phaeum accolunt barbari Amantes et Buliones*: from Theopompus?); Strab. 7.5.8.316C = Posi-don., *FGrHist/BNJ* 87 F 93 = F 235 Edelstein and Kidd = F 46 Theiler; Vitr. 8.3.8–9; Plut. *Sull.* 27; Cass. Dio 41.45; Ampel. 8.1. Cf. also Gal. *SMT* XII p. 375 Kühn. According to Hammond (1967) 234 n. 2, the passages of *Mir.*, Cassius Dio, Plutarch, Aelian, and Pliny derive from Theopompus, while the passages of Strabo and Ampelius depend on Posidonius. On the Nymphaeum and the asphalt mine, cf. Hammond (1967) 231–234; Kidd (1988) 826–828; Morison (2014) on F 316; de Martini (2021) 465–472.

180 Mayhoff (1906) 222.

the war between Apollonia and the Illyrians mentioned by Aelian.[181] Moreover, Pliny's mention of a flame *in Megalopolitanorum agro* cannot but recall the last sentence of the chapter of *Mir.* (Καίεται δὲ συνεχῶς περὶ Λυκίαν καὶ Μεγάλην πόλιν τὴν ἐν Πελοποννήσῳ), which is not reported by Aelian.

We may thus conclude that all the information preserved by *Mir.* and Aelian was reported by Theopompus, who should consequently be regarded as their shared source.[182] The two fragments of Theopompus cited by Pliny, together with the descriptions provided by *Mir.* and Aelian, clearly belonged to one and the same digression in Theopompus' work.[183] According to Morison, Theopompus described this marvel either in his discussion of Epirus in book 39 or in his description of the Adriatic area in book 21 of the *Philippica*.[184]

2.15 Mir. 136

1. Λέγεται δὲ καὶ ἐν Ἰλλυριοῖς τίκτειν τὰ βοσκήματα δὶς τοῦ ἐνιαυτοῦ, καὶ τὰ πλεῖστα διδυμοτοκεῖν, καὶ πολλὰ δὲ τρεῖς ἢ τέσσαρας ἐρίφους τίκτειν, ἔνια δὲ καὶ πέντε καὶ πλείους, ἔτι δὲ καὶ γάλακτος ἀφιέναι ῥαδίως τρία ἡμίχοα. 2. Λέγουσι δὲ καὶ τὰς ἀλεκτορίδας οὐχ ὥσπερ παρὰ τοῖς ἄλλοις ἅπαξ τίκτειν, ἀλλὰ δὶς ἢ τρὶς παρ' αὐτοῖς τῆς ἡμέρας.

1. Cattle in Illyria are said to breed twice during the year, and most commonly of all to have twins, and that goats often bear three or four, and some five or even more; they readily yield three half-pitchers of milk. 2. They also say that hens do not lay once a day, as they do elsewhere, but two or three times.

Part of the first section of this chapter of *Mir.*, which is quoted by Stephanus of Byzantium, A 65 Billerbeck, *s.v.* Ἀδρία,[185] is paralleled in pseudo-Scymnus' description of the peoples living on the Adriatic Sea (369–390), in which the author claims that this area is very fruitful and that animals have twins (379: διδυμητοκεῖν γάρ φασι καὶ τὰ θρέμματα). As noted in Section 1, pseudo-Scymnus explicitly attributes to Theopompus only the first part of his description, which regards the location of the Adriatic Sea, while the ethnographic section that follows (to which v. 379 belongs) is introduced by the verb ἱστοροῦσι. Jacoby considered these verses to be part of Theopompus' fragment since they closely follow those explicitly attributed to the historian (*FGrHist/BNJ* 115 F 130). For this reason, Flashar and others argued that this chapter of *Mir.* depends on the historian.[186] Even though the verbal parallel

---

181 Cf. Jacoby (1930) 395; Dowden (2013) on F 93.

182 Theopompus as a source of *Mir.*: Schrader (1868) 226 n. 11; Müllenhoff (1870) 428; Giannini (1965) 289 ("Ex Theopompo verisimile est"); Flashar (1972) 135; Fraser (1972) II 1080 n. 385; Gómez Espelosín (2008) 239 n. 125. According to Morison (2014) on F 316, "Theopompos was probably the earliest source for the description of this Nymphaion."

183 See Morison (2014) on F 316 and F 320.

184 Morison (2014) on F 316 and F 320.

185 See also *Exc. NA* 2.559 (p. 140.10–15 Lambros).

186 Müllenhoff (1870) 428; Flashar (1972) 135, followed by Vanotti (2007) 201; Vattuone (2000) 25–26.

provided by pseudo-Scymnus is somewhat limited in scope, we may cautiously accept this attribution.

The second part of *Mir*. 136 has parallels in Aristotle, *Historia animalium* 6.1.558b16–21 and *De generatione animalium* 3.1.749b29, according to which Adrianic hens are small in size,[187] but lay every day, while some of those that are bred domestically lay twice a day (or even more often). These Aristotelian passages may be behind the statement that καὶ τὰς ἀλεκτορίδας δὶς τίκτειν τῆς ἡμέρας, τῷ δὲ μεγέθει πάντων εἶναι μικροτέρας τῶν ὀρνίθων ("the hens also lay twice a day, but are smaller than all other fowl"), which Stephanus of Byzantium, A 65 Billerbeck, *s.v.* Ἀδρία, appends to his citation of *Mir.*; but it is uncertain whether the information provided by *Mir*. derives from Aristotle or from the source used in the first part of the chapter, that is, probably Theopompus.

2.16 Mir. 137

Λέγεται δὲ καὶ ἐν Παιονίᾳ τοὺς βοῦς τοὺς ἀγρίους πολὺ μεγίστους ἁπάντων τῶν ἐν τοῖς λοιποῖς ἔθνεσι γίγνεσθαι, καὶ τὰ κέρατα αὐτῶν χωρεῖν τέσσαρας χόας, ἐνίων δὲ καὶ πλεῖον.

It is also said in Paeonia that the wild bulls are bigger than in any of the other nations, and that their horns will hold four pitchers, and some of them even more.

Modern commentaries on *Mir*. usually notice that 137[188] recalls the long description of the *bolinthos* provided in *Mir*. 1, but that animal's horns are said to contain just "more than a half-pitcher."[189] Flashar claimed that the source of this chapter cannot be identified with certainty, but he tentatively attributed it to Theopompus, because he considered it to be connected to the previous one (which he regarded as Theopompan).[190] Flashar was apparently unaware that the information concerning the horns of Paeonian bulls has a parallel in a fragment of Theopompus (*FGrHist/BNJ* 115 F 38) reported by Athenaeus, 11.476d-e:

τοὺς δὲ **Παιόνων** βασιλεῖς φησι Θεόπομπος ἐν δευτέρᾳ Φιλιππικῶν **τῶν βοῶν τῶν παρ᾽ αὐτοῖς γινομένων μεγάλα κέρατα** φυόντων ὡς χωρεῖν **τρεῖς καὶ τέτταρας χόας** ἐκπώματα ποιεῖν ἐξ αὐτῶν τὰ χείλη περιαργυροῦντας καὶ χρυσοῦντας.[191]

But Theopompus in book 2 of the *Philippica* says that because Paeonian bulls produce horns large enough to contain three or four pitchers, their kings make drinking vessels out of them, covering the lips with silver or gold. (trans. Olson, with modifications)

---

187 Cf. also Ath. 7.285d-e.
188 This passage of *Mir*. is quoted by *Exc. NA* 2.444 (pp. 120.25–121.2 Lambros).
189 See, for example, Flashar (1972) 135.
190 Flashar (1972) 135.
191 Ed. Olson (2020) 293. On this fragment (without reference to *Mir*.), see Pédech (1989) 83–84; Flower (1994) 121. Cf. also Ath. 11.468d = Theopomp., *FGrHist/BNJ* 115 F 284.

This passage, which is not mentioned in modern editions of or commentaries on *Mir.*, shows that *Mir.* 137 most likely derives from Theopompus.[192] The passage specifies that the information concerning the Paeonian bulls was originally contained in book 2 of the *Philippica* and reveals that the original context from which the marvel related by *Mir.* was excerpted did not concern marvels of nature as such, but was rather an ethnographic digression about Paeonia, which was part of the historical account of Philip's II war against the Paeonians and the Illyrians in 359/8 BC. If the attribution of *Mir.* 137 to Theopompus is correct, we may infer that the historian also spoke about the large size of the Paeonian wild bulls (τοὺς βοῦς τοὺς ἀγρίους πολὺ μεγίστους ἁπάντων τῶν ἐν τοῖς λοιποῖς ἔθνεσι), a detail which is not explicitly reported by Athenaeus. Moreover, since the two citations are unlikely to derive from a shared intermediate source, we may suspect that at least the words κέρατα . . . χωρεῖν τέσσαρας χόας (which are found in both citations) were found in the original Theopompan text.[193]

2.17 Mir. 138

1. Ἐν Ἰλλυριοῖς δὲ τοῖς Ἀρδιαίοις [ἀρδιαίοις Stephanus: ἀρδίοις BF (cf. *Arduis* φ): σαρδίοις x: ἀρκαδίοις CK] καλουμένοις, παρὰ τὰ μεθόρια τῶν Αὐταριατῶν κἀκείνων, φασὶν ὄρος εἶναι μέγα, τούτου δὲ πλησίον ἄγκος, ὅθεν ὕδωρ ἀναπηδᾶν οὐ πᾶσαν ὥραν, ἀλλὰ τοῦ ἦρος, πολὺ τῷ πλήθει, ὃ λαμβάνοντες τὰς μὲν νύκτας [τὰς μὲν νύκτας . . . τὰς δὲ ἡμέρας Beckmann: τὰς μὲν ἡμέρας . . . τὰς δὲ νύκτας BFx] ἐν τῷ στεγνῷ φυλάττουσι, τὰς δὲ ἡμέρας εἰς τὴν αἰθρίαν τιθέασι. καὶ πέντε ἢ ἓξ ἡμέρας τοῦτο ποιησάντων αὐτῶν πήγνυται τὸ ὕδωρ, καὶ γίνεται κάλλιστον ἅλας, ὃ ἕνεκεν τῶν βοσκημάτων μάλιστα διατηροῦσιν· οὐ γὰρ εἰσάγονται πρὸς αὐτοὺς ἅλες διὰ τὸ κατοικεῖν πόρρω αὐτοὺς θαλάσσης καὶ εἶναι αὐτοὺς ἀμίκτους. 2. Πρὸς οὖν τὰ βοσκήματα πλείστην αὐτοῦ χρείαν ἔχουσιν· ἁλίζουσι γὰρ αὐτὰ δὶς τοῦ ἐνιαυτοῦ. ἐὰν δὲ μὴ ποιήσωσι τοῦτο, συμβαίνει αὐτοῖς ἀπόλλυσθαι τὰ πλεῖστα τῶν βοσκημάτων.

1. Among the Illyrians who are called Ardiaei along the boundary between them and the Autariatae, they say there is a high mountain, and near to it a valley from which the water rises, not at all seasons but in the spring, in considerable quantity, which they take and keep under cover by day, but put in the open at night.[194] After they have done this for five or six days, the water hardens and becomes very fine salt, which they keep especially for the cattle; for salt is not imported to them because they live far from the sea and do not associate with others. 2. Consequently they need it very much for the cattle; for they give them salt twice a year. If they fail to do this, most of the cattle are found to die.

---

192 This silence may be due to the mechanical transposition in β of the phrase λοιποῖς ἔθνεσι γίγνεσθαι, καὶ τὰ κέρατα αὐτῶν χωρεῖν τέσσαρας χόας after the first part of 122: cf. Giacomelli (2021) 51–53. The parallel has been noted only by Geffcken (1892) 89 n. 5 and Gauger and Gauger (2010) 191 (who, however, hesitate between Theopompus and Theophrastus as the source used by *Mir.*).

193 On Athenaeus as a source for reconstructing Theopompus, see Chávez Reino and Ottone (2007).

194 Note that Hett accepts the transmitted form τὰς μὲν νύκτας . . . τὰς δὲ ἡμέρας instead of Beckmann's emendation τὰς δὲ ἡμέρας . . . τὰς μὲν νύκτας, accepted by Giacomelli and others: see Giacomelli (2023) 323. On the form Ἀρδιαίοις, see Giacomelli (2023) 322–323.

This chapter of *Mir.* is paralleled only in a passage from Strabo's description of the Balkan peninsula.[195] With regard to the Illyrian tribe of the Autariatae, Strabo, 7.5.11.317–318C, speaks as follows:

Αὐταριᾰται μὲν οὖν τὸ μέγιστον καὶ ἄριστον τῶν Ἰλλυριῶν ἔθνος ὑπῆρξεν· ὃ πρότερον μὲν πρὸς **Ἀρδιαίους** συνεχῶς ἐπολέμει περὶ ἁλῶν **ἐν μεθορίοις πηγνυμένων ἐξ ὕδατος** ῥέοντος **ὑπὸ ἄγκει** τινὶ **τοῦ ἔαρος** (ἀρυσαμένοις γὰρ καὶ ἀποθεῖσιν **ἡμέρας πέντε ἢ ἓξ ἐπήγνυντο** οἱ ἅλες)· συνέκειτο δὲ παρὰ μέρος χρῆσθαι τῷ ἁλοπηγίῳ, παραβαίνοντες δὲ τὰ συγκείμενα ἐπολέμουν.[196]

Now the Autariatae were once the largest and best tribe of the Illyrians. In earlier times they were continually at war with the Ardiaei over the salt-works on the common frontiers. The salt was made to crystallize out of water which in the springtime flowed at the foot of a certain mountain-glen (for if they drew off the water and stowed it away for five or six days the salt would become crystallized). They would agree to use the saltworks alternately, but would break the agreements and go to war. (trans. Jones, with modifications)

The close similarities between the two passages with regard to content and vocabulary suggest that *Mir.* and Strabo depend on a shared source.[197] It is usually supposed that this source was Theopompus, since the latter is used elsewhere by Strabo in his description of that region,[198] but there may be even stronger reasons to identify Theopompus as the source. Interestingly, we know that Theopompus dealt with the Illyrians in book 2 of his *Philippica* (cf. Section 2.16) and that he specifically recorded the social structure and drinking habits of the Ardiaei as well as a war between them and the Celts (Ath. 10.443a-c = Theopomp., *FGrHist/BNJ* 115 F 39–40; cf. also Ath. 6.271e).[199] Now that we know that the previous chapter of *Mir.* probably derives from the very same book of Theopompus' *Philippica*, it seems all the more probable that 137 and 138 were both excerpted from that book as well.

Both *Mir.* and Strabo can contribute to our knowledge of Theopompus' original account. On the one hand, we can observe that the compiler of *Mir.* selected the material that was relevant to his work, leaving out any reference to the original historical or ethnographic context, which we can recover only thanks to Strabo's

---

195　This chapter is literally quoted by *Exc. NA* 2.560 (p. 140.15–25 Lambros) and perhaps by Leon. Mag. *Therm.* 79–82 = 84–87 Gallavotti.

196　Ed. Radt (2003) 310.

197　See Bearzot (2004) 73; Flashar (1972) 140–141; Radt (2007) 302–303.

198　See Strab. 7.5.9.317C = Theopomp., *FGrHist/BNJ* 115 F 129; Strab. 7.7.5.323–324C = Theopomp., *FGrHist/BNJ* 115 F 382. For the attribution of *Mir.* 138 to Theopompus, see Flashar (1972) 140; Bearzot (2004) 71–73; Vanotti (2007) 209. According to Giannini (1965) 299, followed by Gómez Espelosín (2008) 244 n. 138, *Mir.* 138 may derive from either Timaeus or Theopompus.

199　See Morison (2014) on F 39–40. Polyaenus, *Strat.* 7.42, reports the same stratagem as Athenaeus, but speaks of a war between the Celts and the Autariatae. Mócsy (1972), followed by Shrimpton (1991) 109, 289–290 n. 12, 156, suggested emending "Ardiaei" to "Autariatae" in Athenaeus' text, while Flower (1994) 120 n. 12; Bearzot (2004); Morison (2014) on F 40 argued that this episode concerned the Ardiaei. Cf. also Pédech (1989) 85; Džino (2007) 55; Pajón Leyra (2021).

parallel passage.[200] On the other hand, *Mir.* preserves various details that have been omitted by Strabo but that might have been mentioned by Theopompus (e.g., that salt was not imported to the Ardiaei since they lived far from the sea; that they did not interact with others; and that they needed salt for their cattle). Since *Mir.* and Strabo, moreover, are unlikely to depend on the same intermediate source, it seems probable that their shared vocabulary goes back to Theopompus (cf. τοῦ ἦρος *Mir.* ~ τοῦ ἔαρος Strabo; παρὰ τὰ μεθόρια *Mir.* ~ ἐν μεθορίοις Strabo; ἄγκος *Mir.* ~ ὑπὸ ἄγκει Strabo; πέντε ἢ ἓξ ἡμέρας *Mir.* ~ ἡμέρας πέντε ἢ ἓξ Strabo).

3. Impact on our knowledge of *Mir.* and Theopompus

The following table summarizes the results of the above analysis.

Chapter of Mir.	Source	Location
122	Unknown	Pedasia (Caria)
123	Theopompus: • Ps.-Antig. *Mir.* 136 = Callim. F 407 VIII Pfeiffer = *FGrHist/BNJ* 115 F 268a • *Par. Pal.* 19 = *FGrHist/BNJ* 115 F 268a	Strymon valley (among the Agrieis/Agrianes: ps.-Antig.; among the Sinti and the Maedi: *Mir.*)
124	Unknown (possibly Theopompus: same location as 123)	παρ' αὐτοῖς (i.e., among the Sinti and the Maedi)
125	Theopompus: • Ps.-Antig. *Mir.* 142 = Callim. F 407 XIV Pfeiffer = *FGrHist/BNJ* 115 F 271a • Plin. *HN* 31.17 = *FGrHist/BNJ* 115 F 271b Other passages depending on Theopompus: • *Par. Flor.* 9 = *FGrHist* 1659 F 5	Scotussae (Thessaly [or Thrace?])
126	Aristotle, *Historia animalium* 8(9).36.620a33–620b8	Thrace (Kedropolis? Above Amphipolis?)
127	Theopompus (or Lycus): • Ps.-Antig. *Mir.* 173 = Callim. F 407 XLIV Pfeiffer = *FGrHist/BNJ* 115 F 274a • Ael. *NA* 17.16 = *FGrHist/BNJ* 115 F 274b = *FGrHist/BNJ* 570 F 4	Among the Heneti
128	Theopompus: • Ps.-Antig. *Mir.* 14 = *FGrHist/BNJ* 115 F 266 Other passages possibly depending on Theopompus: • Strab. 7 F 15b Radt • Plin. *HN* 11.99 • Plut. *De tranq. anim.* 15.473e	Near Olynthus

(Continued)

200 Cf. Flashar (1972) 140–141; Bearzot (2004) 73.

(Continued)

Chapter of Mir.	Source	Location
129	Theopompus: • Ps.-Antig. *Mir.* 141 = Callim. F 407 XIII Pfeiffer = *FGrHist/BNJ* 115 F 270a • Plin. *HN* 31.27 = *FGrHist/BNJ* 115 F 270d • *Par. Flor.* 15 = *FGrHist/BNJ* 115 F 270c • *Par. Vat.* 38 = *FGrHist/BNJ* 115 F 270b Other passages possibly depending on Theopompus: • Vitr. 8.3.15	Thrace (Εν δὲ †Κύκλωψι†)
130,1 (Φασὶ δὲ . . . ἥπατα ἔχειν)	Theopompus, book 20 of the *Philippica*: • Ath. 9.401a-b = *FGrHist/BNJ* 115 F 126a • Steph. Byz. B 103 Billerbeck, *s.v.* Βισαλτία = *FGrHist/BNJ* 115 F 126b = Favorin. F 89 Amato • Ael. *NA* 5.27 = *FGrHist/BNJ* 115 F 126a • Ael. *NA* 11.40 • Gell. *NA* 16.15.1 = *FGrHist/BNJ* 115 F 126a	Crestonia, near the country of the Bisaltae
130,1 (καὶ τόπον . . . ἀποθνήσκει)	Unknown (possibly Theopompus: same location as *Mir.* 130.1 [Φασὶ δὲ . . . ἥπατα ἔχειν])	Crestonia, near the country of the Bisaltae
130,2	Unknown (possibly Theopompus: same location as *Mir.* 130.1 [Φασὶ δὲ . . . ἥπατα ἔχειν])	αὐτόθι = Crestonia, near the country of the Bisaltae
131,1	Theopompus (ἐν τοῖς Θαυμασίοις: cf. 131.2) • Ath. 1.34a = *FGrHist/BNJ* 115 F 277 Other passages depending on Theopompus: • Paus. 6.26.1–2	Elis
131,2	Theopompus (ἐν τοῖς Θαυμασίοις): • Apollon. *Mir.* (*FGrHist* 1672) 10 = *FGrHist/BNJ* 115 F 76 Other passages possibly depending on Theopompus: • Paus. 5.14.1 • Ael. *NA* 2.47 • Plin. *HN* 10.28	Elis
132	Unknown	Coroneia (Boeotia)
133	Theopompus: • Ps.-Antig. *Mir.* 137 = Callim. F 407 IX Pfeiffer = *FGrHist/BNJ* 115 F 269	Lusi (Arcadia) and Lampsacus

Chapter of Mir.	Source	Location
134	Theopompus: • Ps.-Antig. *Mir.* 15a = *FGrHist/ BNJ* 115 F 267a • Steph. Byz. K 207 Billerbeck, *s.v.* Κραννών = *FGrHist/BNJ* 115 F 267b = Callim. F 408 Pfeiffer Other passages possibly depending on Theopompus: • Plin. *HN* 10.31	Crannon (Thessaly)
135	Theopompus: • Plin. *HN* 16.59 = *FGrHist/BNJ* 115 F 320 • Plin. *HN* 2.237 = *FGrHist/BNJ* 115 F 316 Other passages depending on Theopompus: • Ael. *VH* 13.16	Apollonia (Illyria)
136,1	Probably Theopompus: • Ps.-Scymn. 369–390 = *FGrHist/ BNJ* 115 F 130	Illyria
136,2	Unknown (possibly Theopompus: same location as *Mir.* 136.1)	Illyria
137	Theopompus, book 2 of the *Philippica*: • Ath. 11.476d-e = *FGrHist/BNJ* 115 F 38	Paeonia
138	Probably Theopompus, book 2 of the *Philippica*: • cf. *Mir.* 137; Ath. 11.476d-e = *FGrHist/BNJ* 115 F 38; Ath. 10.443a-c = *FGrHist/ BNJ* 115 F 39–40 Other passages probably depending on Theopompus: • Strab. 7.5.11 pp. 317–318C	Illyria (among the Ardiaei and the Autariatae)

Before considering Theopompus' role as a source of *Mir.* 122–138, I would like to call attention to an aspect of this section that does not seem to have attracted scholarly attention so far, namely, the geographical order of presentation followed by the compiler. Indeed, he seems to have grouped the reports according to three broad geographical areas:

1. *Thrace: Mir.* 123–124 (Strymon valley); 126 (Kedropolis? Above Amphipolis?); 128 (near Olynthus); 129 (Εν δὲ †Κύκλωψι†); 130 (Crestonia, near the country of the Bisaltae). Two exceptions are represented by the reports concerning Scotussa in Thessaly (125) and the Heneti (127). As noted in Section 2.6, the story concerning the Heneti seems to have been added because it elaborates on the topic treated in the preceding chapter (i.e., the relationship between humans and birds). The presence of 125 is more difficult to explain, unless one assumes

that Theopompus originally spoke of the Thracian Scotussa and that the compiler of *Mir.* added the reference to Thessaly (cf. Section 2.4).

2. Peloponnese and central Greece: *Mir.* 131 (Elis); 132 (Coroneia, Boeotia); 133 (Lusi, Arcadia); 134 (Crannon, Thessaly).
3. Illyria and surrounding regions: *Mir.* 135 (Apollonia); 136 (Illyria); 137 (Paeonia); 138 (Illyria, among the Ardiaei and the Autariatae).

The first chapter of the series, *Mir.* 122, on Pedasia in Caria, seems to be somewhat isolated. However, as a region facing the Aegean Sea, the inclusion of Caria in this section may be less out of place than it might appear at first sight. Be that as it may, we can still conclude that the compiler seems to have structured his account following a geographical order, with a few exceptions.

What then was the role played by Theopompus as a source within this section of *Mir.*? On the basis of the above analysis, we may conclude that Theopompus was probably the source of *Mir.* 123, 125, 128, 129, 130.1, 131.1, 131.2, 133, 134, 135, and 137. To these chapters, we may reasonably add *Mir.* 136.1 and 138. The attribution of *Mir.* 135, 137, and 138 to Theopompus on new textual grounds represents a step forward in our knowledge of *Mir.*'s sources with respect to the results reached by Flashar. *Mir.* 127 probably derives from Theopompus, but we cannot completely exclude the possibility that it is from Lycus. Only *Mir.* 126 is likely to derive from a source other than Theopompus, namely, Aristotle's *Historia animalium* 8(9).

The question now arises as to whether the remaining passages from the section of *Mir.* under discussion can be ascribed to Theopompus, as was argued by Flashar (with the exception of 122). To be sure, the number of parallels between *Mir.* and the named fragments of Theopompus is surprisingly high, which means that we can be virtually certain that also other passages from *Mir.* derive from the historian, since it would be improbable to have parallels for *all* the paragraphs of *Mir.* that tacitly go back to Theopompus. While *Mir.* 122 and 132 concern places that are not treated in the passages that most probably derive from Theopompus, *Mir.* 124, 130.1 (καὶ τόπον . . . ἀποθνήσκει), 130.2, and 136.2 concern places that are dealt with in the Theopompan passages. Their provenance from the *Philippica* is therefore all the more plausible.

However, *Mir.* 126, which interrupts a series of Theopompan passages by inserting Peripatetic material, shows that the *Reihenargument* cannot be systematically applied, since it seems to imply a too rigid, mechanical method of working that is not entirely applicable to *Mir.* We cannot even infer with certainty that sections dealing with one and the same region derive as a rule from the same source, since the Peripatetic *Mir.* 126 concerns a region that is well represented in the Theopompan chapters.[201] Nor can the topics treated in the various chapters be a reliable guide for attribution, since this section of *Mir.* treats a wide range of paradoxographical topics (the amazing characteristics of stones, plants, animals, oracles, and temples) and since we know that Theopompus sometimes dealt with the same topics treated by the Peripatetics (Aristotle and Theophrastus) and other historians (Lycus).[202]

---

201 See *Mir.* 123; 128; 129; 130.1 (Φασὶ δὲ . . . ἥπατα ἔχειν).
202 For the Peripatetics, see *Mir.* 123; 124; 132; 133; 134; for Lycus, see *Mir.* 127.

Therefore, we simply cannot know whether *Mir.* 122, 124, 130.1 (καὶ τόπον . . . ἀποθνῄσκει), 130.2, 132, and 136.2 entirely depend on Theopompus, even though this attribution – I repeat – remains plausible, since the historian represents by far the main source used in the section of *Mir.* under discussion. I am aware that this conclusion might appear to be too cautious, but I think that a clear distinction should be maintained between attributions based on textual evidence, and attributions that are plausible but cannot be proven. One can thus regard the aforementioned chapters of *Mir.* as possibly Theopompan, but we cannot regard them as fragments of Theopompus.

It is difficult to give a clear-cut answer to the question of how the compiler of *Mir.* had access to this Theopompan material, because of the uncertainty surrounding both the chronology of *Mir.* and the transmission of Theopompan reports of a paradoxographical character. Theopompus' interest in marvels of various kinds was well known in antiquity.[203] Some of his amazing stories are said to come from book 8 of the *Philippica* (F 64, 66, 73, 74), while others are attributed to a work entitled *Thaumasia* (F 69, 71, 75, 76). It remains a matter of debate whether the title *Thaumasia* refers to book 8[204] of the *Philippica* – which, just like book 10 (*On the Demagogues*), also circulated separately under its own title[205] – or to a later collection of marvels related by Theopompus throughout the *Philippica* (many of which were in any case originally contained in book 8).[206]

The fragments that Jacoby ascribed to book 8 of the *Philippica* mostly concern prophets or sapiential figures, including the Magi, Zopyrus, Silenus, Epimenides, Pherecydes of Syrus, Zoroaster, and Pythagoras. Only Apollonius, *Historiae mirabiles* (*FGrHist* 1672) 10 (Theopomp., *FGrHist/BNJ* 115 F 76), which is assigned to the *Thaumasia* (ἐν τοῖς Θαυμασίοις), treats the sort of marvels we find in *Mir.*, namely, amazing stories concerning the natural world (in Elis, kites do not touch the meat of the sacred victims). It is worth noting that *Mir.* 131 combines this story with another Theopompan marvel occurring in Elis, namely, the miracle of the wine during a festival in honor of Dionysus (Theopomp., *FGrHist/BNJ* 115 F 277). If Theopompus related the two marvels in the same context, as seems probable, then F 277 may likewise derive from the *Thaumasia*.

According to Apollonius, *Historiae mirabiles* (*FGrHist* 1672) 1, Theopompus in the *Philippica* discussed marvels following a geographical, rather than thematic, pattern (*FGrHist/BNJ* 115 F 67b: Θεόπομπος ἐν ταῖς Ἱστορίαις ἐπιτρέχων τὰ κατὰ

---

203 See, for example, Dion. Hal. *Pomp.* 6.4; 6.11 = Theopomp., *FGrHist/BNJ* 115 T 20a; Cic. *Leg.* 1.5 = Theopomp., *FGrHist/BNJ* 115 T 26a; Ael. *VH* 3.18 = Theopomp., *FGrHist/BNJ* 115 T 26b.

204 And perhaps to book 9 as well: cf. F 77.

205 See Wichers (1829) 29–30; Jacoby (1930) 365; Laqueur (1934) 2212–2213; Fraser (1972) II 1080 n. 385; Pédech (1989) 174–176; Shrimpton (1991) 15–21; Christ (1993) 49; Flower (1994) 34, 163–164; Pownall (2004) 154–155; Morison (2014) Biographical Essay; Ottone (2018) 530.

206 See Ebert (1825) 173–174; Westermann (1839) L-LII; Susemihl (1891) 478–479; Ziegler (1949) 1143–1145; Dušanić (1977); Pajón Leyra (2011) 39 n. 25, 101–102. The two hypotheses, as observed by Giannini (1964) 102–104, are not mutually exclusive. The problem is left open by Flashar (1972) 47–48; Schepens and Delcroix (1996) 380 n. 18; Gómez Espelosín (2008) 291–292; Gauger and Gauger (2010) 195. Cf. also Belousov (2020) Introduction. The story concerning Silenus and Midas is ascribed both to book 8 of the *Philippica* (F 74a) and to the *Thaumasia* (F 75b): on this story, see Zaccaria (2016), with *status quaestionis* and previous literature.

τόπους θαυμάσια). If we identify the *Thaumasia* with book 8 of the *Philippica*, it might be possible to ascribe to this book at least some of the marvels attributed to Theopompus without indication of a book number. However, it is certain that paradoxical stories were not confined to book 8, but were a recurrent topic in the *Philippica*.[207] The reports contained in *Mir.* 137–138 and 130.1, for example, were originally contained in books 2 and 20, respectively. For this reason, and in view of the digressive character of the *Philippica*, we should resist the temptation to assign fragments without a book number to a specific book (even if some hypothetical attributions remain more plausible than others).

We can reasonably assume that later paradoxographers did not regularly take their reports directly from the 58 original books of the *Philippica*, but it is unnecessary to postulate a single intermediate source for the entire paradoxographical tradition either, since the *Philippica* was a well-known work that continued to be read throughout antiquity.[208] The case of pseudo-Antigonus, who drew Theopompan reports from both Callimachus (136 = F 268a; 137 = F 269; 141 = F 270a; 142 = F 271a; 164 = F 278b; 170 = F 273; 173 = 274a) and a different source (directly from Theopompus? 14 = F 266; 15 = F 267a; 119 = F 181b), suggests a rather fluid and open scenario, in which a number of Theopompan reports coming from the *Philippica* entered paradoxographical literature in various ways and freely circulated among paradoxographers and authors interested in natural history.

Still, the fact that *Mir.* contains much well-preserved Theopompan material in a series of more or less contiguous chapters strongly suggests that the compiler did not base these reports on a number of intermediate sources that independently went back to Theopompus, but rather on a single intermediate source which had a good knowledge of Theopompus' work. We cannot even exclude that *Mir.* had a direct knowledge of Theopompus' work, even if the compiler never mentions the historian by name.

Despite its potential value as a source for our knowledge of Theopompus' work, *Mir.* does not seem to have been duly considered in modern studies concerning the historian. *Mir.* is usually cited only with regard to the geographical passages concerning the Adriatic Sea (81, 104–105), of which the Theopompan origin is, however, at best uncertain,[209] while those passages that surely or most probably derive from Theopompus, as far as I can tell, have hardly been taken into account.[210]

---

207 See Wichers (1829) 29–30; Westermann (1839) LI; Schranz (1912) 27; Ziegler (1949) 1144; Giannini (1964) 102–103; Flower (1994) 34, 163–164; Schepens and Delcroix (1996) 380 n. 18. The paradoxographical fragments of Theopompus are collected in Giannini (1965) 365–368.

208 On the transmission of Theopompus' *Philippica*, see Ottone (2018) 109–130. For the papyri possibly containing Theopompan material, see Biagetti (2017); (2019).

209 See especially Shrimpton (1991) 94–101; Vattuone (2000).

210 Shrimpton (1991) 287 n. 29 refers to *FGrHist* 115 F 267b, 268, 269, 271, 274, 277, 316 and to the corresponding passages in *Mir.* only to demonstrate that the compiler of *Mir.* knew Theopompus' work. Occhipinti (2011) claimed that Aristotle (!) used Theopompus' *Philippica*, on the basis of *Mir.* 104–105 and *Mir.* 118 (which was speculatively attributed to Theopompus by Vanotti (1977) 167–168; (2007) 204), without even mentioning the numerous chapters of *Mir.* that are most probably based on Theopompus.

The reason for such neglect is probably that the text of *Mir.* has not been duly exploited in editions of Theopompus' fragments. In their pioneering editions, Wichers and Müller only provided the reader with a few laconic references to *Mir.* (Wichers: F 286; Müller: F 79, 137, 143, 286). Grenfell and Hunt did include some passages of *Mir.* in their Oxford edition, printing them after the parallel passages in which Theopompus is mentioned by name (F 76, 122, 128, 255, 256, 257, 258, 259, 260).[211] However, Jacoby did not follow this practice in *FGrHist* 115 and never mentioned *Mir.* as a source that transmits fragments of Theopompus.

The reason for this choice is Jacoby's general editorial principle not to include passages that are not explicitly attributed to the lost authors. Yet Jacoby did include the description of Tempe related by Aelian, *Varia historia* 3.1 as *FGrHist* 115 F 80, even though Theopompus is not mentioned as a source; Jacoby's reason was that the description has parallels in some named fragments of the historian (F 78–79). Still, Jacoby was well aware that Theopompus was in fact used as a source by the compiler of *Mir.* In his commentary on *FGrHist* 115 F 266, after citing the parallel testimony provided by *Mir.* 128, Jacoby writes as follows: "auch die umgebenden stücke – s. zu F 126; 267–270; 271; 277 – stammen aus Th[eopompus], und vermutlich mehr, als sich nachweisen läßt." The "surrounding chapters" alluded to by Jacoby are *Mir.* 123, 125, 127 (cf. F 274), 129, 130.1, 131.1, 133, and 134. However, Jacoby's list of Theopompan passages is incomplete. Jacoby himself argued elsewhere in his commentary that other sections of *Mir.* depend on Theopompus, namely, *Mir.* 131.2 (cf. F 76), 135 (cf. F 316 and 320) (yet it is unclear whether Jacoby regarded this chapter as Theopompan material), and 136.1 (cf. F 130). The present analysis has confirmed Jacoby's feeling that even more material of *Mir.* probably comes from Theopompus, namely, 137 (cf. F 38) and 138.

Comparison between *Mir.* and the named fragments of Theopompus has also demonstrated the probable Theopompan origin of passages from other authors. Without taking into account passages that may possibly derive from Theopompus (see table above), Jacoby was aware that *Paradoxographus Florentinus* 9 (*FGrHist* 1659 F 5) has parallels in *Mir.* 125 and *FGrHist* 115 F 271; that Pausanias, 6.26.1–2 contains the same report as *Mir.* 131.1 and *FGrHist* 115 F 277; and that Aelian, *Varia historia* 13.16 corresponds to *Mir.* 135 and *FGrHist* 115 F 316 and F 320. We may now add that Strabo, 7.5.11.317–318C, has a parallel in *Mir.* 138, which, together with *Mir.* 137, probably goes back to book 2 of the *Philippica*. Moreover, in the case of *FGrHist* 115 F 126, Jacoby did not include Aelian, *Natura animalium* 11.40, even though Theopompus is explicitly mentioned as a source.

Another problem with Jacoby's treatment of the evidence is that he has not been consistent in his presentation of the material from *Mir.*: sometimes he fully quoted parts of the relevant chapters in his commentary, other times he simply provided the reader with a reference. The consequence of this choice is that the reader does not find in the edition a rich body of material which, even in Jacoby's own opinion,

211 *Mir.* 135, 136.1, 137, and 138, however, are not included by Grenfell and Hunt in their edition.

ultimately derives from Theopompus; the reason for this choice is simply that the compiler of *Mir.* does not mention the historian by name, but it was characteristic of the compiler not to name sources.[212]

The present analysis has shown that *Mir.* 123, 125, 127, 128, 129, 130.1, 131, 133, 134, 135, 136.1, 137, and 138, as well as *Paradoxographus Florentinus* 9 (= *FGrHist* 1659 F 5), Strabo 7.5.11.317–318C, Pausanias 6.26.1–2, and Aelian, *Varia historia* 13.16 ought to be included in an edition of Theopompus' fragments, either in the critical edition itself (perhaps in *petit*, in order to distinguish them from the named fragments) or at least in the commentary. This enlargement of the corpus collected by Jacoby recommends itself not only for the sake of completeness, but also because these neglected passages, as noted above, may contain additional material that could go back to Theopompus (see especially *Mir.* 125, 128, 129, 131.1, 135, 137, and 138). Moreover, they are of major importance for appreciating the influence of Theopompus on paradoxographical literature. In the end, they are as important for our knowledge of Theopompus' work as the named fragments reported by authors such as pseudo-Antigonus, Pliny, and Aelian.

4. Conclusion

The relationship between *Mir.* and Theopompus seems to have played a different role in scholarship of *Mir.* than in studies of Theopompus. One almost gets the impression that the two fields of research traveled on parallel tracks. On the one hand, scholars of *Mir.* traditionally ascribe a large section of this paradoxographical collection to the historian (almost) in its entirety (from 122 or 123 up to 138). On the other hand, this section of *Mir.* is hardly mentioned in editions and studies of Theopompus' *Philippica*. By reassessing the relationship between the two works on a textual level, the present investigation has offered a less rigid and schematic view of Theopompus' role as a source of *Mir.*, on the one hand, and has shown the importance of *Mir.* as a potential source for our knowledge of Theopompus' *Philippica* and its influence on later traditions, on the other.

Theopompus was probably the main source used by the compiler of *Mir.* in his description of paradoxical phenomena concerning Thrace, the Peloponnese and central Greece, as well as Illyria and surrounding regions. The historian can reasonably be regarded as the source of *Mir.* 123, 125, 128, 129, 130.1, 131.1, 131.2, 133, 134, 135, and 137. *Mir.* 136.1 and 138 also probably go back to the historian, while *Mir.* 127 may derive from either Theopompus (as seems more probable) or Lycus. These chapters of *Mir.* ought to find a place in an edition of Theopompus' fragments, together with the corresponding named fragments reported by authors such as pseudo-Antigonus, Pliny, and Aelian, and with other parallel passages that

212 Cf. the critical remarks of Fraser (1972) II 1080 n. 385. Cf. also Shrimpton (1991) 59, 285 n. 3. Jacoby's choice was perhaps also dictated by editorial reasons: his commentary appeared when paper was short in Germany and it was difficult to find the money for the publication of the book. I owe this information to Stefan Schorn.

can be shown to derive from the historian, namely, Strabo, 7.5.11.317–318C; Pausanias, 6.26.1–2; and Aelian, *Varia historia* 13.16.

Sections that do not find parallels in the named fragments of Theopompus may still derive from the historian, but the *Reihenargument*, according to which *Mir.* can be divided into large continuous blocks each deriving from a different source, cannot be safely applied to this text, since 126 derives from a source other than Theopompus, namely, Aristotle's *Historia animalium*. We can therefore consider the possibility that these chapters of *Mir.* go back to Theopompus, but we cannot regard them as fragments of the historian.

Works cited

Amato, E. 2010. *Favorinos d'Arles. Œuvres. Tome III. Fragments. Texte établi, traduit et commenté par E.A.* Paris: Les Belles Lettres.

Ambaglio, D. 2002. "L'Adriatico nei frammenti degli storici greci." *Hesperìa* 15: 95–99.

Amigues, S. 2018. *Théophraste. Les pierres. Texte établi et traduit par S.A.* Paris: Les Belles Lettres.

Amiotti, G. 1994. "Fenomeni naturali della Calcidica." In Biraschi, A.M. ed. *Strabone e la Grecia.* Napoli: Edizioni Scientifiche Italiane. 201–209.

Balme, D.M. 1991. *Aristotle. History of Animals. Books VII-X.* Edited and Translated by Balme, D.M. Prepared for Publication by Gotthelf, A. Cambridge, MA; London: Harvard University Press.

Balme, D.M. 2002. *Aristotle. Historia animalium. Volume I. Books I-X: Text.* Edited by Balme, D.M. Prepared for Publication by Gotthelf, A. Cambridge: Cambridge University Press.

Barigazzi, A. 1966. *Favorino di Arelate. Opere. Introduzione, testo critico e commento a cura di A.B.* Firenze: Felice Le Monnier.

Bearzot, C. 2004. "I Celti in Illiria. A proposito del frg. 40 di Teopompo." In Urso, G. ed. *Dall'Adriatico al Danubio. L'Illirico nell'età greca e romana. Atti del convegno internazionale (Cividale del Friuli, 25–27 settembre 2003).* Pisa: Edizioni ETS. 63–78.

Beckmann, J. 1786. *Aristotelis liber de mirabilibus auscultationibus.* Göttingen: Apud viduam Abrahami Vandenhoek.

Belousov, A. 2020. "Theopompos of Chios (1652)." In Schorn, S. ed. *Jacoby Online. Die Fragmente der Griechischen Historiker Continued.* Part IV. Leiden; Boston: Brill.

Biagetti, C. 2017. "Teopompo e i papiri, Teopompo nei papiri. Acquisizioni, rimaneggiamenti, attribuzioni." In Ottone, G. ed. *"Historiai para doxan". Documenti greci in frammenti: nuove prospettive esegetiche. Atti dell'Incontro Internazionale di Studi (Genova, 10–11 Marzo 2016).* Tivoli (Roma): Tored. 137–177.

Biagetti, C. 2019. *Corpus dei papiri storici greci e latini. Parte A. Storici greci. 2. Testi storici anepigrafi. Vol. 9.1. Frammenti storici attribuiti a Teopompo, a cura di C.B., con un contributo di Valerio, F.* Pisa. Roma: Fabrizio Serra Editore.

Bianchetti, S. 1990. πλωτὰ καὶ πορευτά. *Sulle tracce di una periegesi anonima.* Florence: Università degli Studi di Firenze, Dipartimento di Storia.

Billerbeck, M. 2006. *Stephani Byzantii Ethnica. Volumen I: A-Γ. Recensuit germanice vertit adnotationibus indicibusque instruxit M.B., adiuvantibus Gaertner, J.F., Wyss, B., Zubler, C.* Berlin; New York: Walter de Gruyter.

Billerbeck, M. 2014. *Stephani Byzantii Ethnica. Volumen III: K-O. Recensuit germanice vertit adnotationibus indicibusque instruxit M.B., adiuvantibus Lentini, G., Neumann-Hartmann, A.* Berlin; Boston: Walter de Gruyter.

140   *Pietro Zaccaria*

Boshnakov, K. 2003. *Die Thraker südlich vom Balkan in den Geographika Strabos. Quellen-kritische Untersuchungen.* Wiesbaden: Franz Steiner.

Braccesi, L. 1968. "Statue di Dedalo e Icaro nell'area del delta padano. Nota ad [Aristot.], 836a-b (*Mir.*, 81)." *Studi Romagnoli* 19: 43–48.

Braccesi, L. 1977. *Grecità adriatica. Un capitolo della colonizzazione greca in occidente.* Seconda edizione riveduta e ampliata. Bologna: Pàtron.

Braccesi, L. 2001. *Hellenikòs Kolpos. Supplemento a Grecità adriatica.* Roma: L'Erma di Bretschneider.

Briquel, D. 1984. *Les Pélasges en Italie. Recherches sur l'histoire de la légende.* Roma: L'Erma di Bretschneider.

Caley, E.R. and Richards, J.F.C. 1956. *Theophrastus, On Stones.* Columbus: Ohio State University Press.

Callebat, L. 1973. *Vitruve. De l'architecture. Livre VIII. Texte établi traduit et commenté par L.C.* Paris: Les Belles Lettres.

Chávez Reino, A.L. and Ottone, G. 2007. "Les fragments de Théopompe chez Athénée: un aperçu général." In Lenfant, D. ed. *Actes du Colloque International "Athénée et les frag-ments d'historiens" (Strasbourg, 16–18 juin 2005).* Paris: De Boccard. 139–174.

Christ, M.R. 1993. "Theopompus and Herodotus: A Reassessment." *Classical Quarterly* 43: 47–52.

Cordano, F. 2014. "Dal Mar Nero all'Adriatico: Strabone e le diverse tradizioni." *Rationes Rerum* 4: 13–28.

Cordano, F. 2015. "Strabone e il monte Emo." *I Traci tra geografia e storia. Aristonothos. Scritti per il Mediterraneo antico* 9: 53–65.

de Martini, A. 2021. "Appunti propedeutici a un'edizione del cosiddetto Paradoxographus Palatinus. Parte seconda: i toponimi 'problematici'." *Rivista di Filologia e di Istruzione Classica* 149: 463–498.

De Sensi Sestito, G. 2013. "Lico di Reggio, fra Calcide, Atene e Alessandria. Cultura sto-rica, interessi etnografici, *mirabilia*." In De Sensi Sestito, G. ed. *La Calabria nel Mediter-raneo. Flussi di persone, idee e risorse. Atti del Convegno di Studi (Rende, 3–5 giugno 2013).* Soveria Mannelli: Rubbettino. 83–110.

Dilts, M.R. 1974. *Claudii Aeliani Varia Historia, edidit M.R.D.* Leipzig: Teubner.

Dittenberger, W. 1879. "Ketriporis von Thrakien." *Hermes* 14: 298–303.

Dorandi, T. 2017. "La ricezione del sapere zoologico di Aristotele nella tradizione parados-sografica." In Sassi, M.M., Coda, E. and Feola, G. eds. *La zoologia di Aristotele e la sua ricezione dall'età ellenistica e romana alle culture medievali. Atti della X "Settimana di Formazione" del Centro GrAL, Pisa, 18–20 novembre 2015.* Pisa: Pisa University Press. 60–80.

Dowden, K. 2013. "Poseidonios (87)." In Worthington, I. ed. *Jacoby Online. Brill's New Jacoby.* Part II. Leiden; Boston: Brill.

Dušanić, S. 1977. "On Theopompus' *Philippica* VI-VIII." *Aevum* 51: 27–36.

Džino, D. 2007 [2008]. "The Celts in Illyricum – Whoever They May Be: the Hybridization and Construction of Identities in Southeastern Europe in the Fourth and Third Centuries BC." *Opuscula archaeologica* 31: 49–68.

Ebert, J.F. 1825. *Dissertationes Siculae. Tomus primus.* Königsberg: August Wilhelm Unzer.

Eichholz, D.E. 1965. *Theophrastus. De Lapidibus, Edited with Introduction, Translation and Commentary by D.E.E.* Oxford: Clarendon Press.

Eigler, U. and Wöhrle, G. 1993. *Theophrast. De odoribus. Edition, Übersetzung, Kommen-tar von U.E. – G.W. mit einem botanischen Anhang von Herzhoff, B.* Stuttgart: Teubner.

Eleftheriou, D. 2018a. *Pseudo-Antigonos de Carystos. Collection d'histoires curieuses, 1. Introduction – édition – traduction.* Doctoral thesis. Université Paris Nanterre.

Eleftheriou, D. 2018b. *Pseudo-Antigonos de Carystos. Collection d'histoires curieuses, 2. Commentaire.* Doctoral thesis. Université Paris Nanterre.

Enmann, A. 1880. *Untersuchungen über die Quellen des Pompeius Trogus für die griechische und sicilische Geschichte.* Dorpat: Schnakenburg.

Flashar, H. 1972. *Aristoteles, Mirabilia.* Berlin: Akademie Verlag.

Flensted-Jensen, P. 2004. "Thrace from Axios to Strymon." In Hansen, M.H. and Nielsen, T.H. eds. *An Inventory of Archaic and Classical Poleis.* Oxford: Oxford University Press. 810–853.

Flower, M.A. 1994. *Theopompus of Chios. History and Rhetoric in the Fourth Century BC.* Oxford: Oxford University Press.

Fraser, P.M. 1972. *Ptolemaic Alexandria.* 3 vols. Oxford: Clarendon Press.

García Valdés, M., Llera Fueyo, L.A. and Rodríguez-Noriega Guillén, L. 2009. *Claudius Aelianus. De natura animalium, ediderunt M.G.V. – L.A.L.F. – L.R.-N.G.* Berlin; New York: Teubner.

Gauger, B. and Gauger, J.-D. 2010. *Fragmente der Historiker: Theopomp von Chios (FGrHist 115/116), übersetzt, eingeleitet und kommentiert von B.G. – J.-D.G.* Stuttgart: Hiersemann.

Geffcken, J. 1892. *Timaios' Geographie des Westens.* Berlin: Weidmannsche Buchhandlung.

Giacomelli, C. 2021. *Ps.-Aristotele, De mirabilibus auscultationibus. Indagini sulla storia della tradizione e ricezione del testo.* Berlin; Boston: Walter de Gruyter.

Giacomelli, C. 2023. *Pseudo-Aristotele, De mirabilibus auscultationibus. Edizione critica, traduzione e commento filologico a cura di C.G.* Roma: Bardi edizioni.

Giannini, A. 1964. "Studi sulla paradossografia greca. II. Da Callimaco all'età imperiale: la letteratura paradossografica." *Acme* 17: 99–140.

Giannini, A. 1965. *Paradoxographorum Graecorum Reliquiae, recognovit, brevi adnotatione critica instruxit, latine reddidit A.G.* Milano: Istituto editoriale italiano.

Gómez Espelosín, F.J. 2008. *Paradoxógrafos Griegos. Rareza y maravillas, introducción, traducción y notas de F.J.G.E.* Madrid: Gredos.

Gow, A.S.F. and Scholfield, A.F. 1953. *Nicander. The Poems and Poetical Fragments.* Edited with a Translation and Notes by Jacques, J.-M. Cambridge: Cambridge University Press.

Greene, R. 2022. "1680. Anonymous, Paradoxographus Florentinus." In Schorn, S. ed. *Die Fragmente der Griechischen Historiker Continued. Part IV E. Paradoxography and Antiquities. Fascicle 2. Paradoxographers of the Imperial Period and Undated Authors [Nos. 1667–1693].* Leiden; Boston: Brill. 631–781.

Grenfell, B.P. and Hunt, A.S. 1909. *Hellenica Oxyrhynchia cum Theopompi et Cratippi fragmentis, recognoverunt brevique adnotatione critica instruxerunt B.P.G. – A.S.H.* Oxford: Clarendon Press.

Guenther, P. 1889. *De ea, quae inter Timaeum et Lycophronem intercedit, ratione.* Doctoral thesis. Leipzig: Max Hoffmann.

Habaj, M. 2018. "A Note on the Ancient Idea of a Danube with Two Estuaries." *Graeco-Latina Brunensia* 23: 67–73.

Hammond, N.G.L. 1967. *Epirus. The Geography, the Ancient Remains, the History and the Topography of Epirus and Adjacent Areas.* Oxford: Clarendon Press.

Hammond, N.G.L. and Griffith, G.T. 1979. *A History of Macedonia. Volume II. 550–336 B.C.* Oxford: Clarendon Press.

Hatzimichali, M. forthcoming. "*Mirabilia* 1–15 and *Historia Animalium* VIII (IX)". In Zucker, A., Mayhew, R. and Hellmann, O. eds. *The Aristotelian* Mirabilia *and Early Peripatetic Natural Science*. London; New York: Routledge.

Hett, W.S. 1936. *Aristotle. Minor Works*. London; Cambridge, MA: Loeb Classical Library.

Holford-Strevens, L. 2020. *Auli Gelli Noctes Atticae, ab L.H.-S. recognitae brevique adnotatione critica instructae. Tomus alter libros XI-XX continens*. Oxford: Clarendon Press.

Iliev, J. 2013. "Oracles of Dionysos in Ancient Thrace." *Haemus Journal* 2: 61–70.

Jacoby, F. 1929. *Die Fragmente der Griechischen Historiker. Zweiter Teil. Zeitgeschichte. B. Spezialgeschichten, Autobiographien und Memoiren. Zeittafeln*. Berlin: Weidmann.

Jacoby, F. 1930. *Die Fragmente der Griechischen Historiker. Zweiter Teil. Zeitgeschichte. D. Kommentar zu Nr. 106–261*. Berlin: Weidmann.

Jacoby, F. 1955a. *Die Fragmente der Griechischen Historiker. Dritter Teil. Geschichte von Städten und Völkern (Horographie und Ethnographie). b. Kommentar zu Nr. 297–607*. Vol. I. Leiden: Brill.

Jacoby, F. 1955b. *Die Fragmente der Griechischen Historiker. Dritter Teil. Geschichte von Städten und Völkern (Horographie und Ethnographie). b. Kommentar zu Nr. 297–607*. Vol. II. *Noten*. Leiden: Brill.

Jacques, J.-M. 2002. *Nicandre. Œuvres. Tome II. Les Thériaques. Fragments iologiques antérieurs à Nicandre. Texte établi et traduit par J.-M.J.* Paris: Les Belles Lettres.

Kaibel, G. 1887. *Athenaei Naucratitae Dipnosophistarum libri XV. Vol. I. Libri I-V, recensuit G.K.* Leipzig: Teubner.

Kidd, I.G. 1988. *Posidonius. Vol. II. The Commentary. (ii). Fragments 150–293*. Cambridge: Cambridge University Press.

Korenjak, M. 2003. *Die Welt-Rundreise eines anonymen griechischen Autors („Pseudo-Skymnos"). Einleitung, Text, Übersetzung und Kommentar von M.K. Hildesheim*; Zürich; New York: Weidmannsche Buchhandlung.

Kostov, R.I. 2007. "Notes and Interpretation on the 'Thracian Stone' in Ancient Sources." *Annual of The University of Mining and Geology "St. Ivan Rilski"* 50 Part. I (Geology and Geophysics): 99–102.

Kroll, W. 1940. "Zur Geschichte der aristotelischen Zoologie." *Sitzungsberichte der Wiener Akademie der Wissenschaften, philologisch.-historische Klasse* 218, Band 2: 3–30.

Kullmann, W. 2014. *Aristoteles als Naturwissenschaftler*. Boston; Berlin; München: Walter de Gruyter.

Laqueur, R. 1934. "Theopompos (9)." *Paulys Realencyclopaedie der classischen Altertumswissenschaft* V(A.2): 2176–2223.

Lasserre, F. and Livadaras, N. 1992. *Etymologicum magnum genuinum. Symeonis etymologicum una cum magna grammatica. Etymologicum magnum auctum. Synoptice ediderunt F.L. – N.L.* Athēna: Ekdoseis Philologikou Eullogou Parnassos.

Lindner, K. 1973. *Beiträge zu Vogelfang und Falknerei im Altertum*. Berlin; New York: Walter de Gruyter.

Longo, O. 1986–1987. "'Caccia coi falchi in Tracia?' Atti e memorie dell'Accademia Patavina di Scienze, Lettere ed Arti." *Memorie della Classe di Scienze Morali Lettere ed Arti* 99 Parte III: 39–45.

Loukopoulou, L. 2004. "Thrace from Strymon to Nestos." In Hansen, M.H. and Nielsen, T.H. eds. *An Inventory of Archaic and Classical Poleis*. Oxford: Oxford University Press. 854–869.

Maddoli, G., Nafissi, M. and Saladino, V. 1999. *Pausania. Guida della Grecia. Libro VI. L'Elide e Olimpia. Testo e traduzione a cura di G.M. – M.N., commento a cura di G.M. – M.N. – V.S.* Milano: Mondadori.

Maddoli, G. and Saladino, V. 1995. *Pausania. Guida della Grecia. Libro V. L'Elide e Olimpia. Testo e traduzione a cura di G.M., commento a cura di G.M. – V.S.* Milano: Mondadori.

Marcotte, D. 2000. *Géographes Grecs. Tome I. Introduction générale. Ps.-Scymnos: Circuit de la Terre. Texte établi et traduit par D.M.* Paris: Les Belles Lettres.

Mayhew, R. 2021. "Theophrastus on the Mistletoe in *De causis plantarum* ii 17, and What It Tells Us About Aristotle's *Historia animalium* viii(ix)." *Ancient Philosophy* 41: 463–475.

Mayhew, R. forthcoming. "*Historia animalium* 8(9).5, *De mirabilibus auscultationibus* 5 & 75, and Two of Theophrastus' Lost Works on Animals." In Connell, S. ed. *Philosophical Essays on Aristotle's Historia Animalium.* Leiden; Boston: Brill.

Mayhoff, C. 1892. *C. Plini Secundi Naturalis historiae libri XXXVII. Vol. III. Libri XVI-XXII, post L. Iani obitum recognovit et scripturae discrepantia adiecta edidit C.M.* Stuttgart: Teubner.

Mayhoff, C. 1897. *C. Plini Secundi Naturalis historiae libri XXXVII. Vol. V. Libri XXXI-XXXVII, post L. Iani obitum recognovit et scripturae discrepantia adiecta edidit C.M.* Stuttgart: Teubner.

Mayhoff, C. 1906. *C. Plini Secundi Naturalis historiae libri XXXVII. Vol. I. Libri I-VI, post L. Iani obitum recognovit et scripturae discrepantia adiecta edidit C.M.* Stuttgart: Teubner.

Mayhoff, C. 1909. *C. Plini Secundi Naturalis historiae libri XXXVII. Vol. II. Libri VII-XV, post L. Iani obitum recognovit et scripturae discrepantia adiecta iterum edidit C.M.* Stuttgart: Teubner.

Mili, M. 2015. *Religion and Society in Ancient Thessaly.* Oxford: Oxford University Press.

Mócsy, A. 1972. "Zu Theopompos frg. 39–40." *Rivista storica dell'Antichità* 2: 13–15.

Moraux, P. 1951. *Les listes anciennes des ouvrages d'Aristote.* Louvain: Éditions universitaires.

Morison, W.S. 2014. "Theopompos of Chios (115)." In Worthington, I. ed. *Jacoby Online. Brill's New Jacoby*, Part II. Leiden: Brill.

Müllenhoff, K. 1870. *Deutsche Altertumskunde.* Vol. I. Berlin: Weidmannsche Buchhandlung.

Müller, K. 1841. *Fragmenta historicorum Graecorum.* Vol. I. Paris: Didot.

Musso, O. 1976. "Ps. Aristot. Mir. Ausc. 81." *Rheinisches Museum für Philologie* 119: 369.

Musso, O. 1985. *[Antigonus Carystius]. Rerum mirabilium collectio, edidit O.M.* Naples: Bibliopolis.

Oberhummer, E. 1897. "Binai." *Paulys Realencyclopaedie der classischen Altertumswissenschaft* III(1): 475.

Oberhummer, E. 1927. "Σκότουσσα (2)." *Paulys Realencyclopaedie der classischen Altertumswissenschaft* III(A.1): 617.

Occhipinti, E. 2011. "Aristotele, Teopompo e la politica macedone." *Klio* 93: 291–307.

Öhler, H. 1913. *Paradoxographi Florentini anonymi opusculum de aquis mirabilibus, ad fidem codicum manu scriptorium editum commentario instructum.* Tübingen: Heckenhauer.

Olson, S.D. 2020. *Athenaeus Naucratites Deipnosophistae. Volumen III.A-B. Libri VIII-XI.* Berlin; Boston: Teubner.

Osann, F. 1848. "Ueber Pseudo-Aristot. Ausc. Mirab. 104." *Philologus* 3: 324–330.

Ottone, G. 2004. "Per una nuova edizione dei frammenti di Teopompo di Chio: riflessioni su alcune problematiche teoriche e metodologiche." *Ktèma* 29: 129–143.

Ottone, G. 2018. *Teopompo di Chio, Filippiche (Fozio, Biblioteca, cod. 176), a cura di G.O. Testo critico e Introduzione a cura di Chávez Reino, A.L.* Tivoli (Roma): Tored.

Overduin, F. 2015. *Nicander of Colophon's Theriaca. A Literary Commentary.* Leiden; Boston: Brill.

Pajón Leyra, I. 2011. *Entre ciencia y maravilla: el género literario de la paradoxografía griega*. Zaragoza: Prensas Universitarias de Zaragoza.

Pajón Leyra, I. 2021. "Un frammento di prosa del IV secolo a.C. sugli Autariati: Etnografia, storiografia e movimenti di popolazione in P.Oxy. IV 681." *Rationes Rerum* 17: 85–97.

Papazoglou, F. 1988. *Les villes de Macédoine à l'époque romaine*. Athènes: École française d'Athènes; Paris: Diffusion de Boccard.

Pape, W. and Benseler, G. 1911. *Wörterbuch der griechischen Eigennamen*. 2 vols. Braunschweig: Friedr. Vieweg & Sohn.

Paton, W.R., Pohlenz, M. and Sieveking, W. 1972². *Plutarchi moralia. Vol. III, recensuerunt et emendaverunt W.R.P. – M.P. – W.S.* Leipzig: Teubner.

Pédech, P. 1989. *Trois historiens méconnus. Théopompe – Duris – Phylarque*. Paris: Les Belles Lettres.

Peter, U. 1997. *Die Münzen der thrakischen Dynasten (5.-3. Jahrhundert v. Chr.). Hintergründe ihrer Prägung*. Berlin: Akademie Verlag.

Pownall, F. 2004. *Lessons from the Past. The Moral Use of History in Fourth-Century Prose*. Ann Arbor: University of Michigan Press.

Prandi, L. 2005. *Memorie storiche dei Greci in Claudio Eliano*. Roma: L'Erma di Bretschneider.

Pritchett, W.K. 1999. *Pausanias Periegetes*. Vol. II. Amsterdam: J. C. Gieben.

Prosdocimi, A.L. 1963–1964. "Un frammento di Teopompo sui Veneti." *Atti e memorie dell'Accademia Patavina di Scienze Lettere ed Arti. Memorie della classe di scienze morali lettere ed arti* 76 (parte III): 201–222.

Radt, S. 2003. *Strabons Geographika. Band 2. Buch V-VIII: Text und Übersetzung*. Göttingen: Vandenhoek & Ruprecht.

Radt, S. 2007. *Strabons Geographika. Band 6. Buch V-VIII: Kommentar*. Göttingen: Vandenhoek & Ruprecht.

Regenbogen, O. 1940. "Theophrastos von Eresos." *Paulys Realencyclopaedie der classischen Altertumswissenschaft* VII: 1353–1562.

Rhodes, P.J. and Osborne, R. 2003. *Greek Historical Inscriptions 404–323 BC*. Oxford: Oxford University Press.

Ritschl, F. 1866. *Opuscula philologica. Volumen I. Ad litteras graecas spectantia*. Leipzig: Teubner.

Rivault, J. 2021. *Zeus en Carie. Réflexions sur les paysages onomastiques, iconographiques et cultuels*. Bordeaux: Ausonius.

Rocha-Pereira, M.H. 1977. *Pausaniae Graeciae descriptio. Vol. II. Libri V-VIII, edidit M.H.R.-P.* Leipzig: Teubner.

Rose, V. 1854. *De Aristotelis librorum ordine et auctoritate commentatio*. Berlin: G. Reimer.

Rose, V. 1863. *Aristoteles pseudepigraphus*. Leipzig: Teubner.

Şahin, M.Ç. 1981. *Die Inschriften von Stratonikeia. Teil I. Panamara*. Bonn: Habelt.

Santi Amantini, L. 2005. "A proposito di 'pace' in Teopompo." In Santi Amantini, L. ed. *Dalle parole ai fatti. Relazioni interstatali e comunicazione politica nel mondo antico*. Roma: L'Erma di Bretschneider. 35–59.

Sassi, M.M. 1993. "Mirabilia." In Cambiano, G., Canfora, L. and Lanza, D. eds. *Lo spazio letterario della Grecia antica. Vol. I. La produzione e la circolazione del testo. Tomo II. L'Ellenismo*. Roma: Salerno. 449–468.

Schepens, G. and Delcroix, K. 1996. "Ancient Paradoxography: Origin, Evolution, Production and Reception." In Pecere, O. and Stramaglia, A. eds. *La letteratura di consumo nel mondo greco-latino. Atti del Convegno internazionale. Cassino 14–17 settembre 1994*. Cassino: Università degli Studi di Cassino. 373–460.

Schnieders, S. 2019. *Aristoteles. Historia animalium. Buch VIII und IX, übersetzt und erläutert von S.S.* Berlin; Boston: Walter de Gruyter.

Schrader, H. 1868. "Über die Quellen der pseudoaristotelischen Schrift περὶ θαυμασίων ἀκουσμάτων." *Jahrbücher für Classische Philologie* 14 = *Neue Jahrbücher für Philologie und Paedagogik* 38 (Bd. 97): 217–232.

Schranz, W. 1912. *Theopomps Philippika.* Doctoral thesis. Marburg i. Hessen: Chr. Schaaf.

Sharples, R.W. 1995. *Theophrastus of Eresus. Sources for His Life, Writings, Thought and Influence. Commentary Volume 5. Sources on Biology (Human Physiology, Living Creatures, Botany: Texts 328–435).* Leiden; New York; Köln: Brill.

Sharples, R.W. 1998. *Theophrastus of Eresus. Sources for His Life, Writings, Thought and Influence. Commentary Volume 3.1. Sources on Physics (Texts 137–223), with contributions on the Arabic material by D. Gutas.* Leiden; Boston; Köln: Brill.

Shrimpton, G.S. 1991. *Theopompus the Historian.* Montreal; Kingston; London; Buffalo: McGill-Queen's University Press.

Smith, D.G. 2013. "Lykos (Boutheras) of Rhegium (570)." In Worthington, I. ed. *Jacoby Online. Brill's New Jacoby.* Part III. Leiden: Brill.

Sørensen, S.L. 2020. "Isigonos of Nikaia (1659)." In Schorn, S. ed. *Jacoby Online. Die Fragmente der Griechischen Historiker Continued.* Part IV. Leiden: Brill.

Sørensen, S.L. 2022a. "1679. Paradoxographus Vaticanus." In Schorn, S. ed. *Die Fragmente der Griechischen Historiker Continued. Part IV E. Paradoxography and Antiquities. Fascicle 2. Paradoxographers of the Imperial Period and Undated Authors [Nos. 1667–1693].* Leiden; Boston: Brill. 577–630.

Sørensen, S.L. 2022b. "1681. Paradoxographus Palatinus." In Schorn, S. ed. *Die Fragmente der Griechischen Historiker Continued. Part IV E. Paradoxography and Antiquities. Fascicle 2. Paradoxographers of the Imperial Period and Undated Authors [Nos. 1667–1693].* Leiden; Boston: Brill. 783–827.

Spittler, J. 2022. "1672. Apollonios." In Schorn, S. ed. *Die Fragmente der Griechischen Historiker Continued. Part IV E. Paradoxography and Antiquities. Fascicle 2. Paradoxographers of the Imperial Period and Undated Authors [Nos. 1667–1693].* Leiden; Boston: Brill. 387–519.

Stählin, F. 1927. "Σκότουσσα (1)." *Paulys Realencyclopaedie der classischen Altertumswissenschaft* III(A.1): 613–617.

Steinmetz, P. 1964. *Die Physik des Theophrastos von Eresos, Bad Homburg v.d.H.*; Berlin; Zürich: M. Gehlen.

Stern, J. 2008. "Paradoxographus Vaticanus." In Heilen, S. *et al.* eds. *In Pursuit of Wissenschaft. Festschrift für W.M. Calder III zum 75. Geburtstag.* Hildesheim; Zürich; New York: Olms. 437–466.

Štrmelj, D. 2015. "East Adriatic in Pseudo-Aristotle's *De Mirabilibus Auscultationibus.*" *Electryone* 3(2): 51–74.

Susemihl, F. 1891. *Geschichte der griechischen Litteratur in der Alexandrinerzeit.* Band I. Leipzig: Teubner.

Talbert, R.J.A. 2000. *Barrington Atlas of the Greek and Roman World.* Princeton; Oxford: Princeton University Press.

Terian, A. 1981. *Philonis Alexandrini de animalibus. The Armenian Text with an Introduction, Translation, and Commentary.* Chico: Scholars Press.

Vanotti, G. 1977. "Gerione in Aristot., 830 a, mir. Ausc., 133." *Epigraphica* 39: 161–168.

Vanotti, G. 2007. *Aristotele, Racconti meravigliosi. Introduzione, traduzione, note e apparati di G.V.* Milano: Bompiani.

Vattuone, R. 2000. "Teopompo e l'Adriatico. Ricerche sui frammenti del libro XXI delle Filippiche (*FGrHist* 115 FF 128–136)." *Hesperìa* 10: 11–38.

Voltan, C. 1985. "L'offerta rituale alle 'cornacchie' presso i Veneti." *Archivio Veneto* 125: 5–34.

von Gutschmid, A. 1893. "Ueber Müllenhoff's Deutsche Altertumskunde." *Literarisches Centralblatt* 1871: 521–529; reprinted in von Gutschmid, A. *Kleine Schriften, Vierter Band*, herausgegeben von F. Rühl. Leipzig: Teubner. 123–138.

Westermann, A. 1839. Παραδοξόγραφοι. *Scriptores rerum mirabilium Graeci*. Braunschweig: G. Westermann; London: Black et Armstrong.

Wichers, R.H.E. 1829. *Theopompi Chii fragmenta*. Leiden: S. et J. Luchtmans.

Zaccaria, P. 2016. "Rediscovering Theopompos: Neglected Evidence of the *Meropis*." *Aevum* 90: 51–70.

Zambon, E. 2000. "Filippo II, Alessandro il Grande e la Tracia. Note sulla conquista e l'istituzione della strategia di Tracia." *Anemos* 1: 69–95.

Ziegler, K. 1949. "Paradoxographoi." *Paulys Realencyclopaedie der classischen Altertumswissenschaft* XVIII(3): 1137–1166.

4 *De mirabilibus auscultationibus* and Heraclides of Pontus

Kelly Shannon-Henderson

1. Introduction

My task in this chapter is to assess the relationship between *Mir.* and the now fragmentary historical works of Heraclides of Pontus (ca. 390–320 BCE). Renowned
in antiquity for his elegant and varied yet approachable literary style (Diog. Laert.
5.89), Heraclides took the dialogue form from Plato and made it his own, infusing
philosophy with elements of mythology, religion, storytelling, and the miraculous.
While Heraclides is cited by a variety of authors and earned the admiration of no
less a worthy than Cicero,[1] his brand of history was evidently not for everyone,
and the fabulous elements of his works earned him the scorn of several ancient
readers. Plutarch derides him as "a fabulist and inclined to fiction" (*Cam.* 22.3 = F
102 Wehrli = F 49 Schütrumpf: μυθώδη καὶ πλασματίαν ὄντα τὸν Ἡρακλείδην) for
ascribing the Gallic sack of Rome in 390 BCE to the Hyperboreans, and also seems
to think his work *Abaris* is not particularly serious (Plut. *Quomodo adul.* 1.14e = F
73 Wehrli = F 130 Schütrumpf).[2] Heraclides is also criticized by the Epicurean
speaker in Cicero's *De natura deorum* for "stuff[ing] his books with childish tales"
(*puerilibus fabulis refersit libros, Nat. D.* 1.34 = F 111 Wehrli = 72 Schütrumpf).[3]
Most damning, however, is the criticism of Timaeus of Tauromenium, who, in his
rebuttal of Heraclides' claims about the death of Empedocles, says, "Heraclides is
just this sort of writer of absurdities, saying even that a man has fallen down from
the moon" (ἀλλὰ διὰ παντός ἐστιν Ἡρακλείδης τοιοῦτος παραδοξόλογος, καὶ ἐκ
τῆς σελήνης πεπτωκέναι ἄνθρωπον λέγων, Diog. Laert. 8.72 = F 115 Wehrli = F
94 Schütrumpf).

---

1 Cic. *Tusc.* 5.3.8 = F 88 Wehrli = F 85 Schütrumpf. Cf. Bollansée (1999) 507 n. 232: "Herakleides'
   reputation in antiquity was not one-sidedly negative."
2 The tone here is not as strongly negative as that of the other criticisms, but cf. Bollansée (1999)
   507 n. 232 and Hunter and Russell (2011) 72–73. The text of *Mir.* throughout is that of Ciro Giacomelli (2023), and translations are my own. The text of Heraclides throughout is that of Wehrli
   (1953) except where noted, and translations are those of Jan van Ophuijsen, Susan Prince, and Peter
   Stork published in Schütrumpf (2008) but may contain slight alterations. All other translations are
   my own.
3 Gottschalk (1980) 96–97 attributes the harshness of the criticism to the Epicurean persona of the
   interlocutor. Cicero is otherwise fairly complementary to Heraclides; see above.

DOI: 10.4324/9781003437819-5

Suffice it to say that, while Heraclides was not a paradoxographer *stricto sensu*, his works contain the sort of marvelous details that might have attracted the compiler of *Mir.* Despite this possible affinity, scholars have never investigated Heraclides as a possible source for the pseudo-Aristotelian text. This is perhaps for good reason. *Mir.* never cites Heraclides by name as an authority for any of the anecdotes it reports, nor does *Mir.* repeat anything verbatim from Heraclides or follow any report from Heraclides closely in its details. There are, however, several items in *Mir.* whose subject matter overlaps with surviving fragments of Heraclides. Two topics share a geographical connection to Cumae: the Sibyl of Cumae and/or Erythrae (*Mir.* 95a; Heraclid. Pont. F 130, 131a-c Wehrli = F 119, 120 A-C Schütrumpf) and the unusual properties of Lake Aornos/Avernus (*Mir.* 102a-b; Heraclid. Pont. F 128a-b Wehrli = F 137C Schütrumpf). In these cases, I shall argue, we see in *Mir.* the traces of a polemicizing tradition attacking the claims of Heraclides about both Aornos and the Sibyl as being imprecise or even untrue; so while *Mir.* does not cite Heraclides by name, its author seems to be aware of debates about these topics in which Heraclides was one of the competing voices.[4] Heraclides may have been a source for the information in *Mir.*, but was probably not the only source, and may have been read directly alongside other texts containing geographical material on South Italy, or indirectly through authors by which he was quoted or even criticized.

A different sort of overlap, I believe, is on display in the other items in *Mir.* that show affinity with Heraclides' *On Pleasure*: what I will call "tales of pleasant madness." These are reports of individuals, often (but not always) named and located in a specific geographical location, who experience a period of delusional thinking during which they seem to inhabit a kind of alternative reality accessible only to themselves, but later recover their senses and state that they have had the best time of their lives (*Mir.* 31–32, 178; Heraclid. Pont. F 56 Wehrli = F 40 Schütrumpf). None of *Mir.*'s three examples of this phenomenon have exact parallels in the surviving fragments of Heraclides, nor does the one such tale we know Heraclides told find its way into *Mir.* There are not many other stories of this type preserved in other sources, which may increase the likelihood that these sections of *Mir.* depend on Heraclides, although the picture is complicated by the dialogue form in which Heraclides worked and the possible distortions introduced by Athenaeus when excerpting the *On Pleasure*.

I shall consider each of these points of overlap in turn, and close with some observations about what the similarities between the two can show us. In particular, I want to consider whether thinking about the portions of *Mir.* that seem to lie under the shadow of Heraclides tell us anything we did not already know about what Heraclides' works were like or what they contained.

---

4 Marincola (1997) 221 notes that polemic against predecessors or contemporaries, always a feature of historiography, became more pronounced and more personal beginning in the 4th century BCE.

2. South Italian connections? Sibyls and Aornos

Mir. shows an interest in many Italian phenomena,[5] and it is not surprising that two famous items relating to the area around Cumae – the Cumaean Sibyl, and the infernal Lake Aornos – should find a place among *Mir.*'s παράδοξα. I begin with the Sibyl:

Ἐν τῇ Κύμῃ τῇ περὶ τὴν Ἰταλίαν δείκνυταί τις, ὡς ἔοικε, θάλαμος κατάγειος Σιβύλλης τῆς χρησμολόγου, ἣν πολυχρονιωτάτην γενομένην παρθένον διαμεῖναί φασιν. οὖσαν μὲν Ἐρυθραίαν, ὑπό τινων δὲ τὴν Ἰταλίαν κατοικούντων Κυμαίαν, ὑπὸ δέ τινων Μελάγκραιραν καλουμένην. τοῦτον δὲ τὸν τόπον λέγεται κυριεύεσθαι ὑπὸ Λευκανῶν (*Mir.* 95.1).

In Cumae in Italy, as it seems, an underground room of the soothsayer Sibyl is exhibited; they say that she lived to a very great age but stayed a virgin. They also relate that she was Erythraean, but was called Cumaean by some inhabitants of Italy and Melancraera by others. This place is said to be under the control of the Lucanians.

Parke and Vanotti, who both argue for a 3rd-century BCE date for *Mir.*, claim that *Mir.* is the first source that describes the Cumaean Sibyl as coming from Erythrae. This Hellenistic date would seem to make it more likely that the compiler of *Mir.* was aware of Heraclides, who wrote only a century or so earlier. Giacomelli, however, suggests a much later date, in the early 2nd century CE;[6] on this interpretation, it perhaps becomes less likely that the writer of *Mir.* was using Heraclides directly.

Heraclides, however, is thought to have held an important place in the history of ancient scholarship on the complex traditions about the different prophetesses throughout the Mediterranean world who went under the name of "Sibyl," which he discussed in his treatise *On Oracles* (Περὶ χρησμῶν or Περὶ χρηστηρίων). While there are earlier references to the/a Sibyl in Greek texts of the 5th century BCE, it is widely agreed that Heraclides was the first author "to distinguish individual representatives of this type and try to place each in her historical context."[7] The idea of multiple Sibyls associated with specific places was likely not his invention, but rather an attempt to confront the multiplicity of Sibylline traditions that had independently emerged, and to systematize what he found into a comprehensible catalogue that distinguished the Sibyls by date and geography.[8]

---

5 For example, the strait between Italy and Sicily (*Mir.* 56, 105b, 115[130]), a deadly poison from Mount Circeo (*Mir.* 78), the luxurious garment of Alcisthenes of Sybaris (*Mir.* 96), the "Siren Islands" (*Mir.* 103), and poisonous Italian geckoes (*Mir.* 148a).

6 Parke (1992) 78–79; Vanotti (1998) 168–169; Giacomelli (2021) 20–21.

7 Gottschalk (1980) 130. So also, for example, Parke (1992) 24; Brocca (2011) 19 n. 7. Buitenwerf (2003) 95 (incorrectly, in my view) argues against this interpretation, but still allows that "Heraclides is certainly an ancient witness to Sibylline traditions connected with Asia Minor."

8 Brocca (2011) 25 n. 25. Parke (1992) 26 argues that this was a task Heraclides was prompted to undertake upon finding the evidence for the two Marpessian and Hellespontian Sibyls.

Four fragments of Heraclides' analysis of Sibyls are preserved. One comes to us via Lactantius, who preserves Varro's citation of Heraclides. Heraclides was Varro's authority for the Hellespontian Sibyl, the eighth discussed in Varro's catalogue, and it seems that Heraclides reported her birthplace in the Troad and the fact that she lived in the time of Solon and Cyrus:[9]

ceterum Sibyllas decem numero fuisse (*scil. Varro scripsit*) easque omnes enumeravit sub auctoribus qui de singulis scriptitaverint. . . . octavam Hellespontiam in agro Troiano natam, vico Marmesso circa oppidum Gergithium, quam scribat Heraclides Ponticus Solonis et Cyri fuisse temporibus (Lactant. *Div. inst.* 1.6.8, 12 = Heraclid. Pont. F 131a Wehrli = F 120A Schütrumpf).

(Varro wrote that) furthermore the Sibyls have been ten in number, and he has listed them all under the authors who have written about each of them. . . . The eighth was the Hellespontian Sibyl, born in the Trojan plain in the village Marmessus near the town Gergithium. Heraclides Ponticus wrote that she lived at the time of Solon and Cyrus.

Two further citations from Greek sources convey basically the same information as Lactantius:[10]

ὀγδόη (sc. Σίβυλλα) ἡ Ἑλλησποντία, ἥτις ἐν κώμῃ Μαρμισσῷ τὴν γένεσιν ἔσχεν περὶ τὴν πολίχνην Γεργετίωνα· ὑπὸ τὴν ἐνορίαν δὲ αὕτη τῆς Τροίας ἐτύγχανεν. Ἦν ἐν καιρῷ Σόλωνος καὶ Κύρου, ὡς ἔγραψεν Ἡρακλείδης ὁ Ποντικός (*Schol. in Platonis Phaedrum* 244 B = Heraclid. Pont. F 131b Wehrli = F 120B Schütrumpf).

The eighth (sc. Sibyl) (is) the Hellespontian, who was born in the village Marmissus near the small town Gergetion, which happened to be in the territory of Troy. She lived at the time of Solon and Cyrus, as Heraclides Ponticus wrote.

ὀγδόη (sc. Σίβυλλα) ἡ Ἑλλησποντία τεχθεῖσα ἐν κώμῃ Μαρπησσῷ περὶ τὴν πολίχνην Γεργίθιον, ἥτις ἐνορία ποτὲ τῆς Τρωάδος ἐτύγχανεν ἐν καιροῖς Σόλωνος καὶ Κύρου, ὡς ἔγραψεν Ἡρακλείδης ὁ Ποντικός (*Theosophorum graecorum fragmenta* F 1 Erbse = Heraclid. Pont. F 131c Wehrli = F 120C Schütrumpf).

The eighth (sc. Sibyl), the Hellespontian, was born in the village Marpessus near the small town Gergithium, which once happened to be within the boundaries of the Troad, in the time of Solon and Cyrus, as Heraclides Ponticus has written.

---

9 The detail that Sibyl prophesied to Cyrus while Croesus was on the pyre appears in Nicolaus of Damascus (*FGrHist* 90 F 67–688), for which Heraclides is one possible source; see Parke (1992) 25; Cervelli (1993) 915.

10 They may have used Lactantius directly as a source. For the *Theosophia*, Zaccaria (2021) 754–755 follows the hypothesis of Zingg (2019) that its author mainly relied upon Lactantius, perhaps occasionally supplementing his information from other sources; cf. Lightfoot (2007) 88–89.

A fourth fragment of Heraclides preserved by Clement of Alexandria,[11] however, preserves completely different information: it reveals that Heraclides also discussed *both* a Delphic Sibyl named Artemis, originally from Phrygia, whose verse prophecy is quoted,[12] *and* an Erythraean Sibyl named Herophile:[13]

καὶ οὔτι γε μόνος οὗτος (sc. Μωυσῆς), ἀλλὰ καὶ ἡ Σίβυλλα Ὀρφέως παλαιοτέρα. Λέγονται γὰρ καὶ περὶ τῆς ἐπωνυμίας αὐτῆς καὶ περὶ τῶν χρησμῶν τῶν καταπεφημισμένων ἐκείνης εἶναι λόγοι πλείους, Φρυγίαν τε οὖσαν κεκλῆσθαι Ἄρτεμιν καὶ ταύτην παραγενομένην εἰς Δελφοὺς ᾆσαι

ὦ Δελφοί, θεράποντες ἐκηβόλου Ἀπόλλωνος,
ἦλθον ἐγὼ χρήσουσα Διὸς νόον αἰγιόχοιο,
αὐτοκασιγνήτῳ κεχολωμένη Ἀπόλλωνι.

Ἔστι δὲ καὶ ἄλλη Ἐρυθραία Ἡροφίλη καλουμένη. Μέμνηται τούτων Ἡρακλείδης ὁ Ποντικὸς ἐν τῷ περὶ χρηστηρίων (Clem. Al. *Strom.* 1.21 108.1–3 = Heraclid. Pont. F 130 Wehrli = F 119 Schütrumpf).

But not only he (Moses), but the Sibyl, too, is older than Orpheus. It is said that there are quite a few stories about her name and about the oracles of that woman that were spread abroad, for example, that she was Phrygian and had been called Artemis, and that she arrived in Delphi and sang:

Oh Delphians, servants of far-shooting Apollo,
I have come to pronounce the mind of Zeus the aegis-bearer,
angry at my very own brother Apollo.

There is also another (Sibyl) from Erythrae, called Herophile. Heraclides Ponticus mentions these in his (treatise) *On Oracles.*

None of these fragments mention Cumae in connection with Sibyls; so although Heraclides' treatment of the Erythraean Sibyl is beyond doubt, we have no evidence that he acknowledged the version of Sibylline lore, preserved in *Mir.*, in which she migrated to Italy and became the Cumaean Sibyl. Nevertheless, I suspect it is at least possible that Heraclides' observations on Sibyls might be present indirectly in *Mir.* 95.1. *Mir.*'s claim that the Erythraean Sibyl and the Cumaean Sibyl are considered by some to be the same person shows the shadow of a large and complex discussion about the number and identity of the Sibyls that spanned

11 See Lightfoot (2007) 83 on Clement's attitude to the Sibyls.
12 It is unclear from Clement's wording whether the verse prophecy was also preserved in Heraclides.
13 The plural τούτων in Clement's closural statement μέμνηται τούτων Ἡρακλείδης ὁ Ποντικὸς clearly indicates that *both* Sibyls were discussed by Heraclides. On the (confusing) tradition about the name Herophile, which Paus. 10.12.3 also applies to a Sibyl on Samos, see Parke (1992) 65; Cervelli (1993) 916–918.

many authors and many centuries – a discussion in which Heraclides was likely a significant participant.

There is reason to think that the fragments of Heraclides about Sibyls that we still possess do not give a complete picture of his work on the subject. In the process of excerption by later authors, there seems to have been a "flattening" of what must originally have been a discussion rich with details like dates, names, and geographical locations of the Sibyls. If we had only Lactantius to go on, we might have assumed that Heraclides only discussed the Hellespontian Sibyl; Lactantius also mentions the Delphic Sibyl and the Erythraean Sibyl in his catalog, but cites different authorities for them,[14] and gives no indication that Heraclides also mentioned them. It is only thanks to Clement that we know Heraclides talked about the Delphic and Erythraean Sibyls. Varro has a lot to answer for here; it seems that in his process of distilling the bewildering tradition about a multiplicity of Sibyls – which by then perhaps had multiplied to as many as 30 – down to a canonical 10 sorted into chronological order with one authority cited for each, he had to make some "arbitrary" choices[15] that obscured a lot of the complexities in Heraclides' treatment of Sibyls. Furthermore, Clement by himself also gives only a partial picture of Heraclides' Sibylline works, for he does not tell us about Heraclides' discussion of the Hellespontian Sibyl. If further fragments of Heraclides' *On Oracles* happened to turn up, what further information might they add? We have no way of knowing the full extent of Heraclides' attempts to reconcile and untangle the convoluted Sibylline tradition, but it is clear that each of the authors who cites him gives us only a partial picture. As Parke notes, Heraclides "might have known of other Sibyls," besides these three, "but, if so, our extant authors fail to cite him for them."[16]

Could the Sibyl of Cumae have been one of them? The idea cannot be ruled out. We do not know what other Sibyls already existed in the 4th-century tradition, nor do we know for certain when the Sibyl at Cumae was first mentioned. The earliest evidence for her is Lycophron's *Alexandra*, which mentions some of the same details about her found in *Mir.*: her virginity and her cave (1278–1280), and the name Melancraera (1464). The date of that poem has been much debated; if Hornblower is correct that it was written in the early 2nd century BCE, perhaps around 190, Lycophron himself is too late to be Heraclides' source, but there is good reason to suspect that Lycophron drew on earlier sources.[17] The possibility

---

14 Chrysippus and Apollodorus of Erythrae, respectively, see Cervelli (1993) 917, 926.

15 Nikiprowetzky (1970) 4–5.

16 Parke (1992) 25.

17 Hornblower (2015) 36–39; Hornblower (2018) 3–7. A date earlier in the 3rd century BCE was previously argued, with West (1983) 123 further hypothesizing that the lines dealing with Aeneas in Italy (1226–1280) were a later interpolation. West (1984) 133 has noted that Lycoph. *Alex.* 1226–1280 assumes quite a lot of knowledge of the Aeneas-legend and the topography of Italy on the part of the reader, and this presumed familiarity means that other writers must have discussed this Sibyl before Lycophron. Another writer who seems to have discussed the Cumaean Sibyl is local historian Hyperochus of Cumae (*FGrHist* 576), an author notoriously difficult to date but provisionally assigned by Jacoby to the 3rd century BCE.

most frequently raised is that Timaeus is the source of Lycophron's information[18]; I shall return to the question of Timaeus below. But in any case, if we are willing to suggest that a Cumaean Sibyl was known in the time of Timaeus (working not very much later than Heraclides), there is no particular reason to think that Heraclides could not also have written about her. Additionally, Heraclides may have discussed Aornos (see further below), so he was apparently interested in the region around Cumae. It is hard to imagine that, if he had discovered information about a Cumaean Sibyl in the course of his research on the area, he would have left her out of his Sibylline catalogue in *On Oracles*.

If the above hypothesis is correct, *Mir.*'s claim that some people think the Cumaean Sibyl came from Erythrae reflects a historiographical polemic in which Heraclides may have been one of the multitude of voices. In the second sentence of *Mir.* 95.1 (οὖσαν μὲν Ἐρυθραίαν, ὑπό τινων δὲ τὴν Ἰταλίαν κατοικούντων Κυμαίαν, ὑπὸ δέ τινων Μελάγκραιραν καλουμένην), with μέν . . . δέ . . . δε, the author of *Mir.* draws a contrast between the two geographical epithets: the Sibyl *is* (οὖσαν) actually Erythraean in origin, but she is only *called* (καλουμένην) Cumaean by the Italians, who *also* (second δέ) sometimes refer to her as "Black-Head." In other words, *Mir.* is particularly emphatic about the Erythraean ethnicity of the Sibyl located at Cumae – and it is at least highly plausible that Heraclides was one of the voices, perhaps even an authoritative one, in the complex and probably polemical historiographical conversation lying behind *Mir.*'s claim.[19]

Another place in *Mir.*, also discussing a phenomenon located in South Italy, where Heraclides may be lurking in the background is the discussion of Lake Aornos:

(102.1) Περὶ τὴν Κύμην τὴν ἐν τῇ Ἰταλίᾳ λίμνη ἐστὶν ἡ προσαγορευομένη Ἄορνος, αὐτὴ μέν, ὡς ἔοικεν, οὐκ ἔχουσά τι θαυμαστόν· περικεῖσθαι γὰρ λέγουσι περὶ αὐτὴν λόφους κύκλῳ, τὸ ὕψος οὐκ ἐλάσσους τριῶν σταδίων, καὶ αὐτὴν εἶναι τῷ σχήματι κυκλοτερῆ, τὸ βάθος ἔχουσαν ἀνυπέρβλητον. ἐκεῖνο δὲ θαυμάσιον φαίνεται· ὑπερκειμένων γὰρ αὐτῇ πυκνῶν δένδρων, καί τινων ἐν αὐτῇ κατακεκλιμένων, <u>οὐδὲν ἔστιν ἰδεῖν</u> φύλλον <u>ἐπὶ</u> τοῦ ὕδατος <u>ἐφεστηκός, ἀλλ</u>᾽οὕτω <u>καθαρώτατόν</u> ἐστι τὸ ὕδωρ ὥστε τοὺς θεωμένους θαυμάζειν. (102.2) Περὶ δὲ τὴν ἀπέχουσαν ἤπειρον αὐτῆς οὐ πολὺ θερμὸν ὕδωρ πολλαχόθεν ἐκπίπτει, καὶ ὁ τόπος ἅπας καλεῖται Πυριφλεγέθων. ὅτι δὲ οὐδὲν διίπταται ὄρνεον αὐτήν, ψεῦδος· οἱ γὰρ παραγενόμενοι λέγουσι πλῆθός τι κύκνων ἐν αὐτῇ γίγνεσθαι (*Mir.* 102).

Near Cumae in Italy there is a lake called Aornos, which in itself has (as it seems) nothing marvelous; for they say that hills no less than three stades high lie around it in a circle, and that the lake itself is circular in shape

18 For example, West (1984) 133; Hornblower (2015) 500.

19 For example, Rufus (*FGrHist* 826 T 2) allegedly dealt with the Sibyls in books 4–5 of his now lost *Mousike historia* (Parke [1992] 50 n. 47). Brocca (2011) 25 emphasizes the diversity and complexity of ancient Sibylline scholarship. But the more general point about *Mir.*'s concern with the Sibyl's origin/ethnicity still stands.

and has a depth that is unsurpassable. But this does seem to be marvelous: although thick trees stand above the lake and some even lean over it, it is not possible to see a leaf floating upon the water, but the water is extremely clear, such that those who see it marvel. Around the mainland not very far from the lake, hot water springs up all over the place, and the whole area is called Pyriphlegethon. But it is false that no bird flies over it; for people who have been there say that there is some quantity of swans on it.

Mir.'s comment about birds, which is otherwise a *non sequitur* in a description that focuses mostly on the lake's relationship to the surrounding woods and trees, is a gesture toward the fact that Aornos was famously etymologized in antiquity as ἀ- + ὄρνος, "lacking in birds," because of the notion that vapors from the lake killed birds flying over it,[20] although it is noteworthy that *Mir.* does not actually include this etymology. In trying to refute this claim, the author seems to overextend himself: he says explicitly that the lake has no remarkable properties (οὐκ ἔχουσά τι θαυμαστόν), but later, when giving the detail about its surface remaining clear of leaves, he clearly states that this property *is* marvelous (ἐκεῖνο δὲ θαυμάσιον φαίνεται, . . . ὥστε τοὺς θεωμένους θαυμάζειν). And *Mir.* 81 describes a hot-water lake near the Electrides Islands whose vapors *are* said to be noxious to birds flying overhead,[21] so the author in principle believes that lakes with this property exist, but Aornos in fact is not one of them.

I suggest that *Mir.* is perhaps so emphatic on this point because the author is staking out his own place in a preexisting polemical discussion, one in which Heraclides may have been somehow involved. While *Mir.* 102 cites no explicit source, commentators are generally agreed that the anecdote derives from Timaeus.[22] It is certainly true that Antigonus, who cites Timaeus' name for similar observations on Aornos,[23] uses very similar wording to *Mir.* 102 in his discussion of the failure of leaves to settle on the lake's surface:

Τὴν δὲ ἐν τοῖς Σαρμάταις λίμνην Ἡρακλείδην γράφειν, ὅτι οὐδὲν τῶν ὀρνέων ὑπεραίρειν, τὸ δὲ προσελθὸν ὑπὸ τῆς ὀσμῆς τελευτᾶν. Ὃ δὴ καὶ

---

20 Garnier (2008) suggests that while the ancient etymology is not impossible, it is better to understand it as a toponym formed from a fossilized dialect-word for "inférieur," for which he adduces a parallel in Vedic Sanskrit. Cerasuolo (1987) 125–126 hypothesizes that it is derived from an Oscan word for "water." See *ibid.* 120–121 on the relationship between the Greek Ἄορνος and Latin *Avernus*; 122–123 for a summary of other modern attempts at an etymology.

21 Vanotti (2007) 168 identifies the mephitic swamp as the *Fons Aponi* and compares Claud. *Carm. min.* 26, 28. Flashar (1972) 108 again suggests Timaeus as a possible source for *Mir.* 81. There were a number of bodies of water that were said in antiquity to produce fatal vapors; for examples, see Ogden (2001) 186; Shannon-Henderson (2022a) 30, 44 n. 74.

22 Flashar (1972) 120, quoting Geffcken: "am sichersten timäisch." Vanotti (2007) 182 reaches the same conclusion.

23 Flashar (1972) 120 notes that Antig. Car. and ps.-Arist. *Mir.* both are perhaps quoting Timaeus directly, but we cannot exclude the possibility that both authors were accessing Timaeus through the same intermediary, perhaps a paradoxographical collection. Note that in this section Antigonus purports to be paraphrasing Callimachus.

περὶ τὴν Ἄορνίν τι δοκεῖ γίγνεσθαι καὶ κατίσχυκεν ἡ φήμη παρὰ τοῖς πλείστοις. (2) Ὁ δὲ Τίμαιος τοῦτο μὲν ψεῦδος ἡγεῖται εἶναι· τὰ πλεῖστα γὰρ κατατυχεῖν τῶν εἰθισμένων παρ' αὐτῇ διαιτᾶσθαι· ἐκεῖνο μέντοι λέγει, διότι συνδένδρων τόπων ἐπικειμένων αὐτῇ καὶ πολλῶν κλάδων καὶ φύλλων διὰ πνεύματα τῶν μὲν κατακλωμένων, τῶν δὲ ἀποσειομένων, <u>οὐθέν ἐστιν ἰδεῖν ἐπ'</u> αὐτῇ <u>ἐφεστηκός, ἀλλὰ</u> διαμένειν <u>καθαράν</u> (ps.-Antig. Car. *Mir.* 152 = Timaeus F 57 = Heraclid. Pont. F 128b Wehrli = F 137B Schütrumpf).

[Callimachus says that] Heraclides writes that no bird passes over the lake among the Sarmatians, and if one approaches it dies on account of the vapor. And the same thing seems to happen at Aornis, and the story has prevailed among most people. But Timaeus, for his part, thinks that this is a lie: for most of the [birds] accustomed to live beside it are successful. But he does say this: that although [or because?] the areas near the lake are thickly wooded and many branches and leaves are broken off by winds, and others are shaken off, there is nothing to be seen standing upon the lake's surface, but it remains clean.[24]

Timaeus' discussion of Aornos was evidently polemical in tone, countering the etymological explanation with evidence based upon firsthand observation; the claim of autopsy in Timaeus' refutation is even preserved in *Mir.*'s claim that οἱ γὰρ παραγενόμενοι note that the lake is particularly popular with swans.[25]

As Antigonus notes, Heraclides was also interested in lakes with fatal vapors, and discussed a lake among the Sarmatians whose exhalations kill birds. The same detail is attributed to Heraclides by two other paradoxographical texts:

Ἡρακλείδης φησὶ τὴν ἐν Σαυρομάταις λίμνην οὐδὲν τῶν ὀρνέων ὑπεραίρειν, τὸ δὲ προσελθὸν ὑπὸ τῆς ὀσμῆς τελευτᾶν. Ὁ δὴ καὶ περὶ τὴν Ἄορνιν κατὰ τὴν Ἰταλίαν δοκεῖ γίνεσθαι (*Paradoxographus Vaticanus* 13 = Heraclid. Pont. F 128a Wehrli = F 137A Schütrumpf).

Heraclides says that none of the birds flies over the lake in (the land of the) Sauromatae, but any that approaches is killed by the smell. The same is thought to happen around (Lake) Aornis in Italy.

Ἡρακλείδης ὁ Ποντικὸς λίμνην ἐν Σαυρομάταις φησὶν εἶναι, περὶ ἣν τὰ πετασθέντα τῶν ὀρνέων εἰς αὐτὴν πίπτειν (*Paradoxographus Florentinus* 22 = Heraclid. Pont. F 137C Schütrumpf.)[26]

Heraclides Ponticus says that there is a lake in (the land of the) Sauromatae and that any birds that fly around near to it fall into it.

---

24 My translation.
25 Pearson (1987) 62 n. 36: "If this [Antigonos'] is an accurate quotation, Timaeus must have visited Avernus himself." Cf. Brown (1958) 29; Champion (2010) on Timaeus F 57. Such claims of superior information based on autopsy are familiar among historians; for examples, see Marincola (1997) 63–86.
26 This fragment is not present in Wehrli's edition.

It is important to note that none of the surviving fragments of Heraclides definitively shows that he also mentioned the similar behavior of Aornos. Antigonus also does not put any claims about Aornos in Heraclides' mouth, and in fact changes his sentence structure in his observation about Aornos, moving from the accusative + infinitive construction he used to report (Callimachus' report of) Heraclides' observations about the Sarmatian lake to finite, indicative verbs (δοκεῖ, κατίσχυκεν), suggesting that he does not attribute the information about Aornos to the same source. But if, as I have suggested is at least possible, Heraclides mentioned the Sibyl of Cumae in *On Oracles*, this demonstrates an interest in the geographical area around Cumae that might make it more likely he also discussed Aornos.

A further connection is provided by the idea that Aornos was thought to be the site of an entrance to the underworld and a place of prophecy related to the dead (*nekyomanteion*). The most famous articulation of this concept is Vergil's description of Aeneas descending to the underworld to receive prophecy with the help of the Cumaean Sibyl in *Aeneid* 6, but the idea of an underground oracle at Lake Aornos was much older, and was mentioned first by Ephorus (who is sometimes assumed to be the target of Timaeus' criticism in the passage on Aornos; Strabo 5.4.5.244 = Ephorus F 134a).[27] Aornos also appears at Lycoph. *Alex.* 704 in a way that implies it is intended to be thought of as the site of Odysseus' famous *nekyia* in Homer.[28] Heraclides' hometown of Heraclea Pontica was also the site of an entrance to the underworld – the place where Hercules brought Cerberus up into the world above after kidnapping him from Hades – and there is some suggestion that there was an oracular shrine there.[29] So not only would Aornos potentially have fitted very well into *On Oracles*,[30] where discussion of its birdlessness could have been adduced as evidence of its connection to the underworld, but Heraclides might also have had a special interest in such oracular underworld-portals given the presence of another one near his hometown.

It is true that neither of these examples is very secure; for both Aornos and the Cumaean Sibyl, we know only that Heraclides discussed closely *related* things (the Sarmatian lake and the Erythraean Sibyl), and there is no solid evidence his works included these specific details. But the fact that both have to do with the same geographical area suggests that if Heraclides did discuss one, it is likely he discussed the other. The fact that Timaeus lies in the background of both of these parallels is also

---

27   Heurgon (1987) 158–159; Ogden (2001) 178. On Ephorus and Timaeus, see Geffcken (1892) 31; Flashar (1972) 120; Pearson (1987) 35.

28   West (1984) 140; cf. Edlund-Berry (1987) 44–45. Lycophron describes the lake in sinister terms as λίμνην τ' Ἄορνον ἀμφιτορνηωτὴν βρόχῳ ("lake Aornos encircled by a noose," Hornblower [2015] 291; cf. Gigante Lanzara [2009] 103) and mentions the Coccytus, tributary of the Styx, in the next line.

29   Ogden (2001) 168–171. The only attestation of the oracle at Heraclea being consulted is Plut. *De sera* 555c: Pausanias the Spartan commander allegedly visited in 479/8 BCE after being haunted by the ghost of a girl he killed.

30   The fragments about the Sarmatian lake cannot be assigned securely to any particular one of the known works of Heraclides. Gottschalk (1980) 128 includes the many fragments which are "anecdotes which could have stood somewhere in the philosophical dialogues, although we have no evidence to connect them with any one in particular."

suggestive. As I noted at the beginning of this chapter, Timaeus expressed a particular disdain for Heraclides as παραδοξολόγος (Diog. Laert. 8.72 = Heraclid. Pont. F 115 Wehrli = F 94 Schütrumpf), and Cumae/Aornos was a place where phenomena could be found on which they reported differing details; so, it is possible that the author of *Mir.* became aware of Heraclides' claims about Sibyls and birdless lakes via Timaeus, although in the case of Sibyls Heraclides' fame as an authority on the subject might have convinced the author to take a look at *On Oracles* for himself.

3. Tales of pleasant madness: an argument about ἡδονή?

Finally, let us consider a third area of close overlap between *Mir.* and Heraclides: the motif of pleasant madness. Most descriptions of hallucinatory madness in ancient texts are profoundly negative.[31] But on occasion, hallucinations are harmless and enjoyable,[32] as on two occasions described in *Mir.*:

(31) Λέγεται δέ τινα ἐν Ἀβύδῳ παρακόψαντα τῇ διανοίᾳ καὶ εἰς τὸ θέατρον ἐρχόμενον ἐπὶ πολλὰς ἡμέρας θεωρεῖν, ὡς ὑποκρινομένων τινῶν, καὶ ἐπισημαίνεσθαι· καὶ ὡς κατέστη τῆς παρακοπῆς, <u>ἔφησεν ἐκεῖνον αὐτῷ τὸν χρόνον ἥδιστα βεβιῶσθαι.</u> (32) Καὶ ἐν Τάραντι δέ φασιν οἰνοπώλην τινὰ τὴν μὲν νύκτα μαίνεσθαι, τὴν δ' ἡμέραν οἰνοπωλεῖν. καὶ γὰρ τὸ κλειδίον τοῦ οἰκήματος πρὸς τῷ ζωνίῳ διεφύλαττε, πολλῶν δ' ἐπιχειρούντων παρελέσθαι καὶ λαβεῖν οὐδέποτε ἀπώλεσεν (*Mir.* 31–32).

(31) A certain man in Abydos, so it is said, went out of his mind and went to the theater and for several days watched and applauded as though there were people acting; and when he recovered from his delirium, he said that that was the most pleasant period he had ever lived. (32) And in Tarentum, they say, a certain wine-merchant went mad by night but sold wine by day. And in fact he guarded the key to his storeroom on his belt, and he never lost it, although many people tried to grab it and take it away.

Δημάρατον Τιμαίου τοῦ Λοκροῦ ἀκουστὴν νοσήσαντα ἄφωνόν φασιν ἐπὶ δέκα γενέσθαι ἡμέρας· ἐν δὲ τῇ ἑνδεκάτῃ, ἀνανήψας βραδέως ἐκ τῆς παρακοπῆς, <u>ἔφησεν ἐκεῖνον αὐτῷ τὸν χρόνον ἥδιστα βεβιῶσθαι</u> (*Mir.* 178).

They say that Demaratus, a student of Timaeus the Locrian, got sick and went mute for ten days. But on the eleventh day, having slowly come to his senses again after his delirium, he said that that was the most pleasant period he had ever lived.

---

31 For example, Hippoc. *Int.* 48 states that bile collecting in the liver can cause patients to hallucinate that they are having confrontations with frightening reptiles and hostile soldiers. Kazantzidis (2018a) 231–233 compares these with the hallucinations of Orestes and Heracles in tragedy; cf. Pigeaud (1987) 38; Hoessly (2001) 139–140. Harris (2013) 290 distinguishes such "fictional" examples of madness in tragedy from "texts" like *Mir.* and Heraclides "that claim to describe real events." Cf. O'Brien-Moore (1924) 9–10.

32 Pigeaud (1987) 92–93 posits the existence of a wider tradition of such stories in popular literature. Cf. Kazantzidis (2018a) 237.

Where might these stories have originally come from? One possibility is a medical text, since we do have stories of pleasant madness in a medical context, but these are not numerous.[33]

Another possibility is Heraclides, for all three of *Mir.*'s stories share similarities with a similar report in a fragment of Heraclides' *On Pleasure* quoted by Athenaeus. As we would expect from a text of its genre, *Mir.*'s stories are far less detailed than the one Heraclides reports:

ἐν μανίᾳ δὲ τρυφὴν ἡδίστην γενομένην οὐκ ἀηδῶς ὁ Ποντικὸς Ἡρακλείδης διηγεῖται ἐν τῷ περὶ ἡδονῆς οὕτως γράφων· ὁ Αἰξωνεὺς Θράσυλλος ὁ Πυθοδώρου διετέθη ποτὲ ὑπὸ μανίας τοιαύτης ὡς πάντα τὰ πλοῖα τὰ εἰς τὸν Πειραιᾶ καταγόμενα ὑπολαμβάνειν ἑαυτοῦ εἶναι, καὶ ἀπεγράφετο αὐτὰ καὶ ἀπέστελλε καὶ διῴκει καὶ καταπλέοντα ἀπεδέχετο μετὰ χαρᾶς τοσαύτης ὅσηπερ ἄν τις ἡσθείη τοσούτων χρημάτων κύριος ὤν. Καὶ τῶν μὲν ἀπολομένων οὐδὲν ἐπεζήτει, τοῖς δὲ σῳζομένοις ἔχαιρεν καὶ διῆγεν μετὰ πλείστης ἡδονῆς. Ἐπεὶ δὲ ὁ ἀδελφὸς αὐτοῦ Κρίτων ἐκ Σικελίας ἐπιδημήσας συλλαβὼν αὐτὸν παρέδωκεν ἰατρῷ καὶ τῆς μανίας ἐπαύσατο, διηγεῖτο <πολλάκις περὶ τῆς ἐν μανίᾳ διατριβῆς> οὐδεπώποτε φάσκων κατὰ τὸν βίον ἡσθῆναι πλείονα. Λύπην μὲν γὰρ οὐδ' ἡντινοῦν αὐτῷ παραγίγνεσθαι, τὸ δὲ τῶν ἡδονῶν πλῆθος ὑπερβάλλειν (Ath. 12.554e-f = Heraclid. Pont. F 56 Wehrli = F 40 Schütrumpf).

Heraclides Ponticus narrates not unpleasantly in his (work) *On Pleasure* that in a state of madness luxury becomes most pleasant, writing as follows: "Thrasyllus of the deme Aexone, son of Pythodorus, was once afflicted with a madness of such a kind, with the result that he took all the ships landing at the Peiraeus to be his own. He registered them in his accounts, and sent them out and managed them, and when they returned he received them with such great joy, as one would feel with pleasure in being the owner of so much wealth. He made no search at all for those that were lost, but he rejoiced in those that came back safe, and he lived with the greatest pleasure. But when his brother Crito returned home from Sicily, he (Crito) took hold of him (Thrasyllus) and turned him over to a doctor, and (Thrasyllus) was cured of his madness. Then he <quite often told stories about his life in madness>,

---

33 Hippoc. *Gland.* 12 describes insane patients who experience hallucinations with "grinning laughter" (σεσηρόσι μειδιήμασι), although it is not entirely clear that this is a pleasant experience. Kazantzidis (2018b) 46 cites the *Mir.* passage as a parallel; see also Kazantzidis (2018a) 232. Aretaeus of Cappadocia, writing in the middle of the 1st century CE, describes people experiencing madness as pleasure (καὶ οἷσι μὲν ἡδονὴ ἡ ἡ μανίη) who pose no danger to those around them and seem to have happy hallucinations: for example, they come to the agora wearing garlands because they believe they have won an athletic contest (*On Chronic Diseases* 1.6, p. 41 Hude; see Adams [1856] 301ff. for a text and translation). Sharples (1995) 6 suggests one of Theophrastus' lost medical works as a possible source for *Mir.* On the fraught question of identifying Theophrastean material in *Mir.*, see Giacomelli (2021) 3–4. Geffcken (1892) 88 and Flashar (1972) 40 both suggest Theophrastus' lost Περὶ παραφροσύνης as a possible source for *Mir.* 31.

saying that he had never once enjoyed life more. For not a single sort of pain had befallen him, and the quantity of his pleasures was far greater."

Neither Heraclides nor *Mir.* gives medical details of how the hallucinator regained his sanity; Heraclides mentions that Thrasyllus was healed by a doctor but does not describe the treatment in detail. We would certainly expect *Mir.*, like other para-doxographical texts, to strip out any reference to the medical apparatus of etiology and cure that might have been present in a medical source text.[34] But it may also be significant that the fragment of Heraclides (which Athenaeus purports to be quoting verbatim)[35] does not contain such information either, although we know that Heraclides also wrote a treatise *On Diseases* (F 72–89 Wehrli = 80, 82–95, 130–132 Schütrumpf)[36] and so was liable to have been interested in such matters; in *On Pleasure*, such medical details may not have served Heraclides' purpose and therefore have been omitted. Perhaps, then, the author of *Mir.* was drawn to these stories because Heraclides, their original source, did not diminish their wonder value with medical–scientific explanation.

While the three stories of pleasant madness in *Mir.* share a similar motif with the anecdote in Heraclides, the details of all four stories are different: they have different geographical settings, and different main characters who experience different hallucinations. So if we were to posit that *Mir.* drew on Heraclides, we would have to argue that his treatise *On Pleasure* originally contained more examples of pleasant madness that do not survive. This cannot be ruled out on chronological grounds. None of the anecdotes in *Mir.* contain any details that mean they must have originated later than the 4th century BCE, which would preclude them from having come from Heraclides originally. The stories of the theatergoer of Abydos (*Mir.* 31) and the Tarentine wine merchant (*Mir.* 32) have no chronological details attached. The final story, of Demaratus, the pupil of Timaeus of Locri (*Mir.* 178), is part of the so-called appendix, a later addition to the text of *Mir.* dating from sometime between the 6th and 12th centuries CE, as Giacomelli has shown.[37] (The wording at the end of *Mir.* 178 is identical to that of *Mir.* 31, a phenomenon best explained as a "mechanical problem" resulting from a Byzantine copyist's attempt to "flesh out" a mangled text of *Mir.* and graft the later appendix onto the origi-nal text of *Mir.*)[38] But the original source of the information about Timaeus could have been much more ancient. *Mir.* describes him as being a pupil (ἀκουστής) of the Pythagorean Timaeus, the speaker of Plato's dialogue of the same name, so Heraclides cannot be ruled out as an ultimate source for the story of Demaratus' madness simply based on the late date of the "appendix" of *Mir.*

34 Cf. Jacob (1983) 133; Schepens and Delcroix (1996) 390–392; Pajón Leyra (2011) 31; Kazantzidis (2019) 6.
35 On the quotation of historians by Athenaeus, see Pelling (2000), especially 188–190 on (allegedly) verbatim quotation; Lenfant (2007) 50–53; Gorman and Gorman (2014) 180–183.
36 See further van der Eijk (2009).
37 Giacomelli (2021) 23.
38 Giacomelli (2021) 41–42, 51.

One factor that lends support to the suggestion of Heraclides as a source for *Mir.* 31 and 178 is the fact that the main character of each anecdote, like Heraclides' Thrasyllus, later gives an account of his own madness as the most enjoyable time of his life, a motif that is not present in tales of pleasant madness from medical texts. It may well, however, show the traces of Heraclides' *On Pleasure* as the ultimate source for the anecdotes in *Mir.* The same has been argued by Brink for a passage from Horace's *Epistulae* which describes an unnamed person who believes he is watching plays that no one else can see; Horace's story is very similar to *Mir.* 31, although Horace's version is set in Argos rather than Abydos. It also contains a similar exclamation of the madman after he recovers, thanks to medical treatment: "By god, you've killed me, friends, not saved me, since pleasure has been thus torn away from me, and the most pleasant error of the mind has been taken away by force" (Hor. *Epist.* 2.2.138–40 "*pol me occidistis, amici,* | *non servastis*," ait, "*cui sic extorta voluptas* | *et demptus per vim mentis gravissimus error*").[39] Brink links Horace's anecdote to both *Mir.* and Heraclides, and notes that it "shows some of the argumentative background" of the *On Pleasure* (on which see further below).[40] The same is true of *Mir.* 31: the closing comment about madness being pleasurable would have little place in a purely medical source but makes perfect sense in a philosophical context.[41] So, if any of the stories from *Mir.* originally came from Heraclides' *On Pleasure*, *Mir.* 31 seems the most likely candidate, with *Mir.* 178 also possible but perhaps less likely given its much later date. If we want to assert that Demaratus' story originally came from Heraclides, we would have to assume that whoever composed the later appendix to *Mir.* was reading Heraclides' text independently, found a part of it which we no longer possess, and decided to insert it, with Heraclides' original closing tag about the pleasure of madness being used as a bridge point to connect the appendix with *Mir.* 31.[42]

What role did these tales originally have (or could they have had) in Heraclides' *On Pleasure*? This question is difficult to answer, for the surviving fragments of the treatise strike different tones, and scholars have been divided as to how to interpret them.[43] It has been generally believed that the treatise originally took the form of a dialogue, with some speakers arguing that the pursuit of pure pleasure is destructive both to individuals (such as Deinias the perfume merchant, F 61 Wehrli = F 44 Schütrumpf) and to states (such as Samos or Sybaris, F 57 Wehrli = F 41 Schütrumpf), and others making the case that pleasure can be beneficial for both (such as the striking argument in favor of the luxurious lifestyle of the Persians and Medes, F 55 Wehrli). But it is not clear in the case of any particular fragment which way it was used, since the original context and tone of the anecdotes have been lost in their excerption by Athenaeus, the source of all but one of the fragments of *On*

---

39 See Brink (1968) 355–356 on *me occidistis*: this is "old hyperbole . . . well known from comedy" and suggests "ruin" or "torment" rather than literal murder.

40 Brink (1968) 350.

41 Furthermore, paradoxographical texts like *Mir.* avoid using medical texts as sources, and prefer to use philosophical or historical authors.

42 Cf. Flashar (1972) 61–62 and 154.

43 For an overview of the issue, see Gottschalk (1980) 89–91, Schütrumpf (2009).

Pleasure; indeed, Athenaeus' excerption has obliterated all traces of the treatise's dialogue form, which we only know about thanks to Diogenes Laertius.[44] For the anecdote about Thrasyllus, both arguments have been made. Wehrli interpreted the story as a positive example of pure pleasure with no negative consequences;[45] Thrasyllus' story could thus have served as evidence that certain kinds of madness involving all pleasure and no pain[46] show that pleasure can be an ultimate good. The story in *Mir.* 31 could have been subject to similar interpretation; it is hard to see how the phantom tragedies could have harmed the man from Abydos. Other critics, however, assert that such a positive interpretation of Thrasyllus' hallucinations is impossible because of the idea that appears in various philosophical texts (e.g. Pl. *Grg.* 491c, 492c) that one cannot experience true happiness when he has lost his reason.[47] Certainly Heraclides' anecdote about Deinias the perfume seller (on whom see further below) is most likely meant to serve as a cautionary tale, and something similar may have been true of the Tarentine wine merchant if his story in fact appeared in Heraclides' *On Pleasure*.

Additionally, most of the other surviving fragments of Heraclides' treatise connect pleasure with luxury (τρυφή). This, I believe, can be argued even for the story of Thrasyllus' madness. Wehrli claimed that Thrasyllus is unusual among the fragments of *On Pleasure* in having no connection to τρυφή, since it is concerned instead with pleasure as "seelischer Zustand auf Kosten der Vernunft gepriesen."[48] But since what Thrasyllus' pleasant madness involves is a belief that he owns every cargo ship coming into the Peiraeus, and hence a belief that he is very rich, the anecdote cannot be entirely divorced from the idea of τρυφή.[49] The same is not true, however, of the anecdotes of pleasant madness found in *Mir.*: the phantom plays of the man from Abydos (*Mir.* 31) seem to have nothing to do with luxury, nor do we have any indication of luxury in whatever Demaratus experienced during the ten days of his muteness (*Mir.* 178). The story of the Tarentine wine merchant of *Mir.* 32, if it originally appeared in Heraclides at all, could have been connected to wealth or luxury if we still possessed the complete story; if it was anything like the tale of Deinias the perfume merchant, it might have ended badly, and therefore is at least as likely to have served as an illustration of the dangers of luxury-related madness as it was to have been deployed in its defense. We should be wary, however, of assuming that Heraclides' *On Pleasure* was only about τρυφή-related pleasure. Since most of the fragments come from the book of Athenaeus'

---

44 Brancacci (1999) 101–102; Dillon (2003) 206–207.

45 Wehrli (1953) 79; for a similar argument, see Brancacci (1999) 103; Brancacci (2003) 78–80.

46 Note that Thrasyllus never experiences distress when a ship is lost, but only joy for the ones that do make it safe into harbor.

47 Bignone (1936) 260; Lieberg (1959) 40; Bringmann (1972) 527; Gottschalk (1980) 90–91 ("Thrasyllus' story must have been meant as an argument against accepting pleasure as the ultimate good; it is a strange aberration to regard it as forming part of the hedonist case"); Bosworth (1994) 16. For a different view, see also Repici Cambiano (1977) 235. On the reception of this philosophical debate in Horace, see Citroni (2016).

48 Wehrli (1953) 78–79.

49 Gorman and Gorman (2014) 310 also raise this possibility, only to dismiss it.

Deipnosophistae that focuses on the topic of luxury, the emphasis on τρυφή in the surviving fragments may really be the result of Athenaeus' distortion of Heraclides' original purpose in the work,[50] which might have originally encompassed more different types of pleasure.

Of the pleasant-madness tales from *Mir.*, the anecdote about the Tarentine wine merchant in *Mir.* 32 is the least easy to place in Heraclides' *On Pleasure*. It lacks the telltale closing tag about pleasure, and there also seems to be some bigger piece of the story missing that obscures the story's original context. It is stated that the wine merchant's madness manifested itself only at night, but that during the daytime he continued selling wine as usual (τινὰ τὴν μὲν νύκτα μαίνεσθαι, τὴν δ' ἡμέραν οἰνοπωλεῖν). This motif – an ordinary citizen in all other respects, able to keep his day job, except for madness in an isolated context – is unusual but not unique to *Mir.*[51] But *Mir.* still leaves other questions unanswered. Was the wine merchant's madness even of the pleasant kind, as in the other examples in *Mir.*? Perhaps not; so if the wine merchant appeared in Heraclides, his story might have been put to a very different use. A possible parallel might be Heraclides' story of Deinias the perfume-seller,[52] who "fell into love affairs because of his indulgence in luxury" (διὰ τρυφὴν εἰς ἔρωτας ἐμπεσόντα), spent all his money, and "when he had gotten over his desires" (ὡς ἔξω τῶν ἐπιθυμιῶν ἐγένετο) castrated himself because "he was thrown into turmoil by his grief" (ὑπὸ λύπης ἐκταραχθέντα, F 61 Wehrli = F 44 Schütrumpf). While Deinias' situation is described as love rather than madness, he is an otherwise ordinary merchant not fully in possession of his faculties, not unlike the wine merchant in *Mir.* 32. If Deinias' story ended in tragedy after he regained his senses, it is not impossible that something similar could have happened to the wine merchant in the original ending of the story that has been omitted from *Mir.* If so, the Tarentine wine merchant might have provided a counterexample for the story of Thrasyllus (and of the man from Abydos, if that was originally in Heraclides too). So, on this reading, a place in Heraclides' *On Pleasure* could still be possible for *Mir.* 32.

The connection between the wine merchant's storeroom (οἴκημα) and his madness is also not explained; depending how this detail was originally depicted, the wine merchant's story could have appeared not in the *On Pleasure* but in a different work of Heraclides, and could possibly be understood in the context of Heraclides' interests in south Italy and Pythagoreanism. One other example of madness

---

50 Gorman and Gorman (2014) chapter 3 argue this point extensively; see, however, the salutary criticism of Murray (2017). Bollansée (2008) 411 makes the case for a similar τρυφή-related distortion of Clearchus by Athenaeus.

51 Citroni (2016) 238–239; Kazantzidis (2018a) 238. Galen, for example, mentions the example of the physician Theophilus, who is perfectly rational except that he believes he hears loud flute-players in a certain part of his house (*Symp. diff.* 3 = 7.60–61 Kühn); on the passage, see Nutton (2013) 124–125. Kazantzidis (2019) 2–3 also compares Hippoc. *Epid.* 5.81, the story of Nicanor, who was struck with terror when hearing flute players during the evening at a symposium but was totally unaffected if he heard them in the daytime.

52 Strattis F 34.3–4 Kassel-Austin also refers to a perfume seller named Deinias, whom he claims was Egyptian. Heraclides does not give his nationality.

related to a specific location is known. The medical writer Aretaeus of Cappadocia describes a carpenter who works normally building a house by day, but experiences madness whenever he departs from the job site: "There was this linkage of place and reason" (καὶ ἥδε τοῦ χωρίου καὶ τῆς γνώμης ἡ ξυμβολή, *On Chronic Diseases* 1.6.6 = p. 41 Hude). So again, a medical context for the wine merchant is not impossible, but Aretaeus' narrative is longer and far more detailed than *Mir.*'s story. Other salient details seem to be missing. If the wine merchant's madness took place only behind closed doors within this room, how did anyone know he was mad at all? And why were people trying to steal his keys to get inside the οἴκημα? Possibly the reader is to assume that, since the madness struck only at night, it only manifested itself while the wine merchant was in the room but was perhaps perceptible to those outside via unusual sounds, and that anyone wishing to understand the nature of his madness would have had to look inside for clues. The story might originally have concluded with a revelation scene, where someone finally managed to get inside the wine merchant's οἴκημα and find out the truth; a similar motif is found in folkloric contexts, for example, the revelation that Philinnion is actually a revenant in the famous tale told by Phlegon of Tralles, where her mother and nurse spy on her through a locked door (Phlegon *Mir.* 1.5).[53] If so, these details were not considered important by the author of *Mir.* and were omitted when he excerpted the story from its original source.

But the striking detail that onlookers wanted to break into the wine merchant's room in *Mir.* is very similar to the story of Hermotimus of Clazomenae, apparently first told in Apollonius' *Amazing Stories* (*FGrHist* 1672) and later picked up by other sources. Hermotimus' soul is said to have separated from his body and wandered the earth, until some unnamed men enter the house despite his wife's protestations (Apollonius *Mir.* 3.2 εἰσελθόντες τινὲς εἰς τὴν οἰκίαν)[54] and burn his apparently lifeless body, killing him. While there are no extant fragments of Heraclides preserving this story, we know he was interested in Hermotimus, whose Pythagorean reincarnations he catalogues (F 89 Wehrli = F 86 Schütrumpf), and who seems (along with Empedocles) to have inspired the fictional Empedotimus discussed in Heraclides' treatise *On the Soul* (F 93 Wehrli = F 54 A Schütrumpf). This connection prompted Bremmer to suggest that Heraclides was actually the origin of the story about the fatal destruction of Hermotimus' unattended body.[55] If so, and if *Mir.*'s story of the wine merchant originally contained a similar motif that might explain the detail about the merchant's keys, it is perhaps plausible that Heraclides could have been the source of this anecdote as well. Heraclides' well-documented interest in Pythagoras and Pythagoreans[56] might also provide a link between Heraclides and Demaratus,

---

53 See Shannon-Henderson (2022b) 71.
54 Later versions of the story specify that they were his enemies: Plin. *HN* 7.174, Plut. *De gen.* 592c (who gives his name as Hermodorus), Tert. *De anim.* 44. See further the discussion of Spittler (2022) 439–441.
55 Bremmer (2002) 39: "If anyone, the inventive Heraclides with his preference for fantastic stories must have been the origin of the legend of the seer with his psychic excursion."
56 See, for example, Zhmud (2012) 427–432.

the pupil of Timaeus of Locri, in *Mir.* 178. Timaeus, the title character of Plato's dialogue, is often considered a Pythagorean but was probably not a real person; in fact, this passage of *Mir.* seems to be the only testimony for his existence outside of Plato.[57] It is also not impossible, then, that the stories of the madnesses of Demaratus (*Mir.* 178) and of the Tarentine wine merchant (*Mir.* 32) both came from Heraclides' *On the Soul*, or from one of the other dialogues that are thought to have housed the "various colourful stories" about Pythagoras;[58] in fact, this might make better sense than *On Pleasure*, particularly in the case of the wine merchant, whose madness does not seem to have been pleasurable in the same way as that of Thrasyllus.

4. Conclusions: a Heraclidean mirage?

Now that we have surveyed the evidence for the presence of Heraclides of Pontus in the pseudo-Aristotelian *Mir.*, what we are left with is a suggestive mélange of possibilities, but no certainties. Previous commentators on *Mir.* have also not been confident in identifying a source for any of the chapters I have discussed. Flashar noted that the section of *Mir.* beginning with 78 is the most difficult from a source-critical perspective; the picture of Italy that the text reflects seems to be that of the 3rd century BCE, and Timaeus was certainly the most famous source for Italy in that era, but not the only one.[59] For *Mir.* 95a, on the Sibyl, Flashar seems to have thrown up his hands in despair: "Die Quelle ist nicht sicher bestimmbar." Timaeus is one option. It is also possible that Timaeus cited Heraclides in *his* discussion of the Cumaean Sibyl as part of a polemic against him; if this were the case, Heraclides would have left his stamp on *Mir.* only indirectly. But one also has to consider "daß über die kymaeische Sibylle in ethnographischem Zusammenhang viel gehandelt worden ist."[60] Other possible names have been suggested, including Lycus of Rhegium,[61] Hyperochus of Cumae,[62] or even Apollodorus of Erythrae;[63] and there were probably even more works on Sibyls that we no longer possess.[64] Perhaps because Varro's simplifying version of Sibylline history became so influential, many of these other accounts, just like much of the richness of Heraclides' treatment, seem to have fallen by the wayside and been lost to later

---

57   Marg (1972) 84. On Timaeus' historicity, see, for example, Cornford (1952) 2–3; Muccioli (2002) 342–343; Zhmud (2012) 416, who notes how strange it is that "in order to set out the 'Pythagorean' doctrines, Plato chooses a fictitious character from a city which produced not a single Pythagorean philosopher or scientist, while the actual Italian and Sicilian Pythagoreans are not even mentioned in the dialogue."

58   Dillon (2003) 207.

59   Flashar (1972) 45.

60   Flashar (1972) 115.

61   Flashar (1972) 45; Parke (1992) 78–79; Potter (1994) 41.

62   Vanotti (1998) 269–272; Vanotti (2007) 177.

63   Cervelli (1993) 920.

64   Parke (1992) 50 n. 47. Cf. Brocca (2011) 19 n. 7: "l'interessamento degli eruditi antichi per la Sibilla e le varie tradizioni ad essa legate cominci, *a quanto possiamo vedere*, nella seconda metà del IV sec. a.C., con Eraclide Pontico" (italics mine).

scholarly traditions – the sad fate of relatively unknown Greek writers in the face of a Roman juggernaut. So I suggest that, when we are looking for sources for *Mir.*'s observations on the Cumaean/Erythraean Sibyl, we at least consider Heraclides as one additional possibility. It is likewise possible that a similar erasure has happened with *Mir.*'s Aornos, which has generally been presumed to derive from Timaeus.[65] But if, as I have suggested, Heraclides discussed Aornos too, his observations on the lake's bird-killing properties – which might have found a logical home in a discussion of *nekyomanteia* in the *On Oracles* – seem to lie somewhere in the background of *Mir.*'s polemical assertion that the lake is nontoxic. If you press too hard on any of this evidence, the parallels between *Mir.* and Heraclides vanish. Timaeus is certainly the more famous source for Italy, and the material on both the Sibyl and Aornos could be explained perfectly well via him, without any recourse to Heraclides.

In some senses, we are on surer footing with the tales of pleasant madness. The closing tags about the hallucinator having the time of his life suggest that the theatergoer in *Mir.* 31 and Demaratus in *Mir.* 178 could have been a piece of supporting evidence from a speaker defending the pursuit of pleasure in Heraclides' *On Pleasure*. And depending upon how the story of the Tarentine wine merchant in *Mir.* 32 originally ended, it might have been part of an argument either for or against pleasant madness. But we would then have to assume that Heraclides' original dialogue contained more than one example of pleasant madness. We do not know anything about the length, scope, or arrangement of any of Heraclides' dialogues, and without that information it is hard to know whether it might not have been simply redundant for the stories from *Mir.* to have originally appeared there as well. Perhaps the story of Thrasyllus with his ships was enough to make Heraclides' point all on its own. And Heraclides is not the only author where *Mir.* might have found such anecdotes; he is one possible source for *Mir.* 31–32 and 178, but he is only one possibility among many. We must also consider that there seem to have been a number of contexts for such *exempla* in discussions of pleasure in many Peripatetic writers. We know of treatises on pleasure by Theophrastus, Xenocrates, Eudoxus, Chamaeleon, and others.[66] Indeed, the motif of pleasant madness may once have been more pervasive than we now realize; just like the "Wanderanekdoten" observed in ancient biography that tend to migrate in such a way that the same or similar stories end up attached to a variety of people,[67] a story of someone losing their mind and enjoying it could very well have roamed about freely in ancient literature, with names and other specific details changed to suit a particular context or enhance believability. It might be an accident that Heraclides' story is one of the more famous to have survived, and this need not mean that *Mir.* had to go to him directly to find other such examples.

---

65 Geffcken (1892) 91; Flashar (1972) 120; Vanotti (2007) 182 (who at least mentions Heraclides).

66 Citroni (2016) 234–236. See also Lieberg (1959); Gosling and Taylor (1982) 345; Dillon (1996); Brancacci (2003) 95–96. On Heraclides and the Peripatetic context for such anecdotes, see Repici Cambiano (1977) 216, 235–236. On positive madness in Plato, see Vogt (2013).

67 I owe this point to Gertjan Verhasselt.

Still, it is at the very least possible that Heraclides lies at the back of these stories. I would like to close with a passage that provides an interesting comparandum for the tales of pleasant madness in *Mir.*:

Θράσυλλος ὁ Αἰξωνεὺς παράδοξον καὶ καινὴν ἐνόσησε μανίαν. Ἀπολιπὼν γὰρ τὸ ἄστυ καὶ κατελθὼν εἰς τὸν Πειραιᾶ καὶ ἐνταῦθα οἰκῶν τὰ πλοῖα τὰ καταίροντα ἐν αὐτῷ πάντα ἑαυτοῦ ἐνόμιζεν εἶναι καὶ ἀπεγράφετο αὐτὰ καὶ αὖ πάλιν ἐξέπεμπε καὶ τοῖς περισωζομένοις καὶ εἰσιοῦσιν εἰς τὸν λιμένα ὑπερέχαιρε· χρόνους δὲ διετέλεσε πολλοὺς συνοικῶν τῷ ἀρρωστήματι τούτῳ. Ἐκ Σικελίας δὲ ἀναχθεὶς ὁ ἀδελφὸς αὐτοῦ παρέδωκεν αὐτὸν ἰατρῷ ἰάσασθαι, καὶ ἐπαύσατο τῆς νόσου οὕτως. ἐμέμνητο δὲ πολλάκις τῆς ἐν μανίᾳ διατριβῆς καὶ ἔλεγε μηδέποτε ἡσθῆναι τοσοῦτον ὅσον τότε ἥδετο ἐπὶ ταῖς μηδὲν αὐτῷ προσηκούσαις ναυσὶν ἀποσωζομέναις (Ael. *VH* 4.25).

Thrasyllus of Aexone suffered with an incredible and strange form of madness. For he left the citadel and went down to the Piraeus and lived there, and believed that all the ships putting into port there were his own; he inventoried them and dispatched them out again, and rejoiced exceedingly at those that survived and came into the harbor. He lived with this sickness for a long time. But when his brother returned from Sicily, he gave him to a doctor to be healed, and in this way he ceased from his illness. But he often recalled his period of madness, and said that he never experienced as much pleasure as he felt in those ships that did not belong to him returning safely.

Given the matching details, there can be no doubt that this anecdote derives from Heraclides – but the fact that we are able to say so is a total accident, since Aelian does not cite any author or work as his source. If Athenaeus had not also happened to refer to this story from Heraclides' *On Pleasure*, we would be left to guess at Aelian's source. The wording is not especially close; while there are certainly some close correspondences with the Heraclides fragment (e.g. ἀπεγράφετο, ἐπαύσατο, πολλάκις τῆς ἐν μανίᾳ διατριβῆς), Aelian also seems to have taken pains to vary his vocabulary (e.g., the more unusual ἀρρώστημα instead of νόσος) and phrasing. Some details have been omitted, such as the name of Thrasyllus' brother; notably, the language of τρυφή with which Athenaeus introduces the fragment of Heraclides is also absent from Aelian. Significantly, however, Aelian has preserved, although with slightly different wording (and omitting the idea that Thrasyllus did not grieve at the lost ships) the closing tag about this being the most pleasant time of Thrasyllus' life.

All of this is to say: In Aelian, we have an example of an author working slightly later than Giacomelli has shown the compiler of *Mir.* was working, who without question has also lifted a story about pleasant madness from Heraclides, without attribution but preserving the telltale closing tag about pleasure. It is at least possible that the author of *Mir.* did the same, but with portions of Heraclides' works that we are no longer able to identify.

Works cited

Adams, F. 1856. *The Extant Works of Aretaeus, the Cappadocian.* London: The Sydenham Society.

Bignone, E. 1936. *L'Aristotele perduto e la formazione filosofica di Epicuro.* Florence: La Nuova Italia.

Bollansée, J. 1999. *Felix Jacoby. Die Fragmente der Griechischen Historiker Continued. Part IV. Biography and Antiquarian Literature. A. Biography. Fascicle 3. Hermippos of Smyrna.* Leiden; Boston; Köln: Brill.

Bollansée, J. 2008. "Clearchus' Treatise *On Modes of Life* and the Theme of *Tryphè.*" *Ktèma* 33: 403–411.

Bosworth, A.B. 1994. "Heracleides of Pontus and the Past: Fact or Fiction?" In Worthington, I. ed. *Ventures into Greek History.* Oxford: Clarendon Press. 15–27.

Brancacci, A. 1999. "Le Περὶ Ἡδονῆς d'Héraclide du Pont (Fr. 55 Wehrli)." In Dixsaut, M. and Teisserenc, F. eds. *Le fêlure du plaisir: Études sur le Philèbe de Platon.* Paris: J. Vrin. 99–125.

Brancacci, A. 2003. "Eraclide Pontico e i 'dyschereis' del Filebo." *Wiener Studien* 116: 77–96.

Bremmer, J.N. 2002. *The Rise and Fall of the Afterlife.* London: Routledge.

Bringmann, K. 1972. "Platons Philebos und Herakleides Pontikos' Dialog Περὶ Ἡδονῆς." *Hermes* 100: 523–530.

Brink, C.O. 1968. *Horace on Poetry, Volume III: Epistles Book II: The Letters to Augustus and Florus.* Cambridge: Cambridge University Press.

Brocca, N. 2011. *Lattanzio, Agostino e la Sibylla maga: Ricerche sulla fortuna degli Oracula Sibyllina nell'occidente latino.* Roma: Herder.

Brown, T.S. 1958. *Timaeus of Tauromenium.* Berkeley: University of California Press.

Buitenwerf, R. 2003. *Book III of the Sibylline Oracles and Its Social Setting.* Leiden: Brill.

Cerasuolo, S. 1987. "Il nome del Lago Averno nell'antichità." *Orpheus* 8: 120–126.

Cervelli, I. 1993. "Questioni sibilline." *Studi storici: Rivista trimestrale dell'Istituto Gramsci* 34: 895–1001.

Champion, C.B. 2010. "Timaios (566)." In Worthington, I. ed. *Jacoby Online. Brill's New Jacoby.* Part III. Leiden: Brill.

Citroni, M. 2016. "The Value of Self-Deception: Horace, Aristippus, Heraclides Ponticus, and the Pleasures of the Fool (and of the Poet)." In Hardie, P.R. ed. *Augustan Poetry and the Irrational.* Oxford: Oxford University Press. 221–239.

Cornford, F.M. ³1952. *Plato's Cosmology: The Timaeus of Plato.* New York: The Humanities Press.

Dillon, J.M. 1996. "Speusippus on Pleasure." In Algra, K.A., Van der Horst, P.W. and Runia, D.T. eds. *Polyhistor: Studies in the History and Historiography of Ancient Philosophy.* Leiden: Brill. 99–114.

Dillon, J.M. 2003. *The Heirs of Plato: A Study of the Old Academy, 347–274 B.C.* New York: Oxford University Press.

Edlund-Berry, I.E.M. 1987. "The Sacred Geography of Southern Italy in Lycophron's Alexandra." *Opuscula Romana* 16: 43–49.

van der Eijk, P. 2009. "The Woman Not Breathing." In Fortenbaugh, W.W. and Pender, E.E. eds. *Heraclides of Pontus: Discussion.* New Brunswick; London: Transaction. 237–250.

Flashar, H. 1972. "Aristoteles Mirabilia." In Flashar, H. and Klein, U. eds. *Aristoteles Mirabilia, De Audibilibus.* Berlin: Akademie-Verlag. 5–154.

Garnier, R. 2008. "Sur l'étymologie du nom de l'Averne ('facilis descensus auerno')." *Revue de Philologie, de Littérature et d'Histoire Anciennes* 82: 99–111.

Geffcken, J. 1892. *Timaios' Geographie des Westens*. Berlin: Weidmannsche Buchhandlung.

Giacomelli, C. 2021. *Ps.-Aristotele, De Mirabilibus Auscultationibus: Indagini sulla storia della tradizione e ricezione del testo*. Berlin: De Gruyter.

Giacomelli, C. 2023. *Pseudo-Aristotele, De mirabilibus auscultationibus: Edizione critica, traduzione e commento filologico*. Rome: Accademia Nazionale dei Lincei.

Gigante Lanzara, V. 2009. "Ἔστι Μοι . . . Μυρία Παντᾷ Κέλευθος (Pind. *Isthm*. IV 1)." In Cusset, C. and Prioux, É. eds. *Lycophron: Éclats d'obscurité*. Saint-Étienne: Publications de l'Université de Saint-Étienne. 95–115.

Gorman, R. and Gorman, V.B. 2014. *Corrupting Luxury in Ancient Greek Literature*. Ann Arbor: University of Michigan Press.

Gosling, J.C.B. and Taylor, C.C.W. 1982. *The Greeks on Pleasure*. Oxford: Oxford University Press.

Gottschalk, H.B. 1980. *Heraclides of Pontus*. Oxford: Clarendon Press.

Harris, W.V., ed. 2013. *Mental Disorders in the Classical World*. Leiden: Brill.

Heurgon, J. 1987. "Les deux Sibylles de Cumes, V." In *Filologia e forme letterarie. Studi offerti a Francesco Della Corte*. Urbino: Ed. Quattro Venti. 153–161.

Hoessly, F. 2001. *Katharsis: Reinigung als Heilverfahren: Studien zum Ritual der archaischen und klassischen Zeit sowie zum Corpus Hippocraticum*. Göttingen: Vandenhoeck & Ruprecht.

Hornblower, S. 2015. *Lykophron Alexandra*. Oxford: Oxford University Press.

Hornblower, S. 2018. *Lykophron's Alexandra, Rome, and the Hellenistic World*. Oxford: Oxford University Press.

Hunter, R.L. and Russell, D.A. 2011. *Plutarch, How to Study Poetry*. Cambridge: Cambridge University Press.

Jacob, C. 1983. "De l'art de compiler à la fabrication du merveilleux. Sur la paradoxographie grecque." *Lalies* 2: 121–140.

Kazantzidis, G. 2018a. "Haunted Minds, Haunted Places: Topographies of Insanity in Greek and Roman Paradoxography." In Felton, D. ed. *Landscapes of Dread in Classical Antiquity: Negative Emotion in Natural and Constructed Spaces*. Abingdon: Routledge. 226–258.

Kazantzidis, G. 2018b. "Medicine and the Paradox in the Hippocratic Corpus and Beyond." In Gerolemou, M. ed. *Recognizing Miracles in Antiquity and Beyond*. Berlin: De Gruyter. 31–61.

Kazantzidis, G. 2019. "Introduction: Medicine and Paradoxography in Dialogue." In Kazantzidis, G. ed. *Medicine and Paradoxography in the Ancient World*. Berlin: De Gruyter. 1–40.

Lenfant, D. 2007. "Les 'fragments' d'Hérodote dans les *Deipnosophistes*." In Lenfant, D. ed. *Athénée et les fragments d'historiens*. Paris: De Boccard. 43–72.

Lieberg, G. 1959. *Geist und Lust: Untersuchungen zu Demokrit, Plato, Xenokrates, und Herakleides Pontikus*. Tübingen: Buchdruckerei Bölzle.

Lightfoot, J.L. 2007. *The Sibylline Oracles*. Oxford: Oxford University Press.

Marg, W. 1972. *Timaeus Locrus, De Natura Mundi et Animae: Überlieferung, Testimonia, Text und Übersetzung*. Leiden: Brill.

Marincola, J. 1997. *Authority and Tradition in Ancient Historiography*. Cambridge: Cambridge University Press.

Muccioli, F. 2002. "Pitagora e i pitagorici nella tradizione antica." In Vattuone, R. ed. *Storici greci d'occidente*. Bologna: Il Mulino. 341–409.

Murray, O. 2017. "Review: Greek Luxuries." *Histos* 11: civ–cx.

Nikiprowetzky, V. 1970. *La troisième Sibylle.* Paris: Mouton.

Nutton, V. 2013. "Galenic Madness." In Harris, W.V. ed. *Mental Disorders in the Classical World.* Leiden: Brill. 119–127.

O'Brien-Moore, A. 1924. *Madness in Ancient Literature.* Weimar: Wagner.

Ogden, D. 2001. "The Ancient Greek Oracles of the Dead." *Acta Classica* 44: 167–195.

Pajón Leyra, I. 2011. *Entre ciencia y maravilla: El género literario de la paradoxografía griega.* Zaragoza: Prensas Univ. de Zaragoza.

Parke, H.W. 1992. *Sibyls and Sibylline Prophecy in Classical Antiquity.* London: Routledge.

Pearson, L. 1987. *The Greek Historians of the West. Timaeus and His Predecessors.* Atlanta: Scholars Press.

Pelling, C. 2000. "Fun with Fragments: Athenaeus and the Historians." In Braund, D. and Wilkins, J. eds. *Athenaeus and His World: Reading Greek Culture in the Roman Empire.* Exeter: University of Exeter Press. 171–190.

Pigeaud, J. 1987. *Folie et cures de la folie chez les médecins de l'antiquité gréco-romaine: La manie.* Paris: Belles Lettres.

Potter, D.S. 1994. *Prophets and Emperors: Human and Divine Authority from Augustus to Theodosius.* Cambridge, MA: Harvard University Press.

Repici Cambiano, L. 1977. "Lo sviluppo delle dottrine etiche nel Peripato." In Giannantoni, G. ed. *Scuole socratiche minori e filosofia ellenistica.* Bologna: Il Mulino. 215–243.

Schepens, G. and Delcroix, K. 1996. "Ancient Paradoxography: Origin, Evolution, Production and Reception." In Pecere, O. and Stramaglia, A. eds. *La letteratura di consumo nel mondo greco-latino.* Cassino: Università degli studi di Cassino. 373–460.

Schütrumpf, E.E., ed. 2008. *Heraclides of Pontus: Texts and Translation.* New Brunswick; London: Transaction.

Schütrumpf, E.E. 2009. "Heraclides, *On Pleasure.*" In Fortenbaugh, W.W. and Pender, E.E. eds. *Heraclides of Pontus: Discussion.* New Brunswick; London: Transaction. 69–92.

Shannon-Henderson, K.E. 2022a. "Tacitus and Paradoxography." In McNamara, J. and Pagán, V.E. eds. *Tacitus' Wonders: Empire and Paradox in Ancient Rome.* London: Bloomsbury. 17–51.

Shannon-Henderson, K.E. 2022b. "1667. Phlegon of Tralles." In Schorn, S. ed. *Felix Jacoby. Die Fragmente Der Griechischen Historiker Continued. Part IV. Biography and Antiquarian Literature. E. Paradoxography and Antiquities. Fascicle 2. Paradoxographers of the Imperial Period and Undated Authors [Nos. 1667–1693].* Leiden; Boston: Brill. 9–338.

Sharples, R. 1995. *Theophrastus of Eresus. Commentary Volume 5. Sources on Biology (Human Physiology, Living Creatures, Botany: Texts 328–435).* Leiden; Boston: Brill.

Spittler, J. 2022. "1672. Apollonios." In Schorn, S. ed. *Felix Jacoby. Die Fragmente der Griechischen Historiker Continued. Part IV. Biography and Antiquarian Literature. E. Paradoxography and Antiquities. Fascicle 2. Paradoxographers of the Imperial Period and Undated Authors [Nos. 1667–1693].* Leiden; Boston: Brill. 387–519.

Vanotti, G. 1998. "Riti oracolari a Cuma nella tradizione letteraria di IV e III secolo a. C." In Chirassi Colombo, I. and Seppilli, T. eds. *Sibille e linguaggi oracolari: Mito storia tradizione.* Pisa: Istituti Editoriali e Poligrafici Internazionali. 263–276.

Vanotti, G. 2007. *Aristotele, Racconti meravigliosi.* Milan: Bompiani.

Vogt, K.M. 2013. "Plato on Madness and the Good Life." In Harris, W.V. ed. *Mental Disorders in the Classical World.* Leiden: Brill. 178–192.

Wehrli, F. 1953. *Die Schule des Aristoteles. Heft VII. Herakleides Pontikos.* Basel: Schwabe.

West, S. 1983. "Notes on the Text of Lycophron." *Classical Quarterly* 33: 114–135.

West, S. 1984. "Lycophron Italicised." *Journal of Hellenic Studies* 104: 127–151.

Zaccaria, P. 2021. *Felix Jacoby. Die Fragmente der griechishen Historiker. Continued*. Part IV. *Biography and Antiquarian Literature. A. Biography. Fascicle 5. The First Century BC and Hellenistic Authors of Uncertain Date*. Leiden: Brill.

Zhmud, L. 2012. *Pythagoras and the Early Pythagoreans*. Translated by Kevin Windle and Rosh Ireland. Oxford: Oxford University Press.

Zingg, E. 2019. "Ein anonymer Traktat 'Peri Sibylles' (Iohannes Lydos, Peri Menon 4, 47)." *Segno e Testo* 17: 143–183.

5   Myth, marvels, and *De mirabilibus auscultationibus*

Robin J. Greene

Alongside reports from ps.-Antigonus' paradoxographical compilation preserved in the 9th-century manuscript known as codex Palatinus Heidelberg gr. 398, we find half a dozen instances of the same marginal note: MYΘ.[1] These abbreviations for μῦθος – perhaps made by a schoolteacher marking potential passages to use for progymnasmata, as Eleftheriou suggests[2] — all accompany entries on wonders that include references to mythical legends and events, for example, Leto's birth of Apollo in ps.-Antig. 56 and the death of Nessus the centaur in 117. The anonymous hand's identification of these passages highlights a basic fact about ps.-Antigonus' inclusion of myths, namely, that its infrequency is such that one stops and notices when mythic material appears, but that it happens frequently enough that one may make a list of occurrences. So, in the case of ps.-Antigonus' compilation, such a list as the one included as an appendix to this study would include 12 examples, meaning just about 7% of ps.-Antigonus' entries include mythic material. This modest percentage supports William Hansen's observation that "mythology, despite its fabulous content, never came to play a major role in paradoxography, for the compilers were attracted more to wonders of the contemporary and near-contemporary world than to prehistory."[3]

I agree with Hansen's evaluation. Yet, although mythic material does not play a major role in paradoxography, it does play a minor one in several of the seven surviving or mostly surviving compilations,[4] a supporting role that has historically

---

1 See Eleftheriou (2018) I 100–117 on the codex, with helpful images of the marginalia. The note appears alongside ps.-Antig. 12, 56, 117, 118, 163, and 172. Cf. Eleftheriou (2016) for further discussion.
2 Eleftheriou (2016) 38 and Eleftheriou (2018) I 110–111.
3 Hansen (1996) 8.
4 I refer to this group throughout this study. It includes the compilations of ps.-Antigonus, ps.-Aristotle, Phlegon of Tralleis, Apollonius, and the three anonymous compilations: *Paradoxographus Florentinus, Paradoxographus Palatinus,* and *Paradoxographus Vaticanus.* New texts and commentaries are now available for all but ps.-Antigonus and ps.-Aristotle in Schorn (2022); for ps.-Aristotle, see the new text and philological commentary of Giacomelli (2023), and for both ps.-Aristotle and ps.-Antigonus, see the forthcoming texts and commentaries edited by Schorn. See also Eleftheriou (2018) on ps.-Antigonus, though the commentary is not for the complete compilation. More fragmentary compilations also likely included myth to varying degrees. A tantalizing,

DOI: 10.4324/9781003437819-6

received little critical attention.[5] More importantly, myth has a considerable presence in one particular section of a compilation, 78–121 of *De mirabilibus auscultationibus* (hereafter *Mir.*),[6] with more than 35% of its reports containing at least one reference to a myth. In other words, there are more mythic entries in that section than in the rest of the compilation *and* that of ps.-Antigonus combined. Many of these reports have been studied individually as they relate to a given topic, historical event, or geographical region, but *Mir.*'s myths have not been considered in terms of their general roles in paradoxographical report or in terms of the reasons for and effects of their clustered appearance in a single section.[7] With this study, I aim to begin filling this gap, though there are more examples than can be addressed in a limited space. Rather, in the first section, I consider the typical ways that various paradoxographers use mythic material in their collections, and then in the second section explore how *Mir.*'s incorporation of myth both helps reveal an idiosyncratic conceptual understanding of "wonder" and reflects a sustained interest in the application of Greek myth to regions outside the borders of traditionally Greek areas.

1. Mythic material in paradoxographical compilations

Before turning to specific examples of mythic material in the various collections, I must clarify two points regarding my approach. First, the basic working definition of "mythic material" I use to identify relevant reports only requires that a compiler recounts an action performed by a figure from myth. Thus, for example, I consider *Mir.* 83's claim that there are no aggressive animals on Crete because "Zeus was born there" and ps.-Antig. 111's notice that Heracles had only one daughter as reports that contain mythic material. What this working definition does not include, however, are the many reports set at temples and sanctuaries or those that observe elements related to cult worship unless the reports also include a mythic actor. So the Italic veneration of Sirens in *Mir.* 103 and the strange behavior of a bull set on the altar of Artemis Orthosia in *Mir.* 175 do not constitute "mythic material" for the purposes of this study, whereas *Mir.* 107 on the Sybarites' worship of Philoctetes – which includes some narrative of Philoctetes' actions in Italy – does. A principal drawback of this definition of "mythic material" is that it creates a misleading conceptual rupture between myth and history, which of course were not defined as fundamentally distinct ideas by most Greek authors, who more often than not understood myth as very ancient history. Indeed, we shall see several examples

---

though very damaged, example is *P. Oxy.* II 218 (= *FGrHist* 1682), which combines horror stories, ethnography, and mythology, as Pajón Leyra (2022a) shows in detail.

5 Though see the remarks of Stern (2008) 443–444, who observes that distant (foreign) contemporary *paradoxa* can substitute for the mythic through their reflection of the same sorts of themes, for example, unusual sexual behavior, killing, and metamorphosis.

6 Throughout I use the Greek text, numbering, and reordered arrangement of Giacomelli (2023). I am grateful for the advance copy of the draft.

7 Though see now the recent article by Pajón Leyra (2022b), which was released just before the final review version of this study was due.

in *Mir.* where the compiler traces the history of an area or object from mythic times through historical ones (e.g., 100, 107, 110). Despite my working definition obscuring the nuances of ancient views on myth's relationship to history, it serves the general purposes of this study, though future work on the understudied topic of specific historical references in paradoxographical compilations would also be a welcome addition.[8]

Throughout, I also assume compilers' sources are responsible for the connection of a phenomenon to a myth and not the compilers themselves. *Mir.*'s compiler may, in fact, synthesize different sources together in a single entry, but there is no clear proof that the paradoxographer himself introduced a myth as relevant to a *paradoxon.*[9] Rather, the compilers' responsibility lies in their decision to retain or, as I discuss in the next section, omit mythic material adduced by others.

1.1 Myths unmentioned and mentioned

We often find that paradoxographers do not incorporate mythic references even when other sources explicitly connect a phenomenon to a mythic account. Such is the case with the compiler of *Paradoxographus Florentinus*, who focuses his collection on aquatic *mirabilia*. While several of his reports are linked to mythic material by other authors, *Paradoxographus Florentinus* features only a single mention of a myth. In a report on the spring of Clitor that causes an aversion to wine (*Par. Flor.* 24), the compiler appends his description with an anonymous epigram whose final lines attribute the phenomenon to Melampus' use of the spring as the disposal site for the purificatory materials used to cleanse the Proitids of their madness.[10] We can be sure that the compiler did not include the epigram because of its mythic material, as it appears in a section dedicated to wonders described by epigrams, of which no others contain myths.[11] It is, however, only through the epigram that a reader learns of a mythic *aition* for the spring's effect. Were the epigram omitted, the mythic connection would vanish. Indeed, we see precisely this happen elsewhere in the compilation. *Paradoxographus Florentinus* 12 offers a near-duplicate report on the spring at Clitor but does not include the epigram; there, no mention of Melampus or the Proitids is made.[12] Together, this omission and the absence of other mythic references in the compilation, indicate the compiler's general aversion to or simple disinterest in mythic material. Myth plays no part in his conception of

---

8 Not including Phlegon and Apollonius, who incorporate a number of historical references, reports with historical anecdotes featuring specific historical actors seem by my rough estimation just as rare, if not rarer, than reports with mythic material in the other paradoxographers. Cf., for example *Par. Flor.* 32, *Mir.* 29 and 52, *Par. Vat.* 43.

9 Though see below on Phlegon's pairing of mythic examples of a *paradoxon* with contemporary instances.

10 φεῦγε δ' ἐμὴν πηγὴν μισάμπελον, ἔνθα Μελάμπους/λουσάμενος λύσσης Προιτίδας ἀργαλέης/πάντα καθαρμὸν ἔκρυψεν ἀπόκρυφον· αἳ γὰρ ἀπ' Ἄργους/οὔρεα τρηχείης ἤλυθον Ἀρκαδίης. Text from Greene (2022).

11 See Greene (2022) 724–727 on this report and its parallels.

12 παρὰ Κλειτορίοις ὁ αὐτός (= Isigonus, FGrHist 1660 F 6) φησιν εἶναι κρήνην, ἧς ὅταν τις τοῦ ὕδατος πίῃ, τοῦ οἴνου τὴν ὀσμὴν οὐ φέρει.

paradoxography,[13] just as seems to be the case with Apollonius and *Paradoxographus Palatinus*, whose collections contain no mythic material.

Other paradoxographers also omit mythic material linked to phenomena by various sources, but reasons other than disinterest often seem more probable explanations. This is true of the compiler of *Mir.*, whose comparatively high frequency of mythic material testifies that he was not opposed to including it. Consider, then, the reports on the songless frogs of the Cycladic island Seriphus provided in *Mir.* 70, which does not feature any mythic references, and in ps.-Antig. 4 (along with the preceding two entries), which does:

Mir. 70: Φασὶ δὲ καὶ ἐν Σερίφῳ τοὺς βατράχους οὐκ ᾄδειν· ἐὰν δὲ εἰς ἄλλον τόπον μετενεχθῶσιν, ᾄδουσιν.

They also say that in Seriphus the frogs do not sing, but if they are transferred to another place, they do sing.[14]

ps.-Antig. 2. Καὶ ἄλλο δὲ παρὰ τοῖς Ῥηγίνοις τοιοῦτον ὡς μυθικὸν ἱστορεῖται, ὅτι Ἡρακλῆς ἔν τινι τόπῳ τῆς χώρας κατακοιμηθεὶς καὶ ἐνοχλούμενος ὑπὸ τῶν τεττίγων ηὔξατο αὐτοὺς ἀφώνους γενέσθαι.

ps.-Antig. 3. Καὶ ἐν Κεφαλληνίᾳ δὲ ποταμὸς διείργει, καὶ ἐπίταδε μὲν γίνονται τέττιγες, ἐπέκεινα δὲ οὔ.

ps.-Antig. 4. Οὐδ᾿ ἐν Σερίφῳ δὲ οἱ βάτραχοι φθέγγονται· καὶ μυθῶδες καὶ παρὰ τοῖς Σεριφίοις ἐνίσχυσεν, πλὴν οἱ μὲν περὶ Ἡρακλέους, οἱ δὲ περὶ Περσέως.[15]

2. Another such fabulous thing is recorded at Rhegium: that Heracles, when he was disturbed by cicadas after falling asleep at some spot in the region, prayed that they would become voiceless.

3. In Cephallenia, too, a river partitions the land, and while on the near side there are cicadas, there are none on the far side.

4. Nor do the frogs give voice in Seriphus: and a similar also myth prevails among the Seriphians, except that while the Rhegines tell it about Heracles, the Seriphians tell it about Perseus.

Both *Mir.* 70 and ps.-Antig. 4 belong to a tradition headed by Theophrastus,[16] whose own account of the Seriphian frogs seems to have included the myth, as implied by Aelian's observation at *NA* 3.37 (= Theophrastus F 355A FHS&G): λέγει δὲ Θεόφραστος ἐκβάλλων τὸν μῦθον καὶ Σεριφίους τῆς ἀλαζονείας παραλύων τὴν τοῦ ὕδατος ψυχρότητα αἰτίαν εἶναι τῆς ἀφωνίας τῶν προειρημένων ("But Theophrastus,

---

13 Cf. Greene (2022) 659–660.

14 All translations of *Mir.* are based upon those of Dowdall (1984), with modifications, especially to account for Giacomelli's revisions; throughout I found Robert Mayhew's draft version helpful.

15 I use Musso (1985) for the text of ps.-Antigonus. All translations of ps.-Antigonus are from Hardiman (2021).

16 *Mir.* 70 is part of a cluster of reports (68–70) which all have connections to Theophrastus' *On Differences According to Location*, on which see Flashar (1981) 40–41 and 101–102, Sharples (1995) 51–56, Vanotti (2007) 161.

rejecting the story and freeing the Seriphians from their imposture, says that it is the coldness of the water that is the reason for the dumbness of the afore-mentioned creatures").[17] Given that the compiler of *Mir.* and ps.-Antigonus were working from the same tradition, that they had similar general "paradoxographical" motivations, and that both included more mythic material in their collections than other para-doxographers, why might *Mir.* not include the myth? The simplest explanation may be the likeliest. "Belonging to the same tradition" does not mean that reports were drawn from the same immediate source; *Mir.* 70 may derive from an intermediary that did not report the myth.[18] Another possibility is that the compiler was not inter-ested in this particular myth. Indeed, as will be discussed later, mythic material in *Mir.* occurs more often with what may be classified as topographical, cultural, and ethnographical *paradoxa*, not zoological and biological wonders as in ps.-Antigo-nus' collection. In this case, both paradoxographers have some interest in mythic connections, but apply that interest to different types of phenomena.[19]

Yet questioning why *Mir.* 70 lacks a mythic reference requires us to explain a negative. Rather, a more productive question, both for this example and for the study of myth in paradoxography in general, is why Ps.-Antig. 4 *does* include it. Other reasons for paradoxography's incorporation of mythological material and the effects of that incorporation will be considered throughout the rest of this study, but I suspect that principles of organization exerted some influence over ps.-Antigonus' choice here. His report on the frogs appears in a thematic subsec-tion focused upon melodious and mute animals (Ps.-Antig. 1–8).[20] Of these, three entries treat celebrated figures of myth (2 and 4 on Heracles and Perseus, 5 on Orpheus),[21] while ps.-Antig. 1 recounts a local legend set in the historical past but characterized from the outset as belonging to the realm of μῦθος: λέγεται δέ τι τούτου μυθωδέστερον ("and he [Timaeus] narrates something yet more fabulous than this").[22] Moreover, ps.-Antig. 7 features a verse excerpt from the *Homeric*

---

17 Text and translation from Fortenbaugh *et al.* (1992). Battegazzore (1993) 228–229 interprets Aelian's statement as confirmation that Theophrastus mentioned the myth initially before providing scientific refutation. Sharples (1995) 56–57 is more cautious, deeming Theophrastus' inclusion of the story "uncertain" and suggesting that Aelian or his source introduced the story and then used Theophrastus to reject it. Pliny *HN* 8.277 provides a similar report but mentions no mythic material.

18 See Schorn (this volume, chapter 2) on intermediary sources used elsewhere in *Mir.*

19 Though this general division of interests and mythic references certainly is not absolute. Compare *Mir.* 83 on Zeus' birth as responsible for the lack of aggressive animals on Crete with ps.-Antig. 10, which reports the same phenomenon without mention of the mythic explanation.

20 Jacob (1983) 127. See Lightfoot (2021) 80–82 for further discussion on the arrangement of the section and on the relationship between wonder, music, and musicality in Greek literary traditions. Cf. Eleftheriou (2018) II 9–30 for analysis of the structure of the section and the individual entries; Schepens and Delcroix (1996) 396–398 on ps.-Antigonus' analogical and associative arrangement.

21 For representations of Orpheus in paradoxography and related genres, see Pajón Leyra (2011b).

22 Compare similar terminology in ps.-Antig. 2 (μυθικὸν ἱστορεῖται), 4 (μυθῶδες), and 5 (μυθολογεῖται). Eleftheriou (2016, 2018) II 15–16 understands the terms to reflect not content or judgments regard-ing plausibility, but the reports' origins in an anonymous, often oral source. Her caution in avoid-ing conflating μυθ- terms with more contemporary uses of "myth" is well-placed, though the text also seems to support a basic connection of μυθ- with "myth" as relating to legends (historical or prehistorical) or to stories of gods and heroes. Such terms appear twice more in the collection,

Hymn to Hermes' famous description of the newborn god constructing the first lyre (*H. Merc.* 51) that encourages readers to link the phenomenon of melodious ewe guts with the god's use of them: "One might understand the Poet – who has a real thirst for knowledge and is exceedingly learned – to be referring to this this when he says: 'He [Hermes] strung seven strings of female sheep'" (ὅθεν καὶ τὸν ποιητὴν ὑπολάβοι τις εἰρηκέναι, πολυπράγμονα πανταχοῦ καὶ περιττὸν ὄντα, 'ἑπτὰ δὲ θηλυτέρων ὀίων ἐτανύσσατο χορδάς'). Taken together, the repeated evocations of myth in 1–8 indicate that, although similarity of topic principally governs their grouping, their mythological associations constitute a secondary connective theme that motivated ps.-Antigonus' selection of material.

1.2 Mythic aitia

All of the explicit mentions of myth in ps.-Antig. 1–8 – and nearly all the mythic material throughout that compilation[23] – serve as etiological explanations for the existence of a *paradoxon*. In *Mir.* much the same holds true, with the majority of mythic references functioning as "explanatory" *aitia*. Here the quotation marks are necessary. After all, the fundamental aim of paradoxography targets the cultivation of wonder, hence the genre's famous eschewal of any hint of rationalized discussion or scientific explanation. In the face of a logical accounting for a phenomenon, wonder is lost. As Schepens observed in a seminal article:

> *thauma* is no match for "explanation"; the sense of the marvellous cannot survive on a rational basis. It is imperative for the paradoxographer to concentrate on *historia*, the establishment and recording of facts without explaining them. . . . The deliberate omission of any attempt at rational exegesis must . . . be regarded as an intrinsic part of his endeavor to inspire his readers with a true sense of the marvellous.[24]

---

both explicitly in mythological contexts: 12 (αἰτίαν μυθικῶς) of the tale of Attic crows, Athena, Erichthonius, and the Cecropidae, and 56 (μυθικόν τι) of Leto's labor and its relationship to lupine parturition. Compare also the appearances of μυθ- terms in *Mir.*: 79, 81, 97, 100, 101, 105, 110, all of which except for 101 are used of individuals and events properly deemed "mythic." *Mir.* 101 reports on strange music and noises in cave on Lipara (a phenomenon which may have a connection to Cybele?), so the idea of μυθ- as relating to local, anonymous tales of the supernatural seems to be at work there.

23 The exceptions are 24 and 111 (on which, see p. 180 with n. 34).

24 In Schepens and Delcroix (1996) 391–392. Although Schepens' observations generally hold true for most paradoxographic compilations, recent scholarship has highlighted a few significant counterexamples. In her discussion of Apollonius *Mir.* 21, which admits that a scientific reason for a phenomenon exists but explicitly omits it, Spittler (2022) 420 observes that readers participate in an "inverted 'suspension of disbelief'; that is, instead of suspending judgment on the implausibility of the story reported, the reader willingly foregoes the explanation – which he is assured does indeed exist – in an effort to sustain the sense of implausibility, the sense of wonder." See also Shannon-Henderson (2022) 230 and 241–242 on Phlegon *Mir.* 15, which both adduces an explanation for the difference in size between contemporary humans and ancient heroes and includes a claim to personal autopsy of the evidence.

Mythic *aitia*, however, pose no such threat to wonder, for they likewise find little place in analytical and reasoned discourse on causality.[25] The compiler of *Mir.* may attest that the giant herons of Diomedeia welcome Greeks because the birds descend from the metamorphosed companions of Diomedes (*Mir.* 79), but this hardly constitutes a convincing explanation for animal behavior. Nor does the thunderbolt hurled at Phaethon offer a satisfactory reason for the existence of a boiling, sulfuric lake (*Mir.* 81), or does the claim that "Heracles put it there" truly explain the position of a large Iapygian boulder (*Mir.* 98). Instead, Schepens and Delcroix maintain, such stories "in no way spoil the effect the paradoxographer is trying to achieve," but may even "being mythical, add to the *paradoxon* in [their] own way."[26]

I agree with Schepens and Delcroix's assessment, but the particular way(s) in which mythic *aitia* may add to *paradoxa* – beyond their basic interest value as myths – requires elaboration. Paradoxography is fundamentally a genre of subtraction, with compilers paring down their source texts until only the elements necessary to report and support the wonder remained. That a myth was a myth perhaps may have been enough to snag a compiler's attention so that it survived this editorial process. However, a survey of mythic *aitia* in paradoxographical collections reveals that the vast majority share one attribute other than their essential nature that can help explain their inclusion: relative obscurity.

Calculating the popularity of a myth is a fraught business from the start. The paucity of surviving texts, our inability to account with accuracy for oral and artistic transmission, the comparative fame of a myth of local interest within that location versus without, and many other factors hinder our ability to assess the true popularity of a given story. With these limitations in mind, an attempt at gauging the potential familiarity of mythical *aitia* in paradoxographical compilations by tracing the myths' appearances elsewhere testifies to their relative obscurity as a group. *Paradoxographus Vaticanus* offers the most unique myths. Its four mythical *aitia* are otherwise unattested either in part or entirely, including a "baffling" tale of Marsyas (*Par. Vat.* 18) that omits his famous contest with Apollo and subsequent flaying, and instead recounts his death by drowning in the Phrygian river which bears his name.[27] In contrast, most of ps.-Antigonus' ten mythic *aitia* appear elsewhere, though no more than a few authorities mention any except the most popular (ps.-Antig. 172, the birds of Diomedeia).[28] Moreover, these mentions are typically made in passing or are in works whose authors devote themselves to amassing sundry local stories, obscure trivia, and anecdotes (e.g., Pausanias, Strabo). Likewise, we usually find *Mir.*'s mythic references attested elsewhere in a very small number of sources and, as Vanotti observes, *Mir.*'s versions often include unique elements

---

25 Stern (1996) 7–16 provides a helpful overview of the history of Greek mythological rationalism.

26 Schepens in Schepens and Delcroix (1996) 392, here discussing ps.-Antig. 12.

27 Stern (2008) 455–456 highlights the "baffling" qualities of this "startling" version. On the other three reports with mythic *aitia* (14, 15, 17), see Stern (2008) *ad loc* and Sørensen (2022) *ad loc.*

28 For this report and its myth (which also appears at *Mir.* 79), see p. 200–202.

unseen in other examples.[29] *Mir.*'s propensity for longer reports that combine discrete topics and narratives into a single entry also often results in the obscure existing alongside more familiar material. So in *Mir.* 81, for example, Daedalus' activity in the northern Adriatic and its effects – both of which are otherwise unattested – frame a longer report on the area that includes the better-attested mythic *aition* related to Phaethon.

Mir.'s report on Daedalus inserts the Adriatic episode into the traditional sequence of stories centered on his travels after escaping Crete. Most mythic *aitia* in paradoxographies operate in the same way, offering little-known details incorporated into larger and more famous stories. A few of these connect the well-known experiences of the gods with phenomena in the natural world, as we see in reports about the effects that Zeus' birth (*Mir.* 83) and later dalliance with Europa (ps.-Antig. 163) had on Cretan fauna and topography, or lupine parturition's link to the labor suffered by Leto (ps.-Antig. 56). Far more common, however, are *aitia* integrated with or appended to tales of heroic adventures, travel, and migration, as in the case of Daedalus. Heracles' exploits provide the occasions for the most paradoxographical myths, an unsurprising fact given his ubiquity in the local traditions of areas from Scythia to Iberia as a result of his travels. Ps.-Antig. 117, for example, refers to Myrsilus of Methymna's novel endnote to the traditional tale of the hero's fight with the centaur Nessus. Though other versions conclude with Nessus' death, Myrsilus' account adds that his body polluted the waters of Ozolian Locris and resulted in the area's alleged malodorousness.[30] Similarly, *Mir.* 97 traces a fetid stream of ichor in Iapygia to a version of Heracles' western activities that stages his battle against the Giants in the area. Other entries likewise cast moments from the hero's life and travels – unattested or canonical – into *aitia* for botanical (*Mir.* 51), zoological (ps.-Antig. 2), mineralogical (*Mir.* 58), biological (ps.-Antig. 111), ethnographical (*Mir.* 83), and even art-historical *paradoxa* (*Mir.* 118),[31] and we see much the same in reports that cull recondite details related to the traditional exploits of other mythic heroes.

The relative obscurity shared by nearly all mythic *aitia* in the collections suggests that this was a quality that influenced the paradoxographers' selection of material. Other more ubiquitous stories and myths certainly recount "marvelous" phenomena, but their very popularity may blunt the reader's (and compiler's) interest and dull the wonder. Thus, for example, Schorn suggests that *Mir.* 88 intentionally omits well-known anecdotes about cultural phenomena associated with the Baleares to focus instead upon more unfamiliar material.[32] Obscure stories,

---

29 Vanotti (2007) 37, with unique elements noted throughout the commentary. *Mir.*'s mythic reports most often have parallels in Lycophron (who is famous for his delight in the obscure), Strabo, and Diodorus Siculus, as will be seen in the second half of this chapter.

30 On Myrsilus' unique version, as well as his otherwise unattested variant of the Lemnian women treated in ps.-Antig. 118, see Jackson (1990), esp. 81.

31 See below for further observations about all but ps.-Antig. 2 (discussed above).

32 Schorn, this volume, p. 61. On this report, see also p. 196. Cf. Greene (2022) 690–691 where I suggest that the near absence of anecdotes about healing waters in *Paradoxographus Florentinus*

moreover, not only add novelty to reports, but may stimulate their own sort of fascination, the fascination aroused when learning something new and unfamiliar, especially about that which is otherwise familiar. We find this intellectual attraction to the obscure dramatized in Callimachus' *Aetia*, when the erudite Callimachean narrator describes his response to learning recondite mythic *aitia* for peculiar contemporary cult practices:

ὣ[ς] ἡ μὲν λίπε μῦθον, ἐγὼ δ' ἐπὶ καὶ τ[ὸ πυ]θέσθαι
   ἤ]θελον – ἦ γάρ μοι θάμβος ὑπετρέφ[ε]τρ –,
Κ]ισσούσης παρ' ὕδωρ Θεοδαίσια Κρῆ[σσαν ἑ]ορτὴν
   ἡ] πόλις ἡ Κάδμου κῶς Ἀλίαρτος ἄγ[ει
κ]αὶ στυρὸν ἐν μούνοισι πολίσμασι . . . .ṭ.τῳ

Thus she [the Muse Cleo] ended her story, but I wanted to learn this as well – for, truly, my amazement was fed while she spoke – why near the water of Cissusa the town of Cadmus, Haliartus, celebrates the Theodaesia, a Cretan festival and the earth produces *styrax* only in the towns. (F 43b Harder)

Hearing an unfamiliar *aition* about the origins of a cult in a place he would not expect that cult to exist amplifies the narrator's fascination and whets his appetite to learn other such stories. The mythic *aitia* in paradoxographical compilations can provoke this same response in readers. Such obscure tales add unexpected novelty to what are often well-known mythic traditions, sparking a sense of surprised interest that works in tandem with the central cause of true wonder in any report, the *paradoxon* itself.

1.3 Proving the paradox

Although most mythic material in paradoxographical compilations function etiologically, a handful of reports instead serve to bolster the believability and credibility of a wonder.[33] Mythic *exempla* produced as evidence, a long-standing rhetorical strategy employed even in natural scientific treatises, constitutes one way of doing so. In a report drawn from *Historia animalium* (9(7).6.585b21–24), ps.-Antigonus preserves Aristotle's use of Heracles as a mythic *exemplum* for the claim that some individuals only produce offspring of one sex:

111. Εἶναι δὲ καὶ ἄνδρας καὶ γυναῖκας θηλυγόνους καὶ ἀρρενογόνους, ὃ καὶ περὶ τοῦ Ἡρακλέους ἱστορεῖται· ἐν δύο γὰρ καὶ ἑβδομήκοντα τέκνοις μίαν αὐτὸν γεννῆσαι θυγατέρα.

---

and other compilations stems from such waters' popular renown, rendering them rather mundane *mirabilia*.

33 On believability and credibility as critical components to paradoxography, see Schepens in Schepens and Delcroix (1996) 382–389.

> There are both men and women who have female and male offspring,
> respectively, as is told about Heracles; for out of seventy-two children, he
> begot just one daughter.

Having stripped Aristotle's more substantial discussion on this and similar phenomena down to its bare bones, ps.-Antigonus' report consists of a single statement
of fact with the example of Heracles' progeny as its confirmation. It is no surprise
that the mythic material survived editorial cuts. Not only does Heracles' astounding
propensity for siring sons add interest and wonder in itself, but his example fills out
what would without it be a rather bald and far less appealing claim. Had Aristotle
not made the mythic connection, perhaps ps.-Antigonus would not have included
the report at all.[34]

Whereas ps.-Antigonus only retains a mythical *exemplum* adduced by his
source, Phlegon of Tralleis' *On Marvels* shows the paradoxographer himself
adducing and incorporating myths for suasive rhetorical purposes. He twice pairs
mythological entries with contemporary reports on the same phenomena. His fifth
and sixth entries recount the stories of the changes of sex experienced by Tiresias
and the Lapith Kainis/Kaineus before providing four examples of recent historical androgynes, complete with names, dates, and locations (Phlegon *Mir.* 7–10).
Later, Phlegon concludes a short section focused upon anecdotes regarding contemporary Egyptian women's hyperfertility (28 and 29) with two entries on that
of mythological daughters of the Nile, Euryrhoe (30, with Aegyptus) and Europe
(31, with Danaus). In her recent commentary, Kelly Shannon-Henderson observes
that the examples of the daughters of the Nile reinforce both the validity of the
contemporary reports and the descriptions of Egyptian fertility found in the ancient
natural–scientific texts that could have been Phlegon's sources[35]; the mythic here
serves as support for the accuracy of Phlegon's more modern examples. In turn,
she interprets the contemporary androgynes in 7–10 as "supplementing Phlegon's
accounts of mythical sex changers [in 5 and 6] with four examples drawn from
'real life'."[36] In sum, the myths in *On Marvels* are designed to be comprehended
with contemporary accounts of the same phenomena, and each report helps authen-
ticate the others.[37]

Yet it should also be noted that all four of Phlegon's mythical *paradoxa* enjoyed
considerable popularity from the archaic through the imperial periods. Stories of
the 50 Danaids (and by extension, the Aegyptiads) proliferated, especially during

---

34 Cf. ps.-Antig. 24, which adduces a strange fact about canine behavior to explain Homer's own
    curiousness, there figured as the real *paradoxon* of the report, as evidenced by his treatment of
    Odysseus' response to his dog's reaction to Eumaeus. In this case, a mythic situation serves both as
    an exemplum of a zoological phenomenon and as part of the *paradoxon* itself. Homer's knowledge-
    ability on canine zoology – particularly their possible longevity – elsewhere appears as a discussion
    topic among philosophers and natural scientists, most notably in Aristotle's *History of Animals* and
    possibly in *Homeric Questions*, on which see Mayhew (2019) 55–57.
35 Shannon-Henderson (2022) 282 and 288.
36 Shannon-Henderson (2022) 155.
37 So Doroszewska (2016), 97–98.

the Hellenistic period. Interest in stories of Tiresias and Kainis/Kaineus began, as Phlegon observes at the outset of his fifth entry, with Hesiod and continued into the 2nd century CE, with both being "mentioned together as sex changers *par excellence* several times in Greek imperial literature."[38] We may then question how truly "marvelous" or surprising readers would find *paradoxa* that, in their capacity as myths, had long since become *doxa*. Thus, while I agree that the mythic and contemporary examples mutually support each other's authenticity, I suspect that the mythic in Phlegon's collection carries more weight than the contemporary in terms of facilitating credibility. The more familiar, traditional examples of mythical androgynes and Egyptian fertility encourage readers' acceptance of the unfamiliar modern stories,[39] while underscoring that what may have been long scoffed at by some as "just a myth" in fact have contemporary analogues.

Mir. generally avoids deploying mythic material for its evidentiary value, but one report centers on adducing myth to substantiate a phenomenon and in doing so the compiler adopts a more heavy-handed and contentious approach than that of Phlegon and ps.-Antigonus/Aristotle. Near the beginning of an extended entry centered upon the bifurcated course of the Ister, the report's attention turns to the activity of the Argonauts:

> 105. [1] . . . σημεῖον δὲ οὐ μόνον ἐν τοῖς νῦν καιροῖς ἑωράκαμεν, ἀλλὰ καὶ ἐπὶ τῶν ἀρχαίων μᾶλλον, οἷον τὰ ἐκείνου πλωτὰ εἶναι· καὶ γὰρ Ἰάσονα τὸν μὲν εἴσπλουν κατὰ Κυανέας, τὸν δὲ ἐκ τοῦ Πόντου ἔκπλουν κατὰ τὸν Ἴστρον ποιήσασθαί φασι· καὶ φέρουσιν ἄλλα τε τεκμήρια οὐκ ὀλίγα, καὶ κατὰ μὲν τὴν χώραν βωμοὺς ὑπὸ τοῦ Ἰάσονος ἀνακειμένους δεικνύουσιν, ἐν δὲ μιᾷ τῶν νήσων τῶν ἐν τῷ Ἀδρίᾳ ἱερὸν Ἀρτέμιδος ὑπὸ Μηδείας ἱδρυμένον.

> We have seen a proof not only in present times, but also more fully in antiquity, that the waters are navigable; for they say that Jason sailed into the Pontus by the "Dark Rocks," while he sailed out of it by the Ister, and for this, besides alleging not a few other evidences, they point out altars set up by Jason in the country, and on one of the islands in the Adriatic a temple of Artemis . . . erected by Medea . . . *the remainder of the report largely focuses upon the activity of the Argonauts in the Adriatic and Tyrrhenian seas.*

The Argonauts' return route persisted as a topic of perennial discussion and reinvention among ancient poets, geographers, historians, and other intellectuals for centuries, and this report reflects its entrenchment in the discourse of that debate.[40]

---

38 Shannon-Henderson (2022) 148, noting especially Luc. *Salt.* 57; *Gall.* 19; Ael. *NA* 1.25; Ant. Lib. 17.4–5. Cf. Doroszewska (2013) for a thorough discussion of the many parallels to 5 and 6 and their relationship to Phlegon's truncated report.

39 Compare Aristotle's rhetorical strategy in his discussion of the rare possibility of birthing "twins" who were actually conceived at different times (*HA* 9(7).4.585a12–14), where he first briefly mentions Iphicles and Heracles before introducing three contemporary cases. Here the mythic serves as a familiar bridge to the more unfamiliar.

40 See the detailed overviews of the various Argonautic return routes provided by Vian (1981) 11–46, Vian (1987), and Hunter (2015).

Peppered with the language of argumentation, authority, and autopsy (σημεῖον . . . ἑωράκαμεν, φέρουσιν ἄλλα τε τεκμήρια οὐκ ὀλίγα), the evidence of the Argonautic return via a western route serves to establish the accuracy of the paradoxographer's claims about the Ister. The later introduction of the etiological myth of the "pebbles" of Aethalia (modern Elba), which were reputed to have gained their speckled coloring as a result of being dyed by the oil scraped off the Argonauts' bodies, likewise casts that *paradoxon* as evidence in support of the *Argo*'s western course and thus the opening contention about the Ister: ἄλλα τε δεικνύουσι μνημεῖα τῶν ἀριστέων καὶ τὸ ἐπὶ τῶν ψήφων δὲ λεγόμενον ("they point to other memorials of the heroes, and also what is said regarding the pebbles"). The compiler's approach throughout, especially his explicit characterization of mythic material as evidence for an anecdote's accuracy and his investment in argumentation, runs counter his typical practice of presenting *paradoxa* in a neutral tone with little to no concern for debating the credibility of a report. These differences, combined with the uncharacteristic use of the first person plural at the opening of the excerpt (ἑωράκαμεν),[41] suggest that the report, including its mythic references, was reproduced without significant changes from *Mir.*'s source, which seems to have been directly involved in the "route of the Argonauts" debate.[42] That said, the compiler nonetheless elects to adhere to and replicate his source's argument rather than to temper its presentation to align with the style of reports elsewhere in the collection. As the Argonauts' return route remained in dispute, perhaps the compiler felt such an approach necessary for establishing the believability of the claim about the Ister, especially since readers may well have been aware of the opposing versions of Hesiod, Pindar, and other major authorities.

2. *De mirabilibus auscultationibus* and myth

Thus far I have considered how mythic material may function in paradoxographical compilations and the appeal that myths, particularly obscure ones, could have held for compilers and audiences alike. When considering these and other such questions broadly, it may seem as though ps.-Antigonus, *Paradoxographus Vaticanus*, and *Mir.* adopt fundamentally similar approaches to myth and deploy it in similar ways. A close examination of the mythic reports in *Mir.*, however, demonstrates the contrary is often the case. Beyond sharing the other two compilations' tendency toward including myth for etiological reasons and their taste for relatively unfamiliar stories attached to more familiar mythic sequences, *Mir.*'s myths regularly deviate from those of *Paradoxographus Vaticanus* and ps.-Antigonus in terms

---

41 Compare the strange conclusion of *Mir.* 101, which also breaks from the typical tone of reports and includes a first person reference: τοῦτο μὲν οὖν **ἡμῖν** φαίνεται μυθωδέστερον· ὅμως μέντοι ἔδει μὴ παραλιπεῖν ἀμνημόνευτον αὐτό, τῶν περὶ τὸν τόπον ἐκεῖνον τὴν ἀναγραφὴν ποιούμενον. Flashar (1981) 119, followed by Vanotti (2007) 181–182, attributes this observation to *Mir.*'s source (whom Flashar thinks is Timaeus).

42 For the unidentified source(s), see Flashar (1981) 122–124, who notes the possibility of Müllenhoff's attribution to Lycus or perhaps Theopompus, but rejects the possibility that it derives from Timaeus, as suggested by Geffcken. Cf. Vanotti (2007) 186.

of their presentation, content, and functions within both individual reports and the entire collection.

As a rule, *Paradoxographus Vaticanus* and ps.-Antigonus present mythic *aitia* as simple, unambiguous causes of *paradoxa*: the cicadas of Rhegium are silent because of Heracles' prayer (ps.-Antig. 2), and the Marsyas river reacts differently to flutes and lyres because Marsyas drowned there (*Par. Vat.* 19). The compilers' attention in these and their other mythic reports remains strictly on the stories' causal relationships with the phenomenon. No extraneous details appear, and only a single mythic action or scene receives the spotlight. Even ps.-Antigonus' longest and most narrative entries, 1 on the cicadas and the musical contest of Rhegium and 12 on how the birth of Erichthonius led to the absence of crows on the Athenian acropolis, are confined to the narrative moments that establish the connection between *aition* and *paradoxon*. We do find this restricted focus and unelaborated causal relationship in *Mir.* Yet just as often the compilation's mythic references occur in long entries that amalgamate various combinations of *aitia* linked to a central myth or the phenomenon's location, related miscellanea of mythic prehistory, connections to ancient and contemporary history or cultural practices, and other assorted details. Such reports reach far beyond the immediate context and provide a more expansive view than what we find in *Paradoxographus Vaticanus* or ps.-Antigonus. Gabriella Vanotti stresses *Mir.*'s uniqueness in so blending material indicative of multiple interests, especially myth, history, and ethnography, to the point that an entry's presentation often relegates the *paradoxon* to the background.[43] Indeed, for some entries even identifying the *paradoxon* proves a thorny task.

Mir. also diverges from ps.-Antigonus and *Paradoxographus Vaticanus* in terms of the contexts in which mythic *aitia* occur. The latter almost entirely confine the mythic to *paradoxa* of the natural world, that is, wonders connected to animals, plants, waters, minerals, and other such phenomena upon which human agency has no or negligible influence outside the realm of myth. Roughly half of *Mir.*'s mythic reports are also connected to natural wonders. The remainder, however, turn their attention to human activity. This can come in the form of a mythic *aition* related to a natural phenomenon that instead accounts for a *paradoxon*'s effects on the human world and human culture, as in *Mir.* 51. Here the compiler introduces a peculiar variety of olive and then observes its use in Olympic crowns, a practice etiologically explained as established by Heracles. The hero had nothing to do with the plant's existence or its natural properties, only its integration into Olympic practice.

Even more striking are those reports with mythic material that turn their gaze from the natural world to concentrate upon what we may consider *paradoxa* of history, culture, or human creation. In such reports, myth often moves from an ancillary explanatory position to the foreground as integral to the *paradoxon* itself. As we shall see, *Mir.* operates under a broader conception of what constitutes a "wonder," and my first section below addresses the compiler's incorporation of mythic relics into his working definition of the marvelous. Unlike the "natural"

---

43 Vanotti (2007) 38.

memorials of heroes such as the footprints of Heracles (*Mir.* 97) and the "pebbles" of the Argonauts (*Mir.* 105), *Mir.*'s manufactured relics exist apart from the natural world as products of human (or, in some cases, divine) craftsmanship.

Lastly, *Mir.*'s geographical distribution of mythic material diverges significantly from that of other paradoxographical compilations. Whereas only three of Antigonus' mythic reports and none of *Paradoxographus Vaticanus*' are located in what we may loosely term the "west" – that is, Sicily, the Adriatic, Italy, occidental islands, Iberia, and so on – more than 80% of *Mir.*'s myths have western settings. Of course, *Mir.*'s devotion of a long subsection concentrated largely upon the islands and other territories in the occidental Mediterranean (78–121) explains this discrepancy in part.[44] Nonetheless, the compilation's relative lack of myths for wonders set elsewhere compared to the fact that more than a third of all the reports in *Mir.* 78–121 include mythic material underscores the fundamental significance of Greek myth to *Mir.*'s presentation of western lands. *Mir.*'s west is a very Greek place, mythologically speaking, and in the final section I consider how the compiler's presentation of occidental wonders linked to myths reflects a "Greek" perspective in addressing non-Greek lands and peoples.

2.1 Crafted wonders and cultural relics

Within the larger section devoted to wonders set in the occidental Mediterranean, we find a curious subsection of five sequential reports, *Mir.* 106–110, that form a clear conceptual unit. Not only do they share geographical locations in Italy on the shores of the Adriatic,[45] but also thematic links to Trojan War heroic *nostoi* and southern Italian temples and cults to Hellenic figures. In recent decades, individual reports from this sequence have regularly featured in scholarly discussions in relation to their representations of the activities of the heroic *oikistes* Philoctetes, Epeius, and Diomedes, all famed for founding Greek or, in the case of Diomedes, non-Greek cities in and around Magna Graecia, and for the reports' connections to the archeological remains of Italiote and Italic temples.[46] My interest, however, lies not in the reports as isolated pieces of historical evidence, but in their function as a linked group of paradoxographical reports whose "wonders," at first glance, may not seem all that "wonderful":

106. [1] Ἐν Τάραντι ἐναγίζειν κατά τινας χρόνους φασὶν Ἀτρείδαις καὶ Τυδείδαις καὶ Αἰακίδαις καὶ Λαερτιάδαις, καὶ Ἀγαμεμνονίδαις δὲ χωρὶς θυσίαν ἐπιτελεῖν ἐν ἄλλῃ ἡμέρᾳ ἰδίᾳ, ἐν ᾗ νόμιμον εἶναι ταῖς γυναιξὶ μὴ γεύσασθαι τῶν ἐκείνοις θυομένων. Ἔστι δὲ καὶ Ἀχιλλέως νεὼς παρ᾽ αὐτοῖς. [2] Λέγεται δὲ μετὰ τὸ παραλαβεῖν τοὺς Ταραντίνους Ἡράκλειαν τὸν τόπον

---

44   For the organization and focus of *Mir.* 78–121, see Pajón Leyra (this volume).

45   Flashar (1981) 124–125, who also maintains that they all derive from Timaeus.

46   Though it should be noted that only *Mir.* 107 alludes to settlement or city foundation; the heroic *ktiseis* accounts lie in the background of our reports, but they do not explicitly figure into them. See discussions below for relevant references.

καλεῖσθαι ὃν νῦν κατοικοῦσιν, ἐν δὲ τοῖς ἄνω χρόνοις τῶν Ἰώνων κατεχόντων Πολίειον· ἔτι δὲ ἐκείνων ἔμπροσθεν ὑπὸ τῶν Τρώων τῶν κατασχόντων αὐτὴν †Σίγειον† ὀνομασθῆναι.

106. [1] In Tarentum they say that at certain times people offer sacrifices to the shades of the Atridae, Tydidae, Aeacidae, and Laertiadae, and besides that they celebrate a sacrifice to the Agamemnonidae separately, on another special day, on which it is unlawful for the women to taste the victims offered to those heroes. There is also among them a temple of Achilles. [2] Now it is said that after the Tarentines had taken it, the place which they at present inhabit was called Heraclea, but in earlier times, when the Ionians held it, (it was called) Polieion. Earlier still, before them, when it was held by the Trojans, it was named Sigeum.

107. [1] Παρὰ δὲ τοῖς Συβαρίταις λέγεται Φιλοκτήτην τιμᾶσθαι. κατοικῆσαι γὰρ αὐτὸν ἐκ Τροίας ἀνακομισθέντα τὰ καλούμενα Μάκκαλα τῆς Κροτωνιάτιδος, ἅ φασιν ἀπέχειν ἑκατὸν εἴκοσι σταδίων, καὶ ἀναθεῖναι ἱστοροῦσι τὰ τόξα τὰ Ἡράκλεια αὐτὸν εἰς τὸ τοῦ Ἀπόλλωνος τοῦ Ἀλαίου. Ἐκεῖθεν δέ φασι τοὺς Κροτωνιάτας κατὰ τὴν ἐπικράτειαν ἀναθεῖναι αὐτὰ εἰς τὸ Ἀπολλώνιον τὸ παρ' αὐτοῖς. [2] Λέγεται δὲ καὶ τελευτήσαντα ἐκεῖ κεῖσθαι αὐτὸν παρὰ τὸν ποταμὸν τὸν Σύβαριν, βοηθήσαντα Ῥοδίοις τοῖς μετὰ Τληπολέμου εἰς τοὺς ἐκεῖ τόπους ἀπενεχθεῖσι καὶ μάχην συνάψασι πρὸς τοὺς ἐνοικοῦντας τῶν βαρβάρων ἐκείνην τὴν χώραν.

107. [1] Among the Sybarites Philoctetes is said to be honored; for on his return from Troy he settled in the Crotonian territory in the place called Makkala, which they say is a hundred and twenty stadia away (from Croton); and they relate that he dedicated the bow and arrows of Heracles in the temple of Apollo Alaeus. But from there they say that the Crotonians, during their dominion, took them and dedicated them in the Apollonian temple that is near them. [2] Now it is also said that, when he died, Philoctetes was laid to rest there by the river Sybaris, after he had given help to the Rhodians who, along with Tlepolemus, had been driven off course to those parts, and had engaged in battle with the barbarians who inhabited that area.

108. Περὶ δὲ τῆς Ἰταλίας τὴν καλουμένην †Γαργαρίαν†,[47] ἐγγὺς Μεταποντίου, Ἀθηνᾶς ἱερὸν εἶναί φασιν Εἰλενίας, ἔνθα τὰ τοῦ Ἐπειοῦ λέγουσιν ἀνακεῖσθαι ὄργανα, ἃ εἰς τὸν δούρειον ἵππον ἐποίησεν, ἐκείνη τὴν ἐπωνυμίαν ἐπιθέντος. φανταζομένην γὰρ αὐτῷ τὴν Ἀθηνᾶν κατὰ τὸν ὕπνον ἀξιοῦν ἀναθεῖναι τὰ ὄργανα, καὶ διὰ τοῦτο βραδυτέρας τυγχάνοντα τῆς ἀναγωγῆς εἰλεῖσθαι ἐν τῷ τόπῳ, μὴ δυνάμενον ἐκπλεῦσαι· ὅθεν Εἰλενίας Ἀθηνᾶς τὸ ἱερὸν προσαγορεύεται.

---

47 The reference is certainly to Lagaria, which Lycoph. 930; 946–50 and Strabo 6.26.3 claim Epeius founded. On the text, see Giacomelli (2023) 288–289. For Epeius in Lagaria, see Bérard (1957) 333–337, La Genière (1991), Malkin (1998) 213–214, Genovese (2009) 95–188, who considers archaeological evidence at length, Zachos (2013), who argues that Epeius' role in colonization history was a late development, Hornblower (2015) 348–356, who focuses on Lycophron's presentation.

108. In that part of Italy which is called Gargaria, close to Metapontum, they say that there is a temple of Athena Heilenia, where they state that the tools of Epeius were dedicated, which he made for the construction of the wooden horse; he gave the goddess this title – for (they say that) Athena appeared to him in a dream and desired him to dedicate the tools; and, being delayed in putting out to sea because of this, he was shut up (*heileisthai*) in that place, unable to sail out; hence the temple was called that of Athena Heilenia.

109. [1] Λέγεται περὶ <Λουκερίαν> τὸν ὀνομαζόμενον τῆς Δαυνίας τόπον ἱερὸν εἶναι Ἀθηνᾶς Ἀχαΐας καλούμενον, ἐν ᾧ δὴ πελέκεις χαλκοῦς καὶ ὅπλα τῶν Διομήδους ἑταίρων καὶ αὐτοῦ ἀνακεῖσθαι. [2] Ἐν τούτῳ τῷ τόπῳ φασὶν εἶναι κύνας οἳ τοὺς ἀφικνουμένους τῶν Ἑλλήνων οὐκ ἀδικοῦσιν, ἀλλὰ σαίνουσιν ὥσπερ τοὺς συνηθεστάτους.

109. [1] In the district of Daunia named <Luceria>, there is said to be a temple called that of "Achaean Athena," in which bronze axes and arms of Diomedes and his companions are dedicated. [2] In this place they claim that there are dogs that do no harm to the Greeks who come there, but fawn upon them as though they were most familiar to them.[48]

110. Ἐν δὲ τοῖς Πευκετίοις εἶναί φασιν Ἀρτέμιδος ἱερόν, ἐν ᾧ τὴν διωνομασμένην ἐν ἐκείνοις τοῖς τόποις χαλκῆν ἕλικα ἀνακεῖσθαι λέγουσιν, ἔχουσαν ἐπίγραμμα "Διομήδης Ἀρτέμιδι." Μυθολογεῖται δ' ἐκεῖνον ἐλάφῳ περὶ τὸν τράχηλον περιθεῖναι, τὴν δὲ περιφῦναι, καὶ τοῦτον τὸν τρόπον εὑρισκομένην ὑπὸ Ἀγαθοκλέους ὕστερον τοῦ βασιλέως Σικελιωτῶν εἰς τὸ τοῦ Διὸς ἱερὸν ἀνατεθῆναί φασιν.

110. In the territory of the Peucetians they say there is a temple of Arte-mis, in which, they state, is dedicated the bronze necklace, famous in those parts, bearing the inscription "Diomedes to Artemis." Now the legend relates that he put it around the neck of a stag, and that it remained attached there; and in this way having been discovered later by Agathocles, king of the Sicil-ians, it was, they affirm, dedicated in the temple of Zeus.

Although the reports provide a variety of mytho-historical details, the heroic dedication of storied objects serves as the fundamental theme that defines this subsection, a connection that becomes clear in *Mir.* 108. The first report, whose *paradoxon* revolves around the extensive worship of Trojan War heroes in Taras, only contextualizes the subsection topically within the general dimension of south-ern Italian cults for Greek heroes and temporally within the post-Trojan War period, during which time the western Mediterranean became a key theater for the *nostoi* of Achaeans and the diaspora of Trojans.[49] Next, *Mir.* 107 begins and ends with the

48 *Mir.* 109.3, which relates an ethnographic *paradoxon* connected to Greek myth, is quoted and dis-
    cussed on p. 195.
49 On the cults, see Graf (1982) 161, who views them as artificial and too literary; Malkin (1994) 60
    offers a milder reading of *Mir.* 106 than Graf; Nafissi (1999) 249 advances the compelling theory
    that the cults were introduced in the 4th century as responses to political realities. I count *Mir.* 106
    among those reports that include mythic material due to its reference to the Trojans' foundation of

cultic veneration of Philoctetes himself, but the central section of the report highlights his dedication of Heracles' bow and arrows to a temple either in or around the
Achaean *apoikia* Sybaris.[50] This initiates the theme that persists in the remainder
of the reports. The final three all focus entirely (108 and 110) or initially (109.1)
on offerings of famous mythic objects. Like Philoctetes' bow and arrows, Epeius'
tools (108), used for the construction of the Trojan Horse, were "weapons" instru-
mental to the demise of Troy.[51] The dedications of bronze axes and arms made by
Diomedes and his companions (109.1) do not seem to possess any obvious additional specific mythic significance, though the ὅπλα presumably could include the
armor that Diomedes famously obtained from Glaucus (*Il.* 6.230–326).[52] It should
also be noted *Mir.* does not explicitly link Diomedes' presence in Daunia or his
dedications with the subsequent *paradoxon* of the Greek-loving dogs. Unlike *Mir.*
79, which invokes the death of Diomedes and the metamorphosis of his companions to etiologically explain the similar behavior of birds on the island Diomedeia
(see p. 200), *Mir.* 109 presents the mythic dedication and the canine behavior as
separate anecdotes – each *thaumasios* in its own right – related by their location
within the Leucarian precinct of Athena.[53] In contrast, *Mir.* 110 does feature a zoological curiosity connected with an artifact, an exceptionally long-lived deer in the
precinct of Artemis.[54] Yet the compiler glances over the deer's (implied) extraordinary longevity to focus his attention instead on the otherwise unknown dedicated
necklace – though he assures readers that this too enjoyed some renown, at least
among the locals ("famous in those parts"). In effect, the report relegates to the
background what we would typically expect to be the *paradoxon* so as to concentrate upon the relic itself.

This subsection thus presents four sequential reports linked by their focus upon
nonnatural or "crafted" items with mytho-historical significance offered by heroes
as votives at southern Italian locations. Turning to the rest of the compilation, we
find that *Mir.*'s characterization of relics and crafted items as appropriate subjects
for paradoxographical reports is not an isolated incident. An extravagantly embroi-
dered *himation* dedicated at the Lacinium of Hera by one Alcisthene the Sybarite

---

Sigeum. On Sigeum's relationship to/conflation with Trojan Siris and its mythological history, see
Moscati Castelnuovo (1989) *passim*, esp. 19–27 and 47–55, Malkin (1998) 226–231, Hall (2000),
and Vanotti (2007) 187–189.

50 The location is disputed; see n. 76.

51 Malkin (1998) 214–215.

52 In this scene Homer uses τεύχεα, a rare term in later prose, though it and ὅπλα operate as synonyms
in the *Iliad*; cf. for example, *Il.* 18.614 of Achilles' new arms: Αὐτὰρ ἐπεὶ πάνθ' ὅπλα κάμε κλυτὸς
ἀμφιγυήεις.

53 Though other versions of the tale likely linked the two, on which see Schorn's suggestion that the
Daunian dogs offered a competing version of the Greeks-to-animals tale told in *Mir.* 79, this volume, p. 40–41. For further on *Mir.*'s Hellenophilic and xenophobic dogs, see p. 199–200.

54 On issues related to the location – Campania seems implied, but Apulia makes better sense historically and in terms of the section's itinerary – see Vanotti (2007), who also addresses the theory that
Brindisi is meant because of its derivation from the Messapian term for "deer," and Giacomelli
(2023) 291–292, who also observes, following Beckmann, the tradition of dedicating collars to deer,
as seen in Paus. 8.10.10, where the deer's longevity is also remarked upon.

receives the compiler's full attention in *Mir*. 96,[55] and *Mir*. 155 touts the "imperceptible craftsmanship" (τινος ἀφανοῦς δημιουργίας) displayed by Phidias in his creation of the Acropolis' Athena statue. Elsewhere, *Mir*. observes Heracles' offering of statues made from "mountain-copper" in a digression to a natural wonder (58), while dedications from mythic figures also occur as the principal subjects of entries set in Greece proper (116 and 118, below).[56] A striking example appears in *Mir*. 81. This long report on natural wonders located around the so-called Eridanus river and the Electrides islands has been involved in a number of scholarly discussions due to its representations of the river's geography, a nearby mephitic lake's links to the Phaethon story, and the islands' reputed connection to the ancient amber trade.[57] Yet all these points of interest occur within an entry framed by Daedalus' dedication of statues. So it begins and concludes:

Mir. 81. [1] Ἐν ταῖς Ἠλεκτρίσι νήσοις, αἳ κεῖνται ἐν τῷ μυχῷ τοῦ Ἀδρίου, φασὶν εἶναι δύο ἀνδριάντας ἀνακειμένους, τὸν μὲν κασσιτέρινον τὸν δὲ χαλκοῦν, εἰργασμένους τὸν ἀρχαῖον τρόπον. Λέγεται δὲ τούτους Δαιδάλου εἶναι ἔργα, ὑπόμνημα τῶν πάλαι, ὅτε Μίνω φεύγων ἐκ Σικελίας καὶ Κρήτης εἰς τούτους τοὺς τόπους παρέβαλε . . . [6] εἰς ταύτας οὖν τὰς νήσους Δαίδαλόν φασιν ἐλθεῖν, καὶ κατασχόντα αὐτὰς ἀναθεῖναι ἐν μιᾷ αὐτῶν τὴν αὐτοῦ εἰκόνα, καὶ τὴν τοῦ υἱοῦ Ἰκάρου ἐν τῇ ἑτέρᾳ. Ὕστερον δ' ἐπιπλευσάντων ἐπ' αὐτοὺς Πελασγῶν τῶν ἐκπεσόντων ἐξ Ἄργους φυγεῖν τὸν Δαίδαλον, καὶ ἀφικέσθαι εἰς Ἴκαρον τὴν νῆσον.

[1] In the Electrides Islands, which are situated in the corner of the Adriatic, they say that two statues were erected, the one of tin, the other bronze, wrought in the ancient manner. It is said that these are the work of Daedalus, a memorial of ancient times, when he, fleeing Minos from Sicily and Crete, arrived at these places . . . [*reports of other paradoxa in the area*]. . . . [6] To these islands, therefore, they say that Daedalus came, and, having obtained possession of them, he dedicated on one of them a statute of himself, and on the other that of his son Icarus; but that afterwards, when the Pelasgians, who had been expelled from Argos, sailed against them, Daedalus fled, and arrived at the island of Icarus.

It is tempting to suspect the compiler guilty of significant understatement in his presentation of Daedalus' statues. After all, a robust tradition about the inventor's remarkably life-like – or even truly "living" – works thrived for centuries, to the point that his living statues became popular as plot points on the Attic stage.[58] Such

---

55   On this report, the dedicant's name, and the description of the *himation*, see Giacomelli (2023) 272–275.

56   Given the fundamental memorializing function of all such dedications, we may also note the compiler's broader interest in natural world *paradoxa* as vehicles for commemoration, for example, the "memorials" (μνημόσυνα) of Heracles imprinted on the topography in Iapygia (97) and the "Pebbles" (described as μνημεῖα) of the Argonauts (105).

57   Vanotti (2007) 169–170 provides ample citations to ancient and modern discussions of the area.

58   Frontisi Ducroux (1975) 95–116, Morris (1992) *passim*, Pugliara (2003) 176–207, Mayor (2018) 85–96.

marvelous statues, which could even perform their "own work when ordered or by seeing what to do in advance" (κελευσθὲν ἢ προαισθανόμενον ἀποτελεῖν τὸ αὑτοῦ ἔργον, Arist. *Pol.* 1253b34–35), would neatly align with the sort of wonder cultivated by paradoxographers. However, the compiler's only description of these particular works beyond their materials, that they were "wrought in the ancient manner" (εἰργασμένους τὸν ἀρχαῖον τρόπον), coincides more with Pausanias' sober evaluations of the ancient simplicity of the many statues ascribed to Daedalus that had been dedicated in Greek temples throughout the *oikoumene*. In his description of a Corinthian *xoanon* of Heracles said to be by Daedalus, for example, Pausanias notes that "all the works of this artist, although rather uncouth to look at, are nevertheless distinguished by a kind of inspiration" (Δαίδαλος δὲ ὁπόσα εἰργάσατο, ἀτοπώτερα μέν ἐστιν ἐς τὴν ὄψιν, ἐπιπρέπει δὲ ὅμως τι καὶ ἔνθεον τούτοις, 2.4.5).[59] The compiler of *Mir.*, like Pausanias, betrays no interest in the fantastic popular traditions of Daedalic statuary. Instead, the wonder in this part of the entry lies in the existence and memory of the legendary artisan's crafted dedications, and their surprising (and otherwise unattested) location in the Adriatic.

In *Mir.* 81, readers "discover" two statues in an unexpected location crafted by a marquee name in Greek myth, just as they encounter the Italian afterlives, so to speak, of famous Trojan War relics in 107–109. In *Mir.* 116 and 118, the compiler shines the spotlight on the very process of discovering ancient mythic relics itself. When the Athenians were constructing the temple of Demeter at Eleusis, *Mir.* 116 recounts, they unearthed a stele made from bronze, thus suggesting its great antiquity. The memorializing verse inscription engraved upon it proclaimed that the stele marked the tomb of one Deiope. This prompted a dispute – unresolved, in *Mir.*'s presentation – regarding Deiope's identity as it relates to mythic Attic figures, with some identifying her as the wife of Musaeus, others as the mother of Triptolemus. The wonder here seems rooted in the unexpected discovery of an unusual ancient relic whose mytho-historical significance is clearly felt, though left uncertain. Much the same occurs in *Mir.* 118, except that in this example the mystery is solved. The Aenianes of Hypate once discovered an ancient stele (παλαιά τις στήλη), *Mir.* narrates, whose verse inscription in archaic characters (ἀρχαίοις γράμμασιν) they were unable to decipher. They set out for Athens, hoping to find answers there, only to be diverted to the Theban temple of Ismenian Apollo where, they were told, could be found votives with similar letters that had been translated. Here they learned that the inscription had been made by Heracles to memorialize his dedication of a grove made during his quest to obtain the cattle of Geryon.[60] Throughout the report, *Mir.*'s narrative replicates for readers the mystery which the strange discovery presented to the Aenianes, only disclosing its solution at the

---

59 All translations of Pausanias by Jones (1919 and 1935). Although ἀτοπώτερα could mean "strange" or even "paradoxical," Jones' translation aligns with Pausanias' other descriptions of Daedalic statuary, especially at 9.39–40, on which see Arafat (1996) 67–71.

60 The epigram is almost certainly a forgery whose artificial and often incoherent language, in Giacomelli's estimation, confers upon it a "patina of antiquity." For my purposes, only the compiler's presentation of the epigram as an ancient relic of Heracles matters, not the veracity of his claim. Giacomelli (2023) 302–309 provides a thorough commentary on the piece and synthesize the considerable bibliography regarding it.

conclusion. As with *Mir*. 116, the anecdote's appeal lies in an unexpected confrontation with an inexplicable relic of the past, while the resolution ultimately reveals that the Aenianes were in possession of a true treasure linked to the exploits of a hero. The interest generated by this anecdote may be compounded by the setting at the Theban Ismenium, which boasted a trio of bronze tripods surely meant by *Mir*.'s reference to votives with similar letters.[61] These dedications gained wider exposure thanks to Herodotus, whose fascination was piqued by their inscriptions in ancient "Cadmean letters" (Καδμήια γράμματα), which, once translated, purported the tripods to be dedications made by mythic figures (Amphitryon, Scaeus, and Laodamas).[62] Accordingly, we may see *Mir*.'s report as engaging with two more well-known traditions – Heracles' tenth Labor and the apparent reputation of the Ismenian temple translators – to tell an otherwise "new" or unfamiliar tale.

In total, 8 of the 18 chapters with mythic material in *Mir*. concentrate upon or include votives, and an additional 2 chapters recount historical items. To this number, we may add another (much larger) wonder of human craftsmanship, the famous nuraghes of Sardinia attributed by *Mir*. to Heracles' companion Iolaus (*Mir*. 100, p. 196–198) which are described in terms of their ancient style, using very similar language (Ἑλληνικὸν τρόπον διακείμενα τὸν ἀρχαῖον) to the description of Daedalus' statues in *Mir*. 81 (εἰργασμένους τὸν ἀρχαῖον τρόπον).[63] As a group, these anecdotes testify that, to the compiler, crafted objects and votives with mytho-historical significance can occupy a valid place under the broad conceptual umbrella of "the wondrous."[64] In this, the compiler deviates from what we see in other paradoxographical compilations. Typically, as Delattre has observed, "the paradoxographer points out elements of the world that belong to *phusis* ['what nature offers to us'] but are contrary to *doxa* ['what we know of nature']."[65] Yet despite the crafted wonders and relics of *Mir*. having no foundations in *phusis*, the compiler nonetheless understands them as fit subjects for paradoxographical report. No other surviving paradoxographical compilation offers comparable examples. The few dedications or similar historical objects observed by others all occur as ancillary additions to

---

61   Flashar (1981) 138 and Vanotti (2007) 203 both highlight the link. The report's likely source, Timaeus, would certainly have been aware of the connection between this anecdote and that of Herodotus.

62   Hdt. 5.59–61. The anecdote appears as support for Herodotus' claim for the descent of the Greek language from Phoenician. Scholarship pivots principally on the questionable authenticity of the inscriptions and Herodotus' credulousness, on which see Schachter (1981) 82 n. 2, West (1985) 289–295, Powell (1991) 6, Raubitschek (1991) 256, Nenci (1998) 579–589, Hornblower (2015) 179–180, Harris (2018) 90–91. On the Boeotian cultural values entwined with tripod dedication and their relationship to these examples (forgeries or not), see Papalexandrou (2008).

63   My thanks to Stefan Schorn for highlighting the repeated phrasing in *Mir*. 81 and 100 and its reflection of the compiler's interests. Two other statues are included in the compilation's appendix, though both are said to behave strangely. The Argive statue of Bitys killed the man responsible for Bitys' death in *Mir*. 156, and the bull set upon the altar of Artemis Orthosia bellows when hunters enter the temple in *Mir*. 175; such obviously paradoxical behavior makes these marvels much more in line with what we typically find in paradoxographies.

64   Hunzinger (1995) on Herodotean *thauma* stresses the critical component of individual response to concepts of wonder, as what is wondrous can be anything that causes wonder in that individual.

65   Delattre (2018) 208.

paradoxa drawn from the natural world. So ps.-Antigonus mentions a dedicated bronze cart that inhabitants of Crannon shake when they are desperate for rain in the course of a larger report on the city's peculiar possession of only two ravens (ps.-Antig. 15). Similarly, the single preserved fragment of the paradoxographer Philon of Heracleia recounts Alexander's offering of a Scythian donkey horn of the sort which could, according to the popular claim, hold the water of the Styx.[66] Only *Mir.* includes entries on objects that have no connections to natural phenomena.

Alone among the surviving paradoxographers, then, the compiler of *Mir.* operates under a conception of the marvelous that includes wonders of human craftsmanship and the commemorative relics of both mythic heroes and historical figures. Outside paradoxography proper, however, this perspective resonates with that of others to varying degrees. Several of Posidippus' epigrams on marvelous stones, for example, praise masterful gem carvings as *paradoxa*,[67] and heroic relics and dedications like the Cyzicene anchor (F 108–109a Harder) are commemorated in Callimachus' *Aetia.*[68] Particularly pertinent is Herodotus who, like *Mir.*'s compiler, not only had a special interest in dedications but also included wonders of craftsmanship and human *erga* in his understanding of *thauma*. Herodotus' crafted wonders, however, earn their place among the marvelous because of the ingenuity and quality in their design (e.g., the dikes of Nitocris at 1.185 and the corselet of Amasis at 3.47) or because of their monumental scale or arresting appearance (e.g., the labyrinth near Hawara at 2.148 and the shrine of Leto at Buto at 2.155).[69] While *Mir.*'s reports on Alcisthenes' spectacular *himation* (96), Phidias' "secret craftsmanship" (155), and the Sardinian nuraghes (100) offer fine parallels to both categories in this dimension of Herodotean *thauma*, Herodotus does not characterize dedications or relics as wonders because of their historical value, but only if the grandness of their scale merits it (e.g., the dedications of Croesus).[70] Such objects as the ancient tripods with "Cadmean letters" may be interesting, provide the impetus for narratives, or serve as noteworthy pieces of tangible evidence for historical claims, but they are never explicitly "marvelous."[71] Moreover, but for a few exceptions like the Boeotian tripods from mythic dedicators, Herodotus focuses his attention upon votives and relics that date from historical times and mark more recent occasions rather than items that boast an ancient mythic pedigree.

A better analogue to this aspect of *Mir.*'s conception of wonder is Pausanias. Although he, like Herodotus, rarely applies the specific terminology of wonder to

---

66 *FGrHist* 1676 F 1, with the commentary of Belousov (2022).
67 On which see Krevans (2005) 88–92, Guichard (2006) 121–133, Priestley (2014) 99–104.
68 See Acosta-Hughes and Stephens (2012) 17 on the connections between Callimachus' paradoxographic work and his *Aetia*.
69 Hunzinger (1995) 47 n. 7. See Pajón Leyra (2011a) esp. 175–182 for a full discussion of Herodotean *erga* and wonder.
70 See Priestley (2014) 56–61 and 71 for discussion of wonders in Herodotus unmarked by the typical terminology of wonder but whose wondrous qualities are often intimated through descriptions of size or scale.
71 For the narrative functions and representations of dedications in Herodotus, see Harris (2018) 84–93.

dedications or ancient objects,[72] his consistent concentration on such items directly participates in the central program of the *Periegesis*, the description of the sights (ἴδοι) and wonders (θαύματος) of Greece worthy of being remembered.[73] Moreover, in contrast to Herodotus' inclination toward objects with more recent historical relevance, Pausanias' clear, if unstated, preference for that which is "ancient" manifests in the attention often paid to relics, temples, and assorted cultural detritus ascribed to the mytho-historical past,[74] much like we see in *Mir.*, whose votive wonders are nearly all similarly ancient. At times, the compiler of *Mir.* and Pausanias even adopt comparable approaches to describing such relics. To take just one example, we may compare *Mir.* 107–110 with Pausanias' introduction of the necklace of Eriphyle, an artifact first described in the *Odyssey* (11.327):

> In Cyprus is a city Amathus, in which is an old sanctuary of Adonis and Aphrodite. Here they say is dedicated a necklace of Harmonia, but called the necklace of Eriphyle, because it was the bribe she took to betray her husband. It was dedicated at Delphi by the sons of Phegeus (how they got it I have already related in my story of Arcadia),[75] but it was carried off by the tyrants of Phocis. (9.41.2)

The description begins with the general geographical setting and the relevant temple and then moves on to introducing the relic and its dedication, providing just enough detail to contextualize the object within its mythic sequence, much as in *Mir.* 107–110. From there, Pausanias briefly observes its subsequent movement out of the sanctuary as the result of the Phocian seizure of Delphi during the Third Sacred War, resulting in its eventual re-dedication in Amathus. Pausanias' tendency to provide abbreviated details of post-dedication provenance, which often comprehend major historical events as the background to a relic's movements, is matched in *Mir.* In *Mir.* 107, while describing the Crotonians' transfer of Philoctetes' arrows to their own temple of Apollo,[76] the compiler alludes to a major event – Croton's

---

72 Pausanias tends to dub a dedication or crafted item a θαῦμα only if it exhibits or inspires some strange quality or behavior, that is, a definition in keeping with typical paradoxographical usage. See, for example, the small statues of the Dioscuri on Pellana which cannot be moved by the sea (3.26.3) described in these terms, and also the altar at Olympia from which kites will not take food (5.14.1).

73 Paus. 5.10.1: πολλὰ μὲν δὴ καὶ ἄλλα ἴδοι τις ἂν ἐν Ἕλλησι, τὰ δὲ καὶ ἀκοῦσαι θαύματος ἄξια, with the often-repeated phrase "the things worth seeing" (θέας ἄξια). See Langerwerf (2018). Cf. Porter (2001) for the relationships of wonder, the (fragmented) past, and the sublime in Pausanias; see Delattre (2018) on Pausanias, paradoxography, and aquatic *mirabilia*.

74 Bowie (1974) 188, Habicht (1985) 131–140, Arafat (1996) *passim*, esp. 43–45.

75 Paus. 8.24.10.

76 The sequence of possession in this passage has been a matter of dispute, with the various positions summarized by Malkin (1998) 216–221 and Vanotti (2007) 190–191. Here I generally follow the interpretations of Mele (1984) 36 and Giangiulio (1991), which have Croton seizing dedications from Sybaris, as this seems to best reflect how the compiler's readers would interpret this passage. Ultimately, however, Malkin's view that Philoctetes' worship was originally centered in the "hinterlands" between Croton and Sybaris and that Croton transferred the dedications from this area to

stunning defeat of Sybaris in 510/509 and its aftermath – with the vague κατὰ τὴν ἐπικράτειαν ("during their dominion").[77] Likewise, Agathocles of Sicily's alliance with the Peucetians (295/294), wherein he essentially set them up as pirates to help cement his hold on movement in the Mediterranean,[78] likely lies behind the observation in *Mir.* 110 that Agathocles moved the necklace dedicated to Artemis to a temple of Zeus, presumably on Sicily.[79] Pausanias and the compiler of *Mir.* thus seem to share not only a similar view of ancient relics as being worthy of note, but also a similar feeling as to what information should be included in the descriptions of such items. Of course, if one were to continue reading Pausanias' discussion of Eriphyle's necklace, the similarities wane and the differences between the perigete and the compiler become pronounced. After introducing the necklace, Pausanias launches into a learned authorial discourse aimed at disproving the authenticity of the Cypriotes' claim – theirs is not the *real* necklace of Eriphyle. Such skepticism is a regular feature of the *Periegesis*, where we often find that Pausanias is willing to marvel, but that his wonder must be earned, an attitude that contrasts sharply with the credulity cultivated by *Mir.* and other paradoxographers.[80] Nonetheless, the intersection of Pausanias' and *Mir.*'s interests and basic approaches to describing relics helps us contextualize *Mir.*'s seeming idiosyncratic conception of wonder within a larger tradition beyond the paradoxographical.

2.2 *Myth, Greeks, and the "West" in* **Mir. 78–121**

While the compiler's fascination with crafted wonders and relics stands out as one of his more anomalous qualities, these items comprise just one aspect of his broader interest in wonders linked, etiologically or otherwise, to various mythic traditions associated with occidental lands. As I observed earlier, more than 35% of reports in 78–121 include at least one, sometimes more than one, reference to mythic figures or sequences, resulting in a composite portrait of the "wonderful west" heavily defined

---

their own temple rather than from Sybaris itself seems the most compelling explanation for what may have happened historically.

77 On the various sources observing this defeat, often in the context of Sybaris' infamous *truphe*, and on the creation of a popular tradition of Sybaris' demise as justified punishment, see Ampolo (1993) and del Corno (1993).

78 As recorded by Diod. Sic. 21.4; cf. Plut. *Pyr.* 9.2–3. Cf. Vanotti (2007) 195, with further bibliography. De Lisle (2021) 229–256 offers a helpful updated timeline and discussion of Agathocles' activities in Italy, especially his attempts to gain control of movement and access. The reasoning behind De Lisle's claim at 241 n. 49 that *Mir.* 110 "is set before he became king, probably during his time in Taras as an exile" seems to be based on reading ὕστερον as referring to the time when he assumed the title. On the ambiguity of the adverb, see Giacomelli (2023) 292.

79 Cf. also *Mir.* 98, where Cleonymus the Spartan's sale of Alcisthene's *himation* to the Carthaginians hints not only at Cleonymus' campaigns against southern Italian cities in the early 4th century, but also at his dubious honor of being the first known person to use funds raised from looting temples to support mercenary armies. See Vanotti (2007) 178–179 for further details on the timeline and bibliography; on Cleonymus' use of looted sanctuaries' treasures, see Diod. Sic. 15.14.3, with Miles (2014) 133–34.

80 See Langerwerf (2018) on Pausanias as an arbiter of wonders. For Pausanias' approach to debunking the Cypriots' claim that they house the famous necklace, see Duffy (2013).

relative to the Greek mythic past. In general, *Mir.*'s mythic material traces conventional and expected trajectories for Greek myth set in the west, patterns that had been long established by archaic and classical authors which then flourished even more under Hellenistic-era historians who were invested in comprehending the west within the larger Greek mytho-historical continuum.[81] Accordingly, we see the voyaging Heracles fighting Giants in his capacity as the great "civilizer" who liberates native peoples from disorder and fear (97 and 98).[82] So also do other traveling heroes – the Argonauts (105), Daedalus (81), and Diomedes (78, 109, and 110) – imprint their influence wherever they journey, be it through votive offerings, the foundations of cities and temples, or through changes to the landscape itself. The traces they leave become, in the hands of *Mir.*'s compiler, the stuff of paradoxography. Other reports highlight either the "Greekness" of the west or emphasize its foreignness from a typical Hellenic perspective. Thus, on the one hand, *Mir.* 82 repeats the well-known transposition of the site of Persephone's abduction to Sicily, essentially naturalizing Greek myth in a western setting, while, on the other hand, the ethnographic *paradoxon* of *Mir.* 103 lies in the fact that the Italic people of Seirenusae elevate the Sirens – mythic monsters! – to objects of formalized cult veneration.[83]

That the compiler considered the Italic worship of Sirens strange enough to merit inclusion in the collection illustrates his adoption of an attitude that assumes a Greek perspective – however difficult that can be to define – as the normative position. Indeed, the mythic reports in 78–121 generally look at the wonders of the west and the myths attached to them through Greek eyes. This is especially apparent in the reports threaded throughout the section that dwell on the various interactions of Greeks with non-Greeks, moments whose effects and influence often reverberate into contemporary history or, as he represents it, the compiler's own day. Such reports, examples of which I shall discuss in this section, often comprehend present realities as linked to or resulting from mythic history, or in their telling reflect attitudes about ethnic identity and non-Greeks. Here it must be kept in mind that we cannot be sure how much of the compiler's phrasing and presentation was drawn from his sources and how much is the product of his own hand. Nevertheless, just as the compiler chose what wonders to report on and when to retain mythic material from his sources, so too did he elect to replicate whatever of his presentation also derives from those sources.

Two reports in the section proclaim the impact that Hellenic influences could have upon other peoples by recording ethnographic *paradoxa* concerned with fundamental cultural practices that are then explained through Greek mythic *aitia*. The

81 The bibliography on this subject is considerable; cf., for example, Bérard (1957), Lacroix (1974), Pearson (1975), Lepore (1980), Pearson (1987) esp. 57–90 on Timaeus, Dougherty (1993), Malkin (1994), de Polignac (1997). Malkin (1998), Lamboley (2006), and Briquel (2015), which provides a helpful overview of the various types of tales popularly used of the west.
82 Lacroix (1974).
83 Cf. Strabo 5.4.8 and 1.2.12, and compare Lycoph. 712–737, which considers local worship of the Sirens' tomb sites, with Hornblower (2015) 292–301; cf. Edlund (1987) 45–47. On Italic worship of Sirens at Seirenusae and Naples, see Pais (1905), Breglia Pulci Doria (1987) 65–98, Musti (1999) 33–46, and Bettini and Spina (2007).

third part of *Mir.* 109 is, like the others in that report, set in Daunia, the Apulian region along the coast of the Adriatic populated principally by non-Greek peoples:

109. [3] Πάντες δὲ οἱ Δαύνιοι καὶ οἱ πλησιόχωροι αὐτοῖς μελανειμονοῦσι, καὶ ἄνδρες καὶ γυναῖκες, διὰ ταύτην, ὡς ἔοικε, τὴν αἰτίαν. Τὰς γὰρ Τρῳάδας τὰς ληφθείσας αἰχμαλώτους καὶ εἰς ἐκείνους τοὺς τόπους ἀφικομένας, εὐλαβηθείσας μὴ πικρᾶς δουλείας τύχωσιν ὑπὸ τῶν ἐν ταῖς πατρίσι προϋπαρχουσῶν τοῖς Ἀχαιοῖς γυναικῶν, λέγεται τὰς ναῦς αὐτῶν ἐμπρῆσαι, ἵν' ἅμα μὲν τὴν προςδοκωμένην δουλείαν ἐκφύγωσιν, ἅμα δ' ὅπως μετ' ἐκείνων μένειν ἀναγκασθέντων συναρμοσθεῖσαι κατάσχωσιν αὐτοὺς ἄνδρας. Πάνυ δὲ καὶ τῷ ποιητῇ καλῶς πέφρασται περὶ αὐτῶν· ἑλκεσιπέπλους γὰρ καὶ βαθυκόλπους κἀκείνας, ὡς ἔοικε, ἰδεῖν ἔστιν.

109. [3] Now all the Daunians and the people neighboring them, both men and women, wear black garments, apparently for the following reason – because it is said that the Trojan women, who had been taken as captives, and had come to those parts, fearing that they might experience hard slavery at the hands of the women who already belonged to the Achaeans in their native land, set fire to their ships, in order that they might escape from the expected slavery, and at the same time, that they, being united in marriage with those men, now compelled to stay, might keep the men as husbands. The poet has also very admirably described them; for one may see those women likewise, it seems, "with trailing robes" and "deep-breasted."

Mir. identifies the story of Trojan women burning the ships of their Achaean captors, thus forcing them to stay in Italy as their husbands, as the *aition* (διὰ ταύτην . . . τὴν αἰτίαν) for the claim that Daunians wear black.[84] As presented, the best explanation for the link between the actions of the Trojan women and the local people's adoption of black – that is, mourning – dress is that it reflects the Greeks' sadness at having to permanently remain in Daunia.[85] Lycophron, however, also seems to characterize the ship burning tale as an *aition* for a Daunian ritual observed by girls who, like Cassandra, wished to remain unmarried.[86] This may in turn be a Greek interpretation of a local pre-marriage initiatory rite for prospective brides, as Mari suggests, though by the time it gets to *Mir.*, the details have been thoroughly garbled into "all Daunians wear black, (presumably) all the time."[87] Whatever the original, intended relationship between *aition* and

---

84 The motif of Trojan women burning the ships was popular in *nostoi* set in Italy, for which see Malkin (1998) 197–98, Fowler (2013) 567 n. 150, Hornblower (2015) 387–388, Scheer (2018) 131–132.

85 So Schorn, this volume, p. 39; cf. Geffcken (1892) 22. Hornblower (2015) 362 observes the Trojans' similar attitude of nostalgic mourning upon their arrival at Egesta at Lycoph. 968–969.

86 Lycophron's description of the women dressing like Erinyes (1137) derives from that of Timaeus, whose observation is quoted by the scholiast (= Timaeus, *FGrHist* 566 F 55). See Champion (*BNJ* 566) *ad loc* and Barron (2013) 223 on the Timaeus passage. For Lycophron's treatment of the story, see Hornblower (2015) 402–403.

87 Mari (2009) 423–424. While some have supposed Timaeus to be *Mir.*'s source as well, Schorn's study in this volume, pp. 39–41, offers compelling reasons against this.

paradoxon may have been, *Mir.*'s version of it emphasizes the perceived potency of Greek mytho-history in shaping the daily life of contemporary Daunians. We see much the same in the second half of *Mir.* 88, where the compiler repeats a claim traced to Timaeus that the Iberian inhabitants of the Baleares were forbidden from possessing gold or silver. The alleged reason for the prohibition lay in the distant past: "Heracles made his military expedition against Iberia because of the wealth of the inhabitants" (τὴν στρατείαν Ἡρακλῆς ἐποιήσατο ἐπὶ τὴν Ἰβηρίαν διὰ τοὺς τῶν ἐνοικούντων πλούτους). In this version of the anecdote, as well as Diodorus Siculus' treatment of it (5.17),[88] Heracles' invasion made the islanders so averse to suffering a repeat performance that they transformed the fundamental economic basis of their society in order to make their territory a less attractive target. As a result, *Mir.* observes, the contemporary Balearic mercenaries of the Carthaginians fritter away all their pay on women, since they may not bring it home.

Both mythic *aitia* amplify the reach of Greek impact to absurd extremes, with daily Daunian dress and the entire Balearic economy being determined by ancient mytho-historical encounters. Clearly, these are very much Greek estimations of the culturally-defining power of their influence. Elsewhere in the compilation, however, *Mir.*'s mythic material more often manifests intercultural conflicts with occidental populations and instead expresses early – and not so early – Greek anxieties about the precariousness of their positions abroad. Two of the three reports in the compilation that feature explicit references to the foundations of western *apoikiai*, *Mir.* 100 and 107,[89] both underscore the vulnerability of Greeks or Greek spaces in otherwise alien lands.

In *Mir.* 100's descriptions of the wonders of Sardinia, the mythic colonial past meets more contemporary political realities[90]:

[1] Ἐν τῇ Σαρδοῖ τῇ νήσῳ κατασκευάσματά φασιν εἶναι εἰς τὸν Ἑλληνικὸν τρόπον διακείμενα τὸν ἀρχαῖον ἄλλα τε πολλὰ καὶ καλὰ καὶ θόλους περισσοῖς τοῖς ῥυθμοῖς κατειργασμένους· τούτους δὲ ὑπὸ Ἰολάου τοῦ Ἰφικλέους κατασκευασθῆναι, ὅτε τοὺς Θεσπιάδας, τοὺς ἐξ Ἡρακλέους, παραλαβὼν ἔπλευσεν εἰς ἐκείνους τοὺς τόπους ἐποικήσων, ὡς κατὰ συγγένειαν αὐτῷ τὴν Ἡρακλέους προσήκουσαν, διὰ τὸ πάσης τῆς πρὸς ἑσπέραν κύριον Ἡρακλέα γενέσθαι. Αὕτη δὲ ἡ νῆσος, ὡς ἔοικεν, ἐκαλεῖτο μὲν πρότερον Ἰχνοῦσσα, διὰ τὸ ἐσχηματίσθαι τῇ περιμέτρῳ ὁμοιότατα ἀνθρώπων ἴχνει. [2] Εὐδαίμων δὲ καὶ πάμφορος ἔμπροσθεν λέγεται εἶναι· τὸν γὰρ Ἀρισταῖον, ὅν φασι γεωργικώτατον εἶναι ἐπὶ τῶν ἀρχαίων, τοῦτον αὐτῆς ἄρξαι μυθολογοῦσιν, ὑπὸ μεγάλων ὀρνέων ἔμπροσθεν καὶ πολλῶν κατεχομένης. Νῦν μὲν οὖν οὐκέτι φέρει τοιοῦτον οὐδέν, διὰ τὸ κυριευθεῖσαν ὑπὸ Καρχηδονίων ἐκκοπῆναι πάντας τοὺς χρησίμους εἰς προσφορὰν καρπούς, καὶ θάνατον τὴν ζημίαν τοῖς ἐγχωρίοις τετάχθαι, ἐάν τις τῶν τοιούτων τι ἀναφυτεύῃ.

---

88 See Schorn, this volume, pp. 59–63, for a comparison of the accounts of *Mir.* and Diodorus and their relationship to Timaeus.

89 The third is *Mir.* 106 on Tarentine Heraclea.

90 The following discussion relies heavily (and gratefully) on that of Schorn (this volume, pp. 54–59).

[1] On the island of Sardinia they say there are many beautiful build-ings constructed in the ancient Greek style, and, among them, domes built in remarkable proportions. It is said that these were built by Iolaus, son of Iphi-cles, when he, having taken with him the Thespiadae, the sons of Heracles, sailed to these parts with the intention of making settlements there, consid-ering that they belonged to him through his kinship with Heracles, because Heracles was lord of everything to the west. This island, as it appears, was formerly called Ichnussa, because it was shaped in its outline very simi-larly to a human footprint. [2] It is said to have been previously fertile and productive; for the legend states that Aristaeus, whom they assert to have been the most skillful in agriculture among the ancients, ruled over these parts, which were formerly occupied by many large birds. At present, how-ever, it no longer produces crops like it used to, because when it was ruled by the Carthaginians it had all its trees that were useful for producing fruit destroyed, and death was fixed as the penalty for the inhabitants if any one should plant again anything of the kind.

Mir.'s report traces three different phases in the occupation of Sardinia, each with its own "wonder." Chronologically, the arrival of the civilizing deity/hero Aristaeus comes first, who, according to *Mir.*, found the island unpopulated but for birds. This initial phase witnessed Aristaeus' application of his agricultural expertise (certainly a sort of human *technē*) to cause the island's great fecundity.[91] The immigration of Iolaus and the Thespiadae followed, during which time *Mir.* claims that Iolaus constructed the island's famous nuraghes, an example of crafted wonders on a much larger scale than what we find elsewhere in the compilation.[92] Both sequences emphasize the instrumentality of Greek hands in producing Sar-dinia's most noteworthy qualities, effectively erasing, at least in the case of the nuraghes, the achievements of those non-Greeks already occupying the island whose existence *Mir.* never mentions. With the hostile occupation of the Carthag-inians in the third phase of *Mir.*'s chronology, these (purported) testaments to Hel-lenic skills in agriculture and architecture either fall into the hands of others or are erased by them. The ultimate *paradoxon* of the report, as Schorn notes, "consists in the fact that Greeks seem to have lived on the island in early times while it is now inhabited by barbarians," though I would add that the elimination of what once had been remarkable – the fecundity caused by Aristaeus – also contributes to the wonder.

As *Mir.* presents it, the anecdote summarizes the rise and fall of an early *Greek* territory, with its barbarization implied as a recent consequence of Carthaginian dominance. In contrast, Diodorus maintains in one of his reports on Iolaus' colo-nization of Sardinia that "the people of the colony in the long course of time came

---

91 For the myth and worship of Aristaeus on Sardinia, see Pianu (2004) and Sanna (2004); for Dio-dorus' version of Aristaeus' Sardinian excursion, see Bernardini (2004) 56–58.

92 On Iolaus' expedition, see Mastino (2004) and Bernardini (2004). For the nuraghes and Nuragic Sardinia, see most recently Webster (2015).

to be barbarized, since the barbarians who took part in the colony outnumbered them [the Greeks]" (οἱ μὲν γὰρ λαοὶ διὰ τὸ πλῆθος τοῦ χρόνου, πλειόνων τῶν βαρβάρων ὄντων τῶν μετεσχηκότων τῆς ἀποικίας, ἐξεβαρβαρώθησαν).[93] Here, Sardinian settlements had never been exclusively Greek. Gradual acculturation with non-Greeks, figured by Diodorus in terms of degeneration, had been a factor in Sardinian life since the days of Iolaus. We cannot know if *Mir.*'s source included information about the island's non-Greek populations, so it may be that the compiler was unaware of this aspect of Sardinian history rather than that he intentionally omitted it. Regardless, the contrast between *Mir.* 100 and Diodorus' version throws *Mir.*'s simplification of the island's residents to just Greeks and Carthaginians into sharp relief. Assuming that Pausanias was correct in his observation that "Greeks are very ignorant" about Sardinia,[94] it seems unlikely that *Mir.*'s readers would see past the uncomplicated cultural binary that *Mir.* 100 constructs.

The vicissitudes of colonization also lie in the background of *Mir.* 107 (quoted on p. 185), where conflicts occur not with other outsiders like the Carthaginians but rather with the indigenous population. *Mir.* relates that Philoctetes had settled in Makkala, a remote area between the Achaean *apoikiai* Sybaris and Croton that was likely populated by Oenotri or Chones.[95] Later, the hero perished in battle after coming to the aid of a contingent of Rhodians who, according to Lycophron, had been blown off course to the area while journeying to found their own colony.[96] The enemies with whom they clashed are identified by *Mir.* only as "the barbarians who inhabited that area" (τοὺς ἐνοικοῦντας τῶν βαρβάρων ἐκείνην τὴν χώραν). This disagrees with Lycophron's assertion that Philoctetes and his allies fought near the river Sybaris against "Ausonians from Pellene" (Αὔσονες Πελλήνιοι), that is, residents of the Achaean colony Pellene just inland from the

---

93 4.30.5; translation by Oldfather (1935). Compare Diodorus' second account of the Sardinian expedition, 5.15.1, which observes that Heracles sent along with Iolaus and the Thespiadae a "noteworthy contingent of both Greeks and barbarians" (δύναμιν ἀξιόλογον Ἑλλήνων τε καὶ βαρβάρων). See Schorn, this volume, on the relationships between Diodorus' accounts, Timaeus, and *Mir.* On the terminology and representation of the "barbarization" of Sardinia, see Mastino (2004) esp. 20–22 and Galvagno (2004); on Greek views of the "barbarization" of *apoikiai* in general, see Bowersock (1995).

94 Paus. 10.17.13: . . . ἥκιστα καὶ ἐς ταύτην οἵ Ἕλληνες τὴν νῆσον ἀνηκόως εἶχον.

95 For Philoctetes as the founder of Petelia, a city of the Chones, and nearby Crimissa, see Strabo 6.3 with the commentary of Musti (1988) *ad loc*, Giangiulio (1991), and Malkin (1998) esp. 215–221 and 226–227 on Philoctetes and Siris. *Mir.* says nothing of his potential role as an oikist, and the representation of his occupation of Makkala here aligns with the interpretation of Malkin (1998) of Philoctetes as a hero connected initially with the "middle ground" between cultures. On the local Italic peoples in the area, see Lycoph. 913–921, Strab. 6.1.3, Verg. *Aen.* 3.402 with Servius *ad loc*, Steph. Byz. χ 64, s.v. *Chone*, though *Mir.* displays no knowledge of either people. Genovese (2009) 25–94 on the "territory of Philoctetes" and the material cultural remains related to his presence in southern Italy is essential. Vanotti (2007) 189–192 provides an overview of the geography and toponomastics of the area.

96 Lycoph. 923–926. On the question of possible Rhodian colonization in Italy, see most recently Hornblower (2018) 81–105, who also reviews the issue of *Mir.*'s inclusion of Tlepolemus in this group, a Rhodian Homeric hero whose death was featured in the *Iliad*.

Corinthian Gulf in Magna Graecia.[97] For Lycophron, the dispute occurs between two different groups of Hellenic settlers. Accordingly, *Mir.*'s identification of the opponents as βάρβαροι has been deemed an error not present in his source,[98] or a confused reference to these Ausonian Achaeans.[99] Either explanation may be correct, though the compiler instead could have worked from an intermediary which attested that the opponents were a local Italic group. Whatever the reason, *Mir.*'s report unequivocally casts "the local barbarians" as Philoctetes' and the Rhodians' adversaries, and in so doing inserts a vein of barbaric alterity into both the narrative of Philoctetes' death and its representation of southern Italy as a whole. With the exception of *Mir.* 79 (which offers a somewhat comparable context to *Mir.* 107, see pp. 200–201), the compiler elsewhere avoids using the simplified βάρβαροι in favor of retaining specific ethnic identifiers applied by the Greeks to the various native populations of Italy and further west. At the same time, however, we regularly see the Greeks themselves referred to as homogeneous Ἕλληνες, which indirectly perpetuates the basic dichotomy perceived between Greeks and "others."

So we find in the second *paradoxon* reported by *Mir.* 109 (quoted on p. 186), the Hellenophilic behavior of dogs that live around the temple of Athena in Daunia, a region with which Diomedes, as here, is regularly linked, particularly in his capacity as a founder of non-Greek cities in Apulia.[100] The dogs' inexplicable ability to recognize a person's Greekness and their subsequent fawning over those Greeks who visit the temple (κύνας οἳ τοὺς ἀφικνουμένους τῶν Ἑλλήνων) has been interpreted by Malkin as drawing on folk motifs "to express the insecurity of a Greek in a foreign land, where the presence of the Greek would be an exception."[101] Through the behavior of the dogs at the Daunian "temple called that of Achaean Athena" (ἱερὸν . . . Ἀθηνᾶς Ἀχαΐας καλούμενον) – a moniker that underscores the temple's location *outside* the ambit of Greek control – nature itself privileges Greekness and offers Greek visitors a special welcome and safety from the potential vagaries of interactions with non-Greeks. Erskine offers a somewhat different reading of the story that relies not so much on Greek fears as on contemporary Daunian responses to Greek tourists. Drawing on other examples of Italic peoples' integration of myth within their own communities and their use of myth as a vehicle for

---

97 Lycoph. 922, with Giangiulio (1991) 47–48. Cf. Hornblower (2015) 347, who also observes that "'Ausonian' means 'of Magna Graecia'" and that the name of Pellene likely serves to indicate the Achaean region of colonization in southern Italy as a whole. Compare Malkin (1998) 222, who suggests they are instead Achaean Greeks living in the mixed Greek/indigenous communities on the periphery of Croton.

98 Van Gelder *apud* Hornblower (2015) 92.

99 Bérard (1957) 347.

100 Diomedes' activities in Italy and the Adriatic (as in *Mir.* 79, below) have been the subject of considerable discussion, particularly since the Adriatic Island Project unearthed remains of his cult worship at Cape Ploča and Palagruža in the 1990s. See Coppola (1988), Malkin (1998) 234–257, Braccesi (2004), Giangiulio (2006), Castiglioni (2008), Genovese (2009) 189–266, Hornblower (2015) 256–268; Hornblower (2018) 77–79 offers a short, updated summary of Daunia and its relationship to Diomedes.

101 Malkin (1998) 238–239.

intercultural discourse, Erskine understands the tale as offering a "Greek-friendly" story designed to please Greek visitors to the temple.[102] While Erskine's argument from the Daunian perspective is compelling, I suspect that the sense of "safety from the dangers outside the Greek world" identified by Malkin lies at the heart of *Mir.*'s use and understanding of the story, especially given the preceding account of Philoctetes' death.

Intercultural conflict in the context of Greek arrival in southern Italy also defines the tale of the xenophobic birds of Diomedeia (*Mir.* 79), a sibling *paradoxon*, so to speak, of the Hellenophilic dogs of Daunia and one of the only relatively well-known myths in the compilation[103]:

[1] Ἐν τῇ Διομηδείᾳ νήσῳ, ἣ κεῖται ἐν τῷ Ἀδρίᾳ, φασὶν ἱερόν τι εἶναι τοῦ Διομήδους θαυμαστόν τε καὶ ἅγιον, περὶ δὲ τὸ ἱερὸν κύκλῳ περικαθῆσθαι ὄρνιθας μεγάλους τοῖς μεγέθεσι, καὶ ῥύγχη ἔχοντας μεγάλα καὶ σκληρά. τούτους λέγουσιν, ἐὰν μὲν Ἕλληνες ἀποβαίνωσιν εἰς τὸν τόπον, ἡσυχίαν ἔχειν, ἐὰν δὲ τῶν βαρβάρων τινὲς τῶν περιοίκων, ἀνίπτασθαι καὶ αἰωρουμένους καταράσσειν αὐτοὺς εἰς τὰς κεφαλὰς αὐτῶν, καὶ τοῖς ῥύγχεσι τιτρώσκοντας ἀποκτείνειν. [2] Μυθεύεται δὲ τούτους γενέσθαι ἐκ τῶν ἑταίρων τῶν τοῦ Διομήδους, ναυαγησάντων μὲν αὐτῶν περὶ τὴν νῆσον, τοῦ δὲ Διομήδους δολοφονηθέντος ὑπὸ τοῦ Δαύνου[104] τοῦ τότε βασιλέως τῶν τόπων ἐκείνων γενομένου.

[1] On the island of Diomedeia, which lies in the Adriatic, they say there is a temple of Diomedes, wonderful and holy, and around the temple there sit in a circle birds of a large size, having great hard beaks. They say that if Greeks disembark at the place, these birds keep quiet, but if any of the barbarians who live around the area approach, they fly up, and soaring in the air swoop down upon their heads, and, wounding them with their beaks, kill them. [2] The story goes that the companions of Diomedes were metamorphosed into these, when they had been shipwrecked off the island and Diomedes was treacherously slain by Daunus, who was then king of those regions.

Like the Daunian dogs, the birds of Diomedeia possess the uncanny ability to detect ethnicity and respond accordingly, though their reaction – no quarter for non-Greeks – inverts the fawning welcome accorded Greeks in *Mir.* 109. The mythic *aition* provided for their behavior, the only report in *Mir.* that includes

---

102 Erskine (2005) 122–126, who emphasizes Daunian towns' positive incorporation of Diomedes into their local history. See also Schorn, this volume p. 40–41, for discussion of this story as stemming from a tradition in which Diomedes spent the remainder of this life ruling in the area and married to a local princess.

103 Versions of this tale or references to the wonder occur at Mimnermus ap. *Schol. Lycoph.* 610, Timaeus and Lycus of Rhegium ap. *Schol. Lycoph.* 615, Lycoph. 592–632; ps.-Scymn. 431–433; Verg. *Aen.* 11.271–274; Ov. *Met.*14. 483–509; Strabo 6.3.9; Plin. *HN* 3.30.1, 10.61.2, 12.3.1; Ael. *NA* 1.1, Ant. Lib. 37; August. *De civ. D.* 18.16.

104 I follow the reading of the name as Daunus, see Giacomelli (2023) 259–260 for discussion of the text and compare Vanotti (2007) 166, who reads the name as Aeneas.

human metamorphoses, draws on the conclusion of what was an old tale of "native perfidy,"[105] though *Mir.* does not elaborate why Daunus' actions amounted to treachery.[106] Other versions fill in the gaps in our report. After being forced to leave Argos, Diomedes arrived in southern Italy where he founded a number of non-Greek colonies, including Agyrippa in Apulia. Meanwhile, Daunus was beset by local enemies, so Diomedes allied with him on the condition that portions of his lands would be granted to the Greeks. Later, rather than abide by their agreement, Daunus had Diomedes killed, and his companions then were transformed into the birds that guarded his tomb.[107] The potential anxieties about intercultural relationships implied in *Mir.* 109 are thus realized in the failed alliance and cautionary example of Diomedes and Daunus. *Mir.*'s unconventional language further stresses this point. Here appears his only other description of non-Greeks as "local barbarians" (τῶν βαρβάρων τινὲς τῶν περιοίκων), which we saw used of the opponents against whom Philoctetes fought and died in *Mir.* 107. These are the only two reports in the section that feature explicit deaths of specific Greek individuals[108] – heroes' deaths, at that – which suggests *Mir.* may restrict the term for particularly adversarial or momentous contexts. No less noteworthy is the term applied to Daunus' dispatchment of Diomedes, δολοφονηθέντος. Common enough among ancient historians as an expression for "assassination" or any sort of covert murder, the term can carry with it a sense of moral indignation or imply, as in this example, a negative idea of deviousness and nonheroic action. Here, the great Homeric hero suffers an unheroic death. *Mir.* 79's presentation of "barbarians" and a "treacherously murdering" Apulian king, whether *Mir.*'s language or language retained from his source, combines with the content of both *paradoxon* and *aition* to offer a distinct expression of Greek insecurities regarding their positions abroad.

We feel this undercurrent all the more when we compare ps.-Antigonus' report on the same phenomenon[109]:

172. Περὶ δὲ τῶν ζῴων Λύκον μὲν ἐν τῇ Διομηδείᾳ τῇ νήσῳ φησὶν ἱστορεῖν τοὺς ἐρωδιοὺς ὑπὸ μὲν τῶν Ἑλλήνων, ὅταν παραβάλλῃ τις εἰς τοὺς τόπους, οὐ μόνον ψαυομένους ὑπομένειν, ἀλλὰ καὶ προσπετομένους εἰς τοὺς κόλπους ἐνδύνειν καὶ σαίνειν φιλοφρόνως. . . . Λέγεσθαι δέ τι τοιοῦτον ὑπὸ τῶν ἐγχωρίων, ὡς τῶν τοῦ Διομήδους ἑταίρων εἰς τὴν τῶν ὀρνέων τούτων φύσιν μετασχηματισθέντων.

He [Callimachus] says that Lycus relates, concerning animals, that in the island of Diomedeia the herons not only endure the touch of any Greek that approaches their haunts but even fly towards him and sink into his breast and fawn in a friendly manner. . . . And [Lycus says] that something like the

---

105 Cf. Briquel (2015) 15–16 on tales of this type.
106 Given the popularity of the tale, elaboration may have been felt unnecessary.
107 On this version, as well as variants (especially that of Liberalis 37, who instead assigns guilt for Diomedes's death to Illyrians), see Bérard (1957) 368–373 and Malkin (1998) 237–249.
108 Though *Mir.* 81 mentions the fall and thus the implied death of Phaethon.
109 *Mir.* 79 is the only mythic report in the collection which has a parallel in another compilation that also includes the mythic material.

following is said by the natives, that the comrades of Diomedes have been metamorphosed into the form of these birds.

Three key differences between *Mir.*'s version and that of ps.-Antigonus are worth noting. First, the behavior of the birds coincides more with the behavior of the Daunian dogs in *Mir.* 109, that is, fawning over Greeks, whereas in *Mir.* they simply keep quiet and, more importantly, attack non-Greeks, a reaction about which ps.-Antigonus says nothing. Moreover, while both compilers link the birds to the metamorphoses of Diomedes' comrades, ps.-Antigonus omits any mention of the reason for their transformation or the murder of Diomedes; nothing betrays the *aition*'s origins in a tale focused upon the treacherousness of the Daunians. What's more, the *aition* is said to have been told by the non-Greek natives of the area themselves (ὑπὸ τῶν ἐγχωρίων), and we may well imagine Erskine's scenario of non-Greeks developing and promulgating "Greek-friendly" myths for tourists reflected here. In *Mir.*'s case, however, no specific authority appears, and the only image readers receive of the locals relates to Diomedes' murder. Taken together, *Mir.*'s report stages intercultural conflict and manifests tensions between Greek and non-Greek in ways largely absent from ps.-Antigonus' version.

The placement of *Mir.* 79 in the compilation amplifies this message. The second report in the subsection (*Mir.* 78–121), 79 follows an anecdote about an Italian plant-based poison with which, at first glance, it may seem to have no substantial connection beyond its possible progression along a hodological route.[110] Instead, the various details in *Mir.* 78 align with critical points in 79 to provide a contemporary historical analogue to the mytho-historical narrative of Diomedes and Daunus:

78. [1] Λέγεται δὲ περὶ τὴν Ἰταλίαν ἐν τῷ Κιρκαίῳ ὄρει φάρμακόν τι φύεσθαι θανάσιμον, ὃ τοιαύτην ἔχει τὴν δύναμιν ὥστε, ἂν προσρανθῇ τινι, παραχρῆμα πίπτειν ποιεῖ, καὶ τὰς τρίχας τὰς ἐν τῷ σώματι ἀπομαδᾶν, καὶ τὸ σύνολον τοῦ σώματος διαρρεῖν τὰ μέλη, ὥστε τὴν ἐπιφάνειαν τοῦ σώματος εἶναι τῶν ἀπολλυμένων ἐλεεινήν. [2] Τοῦτο δέ φασι μέλλοντας διδόναι Κλεωνύμῳ τῷ Σπαρτιάτῃ Αὖλον τὸν †Πευκέστιον† καὶ Γάϊον φωραθῆναι, καὶ ἐξετασθέντας ὑπὸ Ταραντίνων θανατωθῆναι.

[1] It is said that on the Circaean mountain in Italy there grows a deadly poison, which is so potent that, if sprinkled on anyone, it straightaway causes him to fall, and the hair on his body drops off, and generally the limbs of his body waste away, so that the surface of the body of those who are dying is a pitiable sight. [2] They say too that Aulus the Peucestian (?) and Gaius were caught when they were about to administer this poison to Cleonymus the Spartan, and that after being interrogated they were put to death by the Tarentines.

---

110   So Flashar (1981) 46 on *Mir.* 78–81: "Als zusammengehörig erweist sich zunächst 79–81, indem hier in süd-nördlicher Richtung der Osten Italiens behandelt wird." This could work, provided the compiler was focused upon the location of the anecdote (Taras), not the location of the *paradoxon* (Mount Circeo, in southern Latium on the coast of the Tyrrhenian Sea).

The first half of the report demonstrates *Mir.* not integrating mythic material even when he likely could have done so. The name of the mountain and its deadly φάρμακον naturally calls to mind Circe and her own famous φάρμακον that she used on Greek visitors in *Odyssey* 10. In fact, the mountain that dominates a promontory in southern Latium had long been advertised by locals as the "real" location of Circe's island, complete with a temple and a guided tour that touted alleged Odyssean relics.[111] Ignoring or unaware of the mythic associations, *Mir.* instead fills out his report with a reference to the plant's use in an attempted assassination drawn from more recent contemporary history. The enigmatic tale of the foiled murder plot against Cleonymus the Spartan appears nowhere else in our sources, though we can sketch the essential context.[112] Diodorus relates that Cleonymus, who did not succeed his father as king due to his uncontrollable temperament, went, with the backing of Sparta, to aid Taras in their conflict with the encroaching Lucani in 303/302. Gaining the support of many Greek *apoikiai*, as well as the Messapians, he amassed such a large mercenary army, mainly drawn from southern Italy, that the Lucani abandoned aggressions against Taras, turning their attention to Metapontum instead (Diod. Sic. 17.104). The attempt on Cleonymus' life may have occurred upon his arrival in Italy while he was recruiting his mercenary force and before the resolution with the Lucani. The specific identities and motivations of the would-be assassins remain mysteries, though the names *Mir.* provides and the problematic Πευκέστιον make their non-Greek ethnicity clear.[113]

As presented by *Mir.*, the experience of Cleonymus maps on to that of Diomedes and his companions in key ways. Both anecdotes recount plots against the life of a visiting Greek leader who had entered into an alliance with a southern Italian power against that power's local enemies. The perpetrators in both are non-Greeks, who either themselves earn death sentences for their actions, as in *Mir.* 78, or whose crime constitutes the etiological basis for the capital penalty imposed on others, that is, the "death by birds" suffered by non-Greeks who visit the temple on Diomedeia. The critical difference between the two anecdotes lies in the part played by the Tarentines. Unlike Diomedes, who partnered with a non-Greek king, Cleonymus allied principally with the local Greeks, particularly those of the Spartan *apoikia* Taras. Italic treachery threatened the lives of both Diomedes and Cleonymus, but the latter's Greek allies not only thwarted the assassination plot but also meted out justice for the attempt, whereas Diomedes' comrades could only avenge his death through their subsequent avian attacks on non-Greeks. The many points of convergence between the two reports are such that their placement together could hardly have been dictated by a vague hodological route. Rather, *Mir.* 78 and 79 form an analogical diptych – historical on one side, mythical on the other – that traces threats faced by Greeks when entering into political relationships in Italy.

---

111 Strabo 5.3.6, Thphr. *Hist. Pl.* 5.8.3, Pliny *HN* 15.119, Cic. *Nat. D.* 3.48; cf. Erskine (2005) 121–22.
112 See Vanotti (2007) for an overview of what little we can say about the particulars and relevant bibliography.
113 On the names and descriptions, see Giacomelli (2023) 258–259.

In light of this diptych's placement at the beginning of a long section largely devoted to wonders in the west, it may be tempting to see in it a broad programmatic or thematic statement, particularly as the reports stand in the shadow, so to speak, of Circe (hardly an ideal representation of *xenia*). I suspect doing so would be mistaken, for two main reasons. First, unlike ps.-Antigonus, who includes two clear section prefaces (ps.-Antig. 26 and 129), the compiler of *Mir.* elsewhere does not. No textual evidence beyond his creation of a curious linked pair supports investing *Mir.* 78 and 79 with any sort of introductory significance. Second, although many of *Mir.*'s anecdotes communicate anxieties about the dangers to Greeks abroad, the section as a whole constructs a composite portrait of the west and its wonders that includes such concerns but certainly is not defined by them, as we see even in just the considerable variety of reports with mythic material. Rather, the value in identifying this diptych lies in what it may tell us about *Mir.*'s approach elsewhere. The compiler's pairing of two reports on different types of *paradoxa* that feature anecdotal and etiological material with corresponding themes reveals that, at least here, *Mir.*'s organization is influenced not only by geography and itinerary, but also by associative and analogical connections formed between the secondary mythical and historical details that the compiler so seems to enjoy including in his entries.[114] In *Mir.* 78 and 79, it is the "unnecessary" but fascinating additions from myth and history that defines the reports' relationship.

3. General conclusions

While William Hansen's claim that myth never played a major role in paradoxography still rings true, my purpose here has been to demonstrate that consideration of myth's appearances in supporting roles, so to speak, can shed light on paradoxographers' processes, perspectives, and interests. Compilers of collections of wonders are invested in presenting readers with the strange and unbelievable, and this commitment in turn influences their incorporation of mythic material, as I showed in my first section. The single critical feature shared by nearly all myths in paradoxographical compilations – relative obscurity – aligns with and even contributes to paradoxography's central goal of cultivating wonder. Certainly, many traditional etiological myths recount the fabulous origins of a natural phenomenon, but those with widespread familiarity hardly inspire fascination. Consequently, these are not tales that we find in paradoxographical compilations. Rather, paradoxographers direct their attention to unfamiliar mythic material, often culled from lesser-known works like that of Myrsilus of Methymna, which retained the power to surprise audiences. *Mir.*'s sustained focus upon western marvels in 78–121 especially afforded the opportunity to include mythic material unfamiliar to a readership (and perhaps the compiler himself) centered in the eastern environs of Greece proper.

114 Cf. Rosenkranz (1966) on associative organization in parenetical speeches and Schepens in Schepens and Delcroix (1996) 397–398 on the importance of analogy and association in the organization of ps.-Antigonus' compilation.

In contrast to other compilers, Phlegon pairs more ubiquitous mythic *paradoxa* with likely–unfamiliar contemporary instances of the same phenomenon for suasive purposes. In doing so, he provides a helpful example of the gulf that can exist between the approaches and practices of paradoxographers, who often are loosely comprehended in modern scholarship as an undifferentiated group. The second half of my study in turn identified critical differences in *Mir.*'s use of mythic material as compared to that of ps.-Antigonus and the compiler of *Paradoxographus Vaticanus*. Although *Mir.*'s inclusion of obscure myths as etiological explanations for wonders corresponds with what we find in other compilations, *Mir.*'s penchant for more elaborate reports set in the west that integrate elements drawn from myth, history, and ethnography finds few analogues elsewhere. Most importantly, *Mir.* often moves mythic material from the background to the foreground, especially in anecdotes about dedications made by mythic figures. The compiler's interest in the cultural relics of the ancient and mythic past and his representation of such relics as wonders in their own right reflect a fundamental discrepancy between his interpretation of what constitutes a marvel and that of other paradoxographers.

Mir.'s short subsection topically devoted to cults and mythic votives in southern Italy (106–110) also underscores the role that mythic material can play in the organization of a compilation. While the surviving collections typically are structured around criteria such as geography, hodology, type of wonder, and/or the compiler's source,[115] we have seen the inclusion of mythic material as a secondary or supporting consideration for the arrangement of sections within works broadly organized under other criteria. So ps.-Antigonus' first section on noteworthy animal sounds introduces mythic links as an additional connective theme, while Phlegon's subsections on spontaneous changes of sex and multiple births rely on the union of mythic and contemporary exempla. The pairing of *Mir.* 78 and 79, meanwhile, depends on the close correspondence of the historic and mythic anecdotes which fill out reports on otherwise unconnected botanical and zoological *paradoxa*. Such examples testify to the potential benefits of continued exploration of the ways that a variety of secondary elements may interact with each other between reports, and also how such elements may participate in or contribute to a compilation's organization.[116]

While questions of organization concentrate on the motivations and decisions of the compiler, we may also consider the ways that these decisions could cumulatively affect the reader. In my final section on *Mir.* 78–121, I highlighted the repeated representation of occidental lands as the stomping grounds of heroes, the "Hellenized" locations of cults to Greek figures, the sites of foreign exotica experienced and sometimes influenced by Greeks, and even as fundamental threats to Hellenes, be they mythical or more contemporary. Whether the compiler intended it or not, the quantity and quality of mythic material in *Mir.* 78–121 ultimately communicates to readers a composite portrait of the "wonderful west" that is

---

115 For remarks on the various types of organization schemes used, see, for example, Schepens in Schepens and Delcroix (1996) 395–398, Geus (2016), and Greene (2022) 657–658.

116 The introductions to many of the editions of imperial age and anonymous paradoxographers in Schorn (2022) offer very helpful thoughts on this general topic.

consistently defined by Greek perspectives and reflective of Greek interests. In other words, *Mir.* doesn't just inspire fascination in its readers, it creates a cumulative impression of cultural relationships with western lands. Scholarship on paradoxography's effects on readers usually – and understandably – has centered upon the dynamics of the cultivation of wonder. However, this reading of *Mir.* 78–121 suggests the possible value of further study of the ways that a given compilation's areas of focus and presentation of reports may foster particular impressions or reactions in readers beyond just wonder.[117]

Appendix: myths in paradoxographical compilations

The following list includes all reports with explicit mythic material in the seven surviving compilations. As I discuss in the introduction to the first section, a clear mention of a mythic actor (e.g., Heracles) performing a mythic action (e.g., Trojans founding Sigeum in *Mir.* 106) constitutes my criteria for "explicit mythic material."

Those entries preceded by an asterisk (*) were not included by Giannini in his (1966) appendix's list of reports περὶ μύθων (p. 430). Giannini's appendix focused more upon the quality of a *paradoxon*, and thus many reports with mythic material appear in his appendix under headings for various types of *paradoxa*. I focus instead only upon the inclusion of mythic material regardless of the type of *paradoxon*, though this information is provided below for each report.

ps.-Antigonus

1. Local legend of singing contest and silent cicadas at Rhegium (*zoological paradoxon; legendary aition; no figures from myth*)[118]
2. Mute cicadas of Rhegium (*zoological paradoxon; mythic aition; Heracles*)
4. Mute Seriphian frogs (*zoological paradoxon; mythic aition; Pericles*)
5. Melodious nightingales of Antissa (*zoological paradoxon; mythic aition; Orpheus*)
12. Absence of Attic crows on the Acropolis (*zoological paradoxon; mythic aition; Athena, Erichthonius, Cecropidae*)
* 24. Canine behavior in the *Odyssey* (*zoological paradoxon; mythic example*)
56. Parturition of wolves (*zoological paradoxon; mythic aition; Leto*)
111. Idiosyncratic distribution of sex in human offspring (*human paradoxon; mythic example; Heracles*)
117. Malodorousness of Ozolian Locris (*human/topographical paradoxon; mythic aition; Heracles and Nessus*)

---

117   My thanks to Robert Mayhew and Stefan Schorn both for their many helpful insights and for inviting me to contribute this study despite being unable to attend the conference.
118   I hesitate to include this report since it is a local legend and no figures from myth appear (see p. 175). However, I follow Giannini, who also included it in his list.

118. Malodorousness of Lemnian women (*human paradoxon; mythic aition; Medea and Jason*)
* 163. Cretan rivulet unaffected by rain (*topographical paradoxon; mythic aition; Europa and Zeus*)
* 172. Hellenophilic herons of Diomedeia (*zoological paradoxon; mythic aition; Diomedes and companions*)

ps.-Aristotle

* 51. Olives of Pantheion and Olympia (*botanical paradoxon; mythic aition; Heracles*)
* 58. Metals of Demonesus (*mineralogical paradoxon; mythic example; Heracles*)
* 79. Hellenophilic herons of Diomedeia (*zoological paradoxon; mythic aition; Diomedes and companions*)
81. Wonders of the Electrides Islands and Eridanus (*1. mythic/cultural paradoxon; Daedalus; 2. aquatic paradoxon; mythic aition; Phaethon*)
82. Wonders of Sicilian Enna (*topographical paradoxa; mythic aitia; Kore, Pluto, Demeter*)
83. Absence of dangerous animals on Crete (*zoological paradoxon, mythic aition; Zeus*)
* 88. Licentiousness of Gymnesian men and prohibitions against their use of gold (*ethnographical paradoxon; mythic aition; Heracles*)
97. Iapygian ichor stream and giant footprints (*aquatic and topographical paradoxa; mythic aition; Heracles and Giants*)
98. Large stone Iapygian stone (*topographical paradoxon, mythic aition; Heracles*)
100. Ancient temples and former fecundity of Sardinia (*1. cultural paradoxon; mythic aition; Iolaus and Heracles; 2. topographical paradoxon; mythic aition; Aristaeus*)
105. Course of the Ister and phenomena associated with that area (*1. aquatic paradoxon; mythic example; Argonauts; 2. topographical paradoxon (pebbles of Elba); mythic aition; Argonauts*)
* 106. Tarentine cult worship of Greek heroes and Trojan settlement (*ethnographic/cultural paradoxon*)
107. Philoctetes at Sybaris (*ethnographic/cultural paradoxon; Philoctetes*)
108. The tools of Epeius at "Gargargia" [Lagaria] (*cultural paradoxon; Epeius and Athena*)
109.1. Weapons of Diomedes in Daunia; 2. Hellenophilic dogs at temple of Athena in Daunia; 3. Daunian clothing (*1. cultural paradoxon; Diomedes and companions; 2. zoological paradoxon; 3. ethnographic paradoxon; mythic aition; Trojans and Achaeans*)
110. Dedication history of bronze necklet in the territory of the Peucetini (*zoological/cultural paradoxon; Diomedes*)
116. Inscription discovered at Athens (*cultural paradoxon; Deiope*)
118. Inscription of Heracles on a dedicated pillar (*cultural paradoxon; Heracles, Cattle of Geryon*)

Paradoxographus Florentinus

* 24. Spring of Clitor (*aquatic paradoxon; mythic aition; Melampus and Proitids*)

Paradoxographus Vaticanus

14. Bay leaves and stones (*botanical paradoxon; mythic aition; Apollo and Daphne*)
15. Winds roused by touching of willow-like trees (*botanical paradoxon; mythic aition; Boreas*)
18. Perinthus river causes intestinal swelling (*aquatic paradoxon; mythic aition; Perseus and Medusa*)
19. Marsyas River responds to music (*aquatic paradoxon; mythic aition; Marsyas*)
31. Change of sex: Tiresias (*mythic/human paradoxon; Tiresias*)
32. Mythological shapeshifters (*mythic/human paradoxon; Proteus, Nereus, Mestra*)
* 58. Cretan law and child-rearing practices (*ethnographic paradoxon; Minos*)

Phlegon of Tralleis **(this work was not included in Giannini 1966)**

* 4. Changes of sex: Tiresias (*human/mythic paradoxon*)
* 5. Changes of sex: Kainis/Kaineus (*human/mythic paradoxon*)
* 11. Bones of Idas (*human/mythic paradoxon*)
* 30. Hyperfertility: Aegyptus and Euryrrhoe (*human/mythic paradoxon*)
* 31. Hyperfertility: Danaus and Europe (*human/mythic paradoxon*)

Apollonius[119] *and* **Paradoxographus Palatinus:** no explicit mythic material

Bibliography

Acosta-Hughes, B. and Stephens, S. 2012. *Callimachus in Context*. Cambridge: Cambridge University Press.

Ampolo, C. 1993. "La città dell'eccesso: per la storia di Sibari fino al 510 a.C." In Stazio, A. and Ceccoli, S. eds. *Sibari e la Sibaritide. Atti del trentaduesimo Convegno di studi sulla Magna Grecia, Taranto-Sibari, 7–12 Ottobre 1992*. Taranto: Istituto per la storia e l'archeologia della Magna Grecia. 213–254.

Arafat, K. 1996. *Pausanias' Greece. Ancient Artists and Roman Rulers*. Cambridge: Cambridge University Press.

Barron, C. 2013. *Timaeus of Tauromenium and Hellenistic Historiography*. Cambridge: Cambridge University Press.

Battegazore, A. 1993. "Communis opinio e contro-argomentazione nella fisica e nella scienza naturale di Teofrasto." In Battegazzore, A. ed. *Dimostrazione, argomentazione dialettica e argomentazione retorica nel pensiero antico*. Genoa: Sagep Editrice. 201–261.

---

119 Giannini (1966) 430 includes 24 and 36 from this compilation in his list περὶ μύθων. I am unsure of his reasoning for doing so, but they contain no mythic material that meets my criteria. On the reports, see the commentary of Spittler (2022).

Belousov, A. 2022. "1676. Philon of Herakleia." In Schorn, S. ed. *Felix Jacoby. Die Fragmente der Griechischen Historiker Continued. Part IV. Biography and Anti- quarian Literature. E. Paradoxography and Antiquities. Fasc. 2. Paradoxographers of Imperial Times and Undated Authors [Nos. 1667–1693].* Leiden; Boston: Brill. 543–550.

Bérard, J. 1957. *La colonisation grecque d'Italie méridionale et de la Sicile dans l'antiquité: Hisotoire et légende.* 2nd ed. Paris: Presses Universitaires de France.

Bernardini, P. 2004. "Gli eroi e le fonti." In Zucca, R., ed. *Logos peri tēs Sardous: le fonti classiche e la Sardegna. Atti del convegno di studi, Lanusei, 29 dicembre 1998.* Rome: Carocci. 39–62.

Bettini, M. and Spina, L. 2007. *Il mito delle Sirene: immagini e racconti dalla Grecia a oggi.* Turin: Einaudi.

Bowersock, G. 1995. "The Barbarism of the Greeks." *Harvard Studies in Classical Philol- ogy* 97: 3–14.

Bowie, E.L. 1974. "The Greeks and their past in the second sophistic." In Finley, M. ed. *Studies in Ancient Society.* London: Routledge. 166–209.

Braccesi, L. 2004. "The Greeks on the Venetian Lagoon." In Lomas, K. ed. *Greek Identity in the Western Mediterranean. Papers in Honour of Brian Shefton.* Mnemosyne Supple- ments 246. Leiden: Brill. 349–361.

Breglia Pulci Doria, L. 1987. "Le Sirene, il canto, la morte, la polis." *Aion* 9: 5–98.

Briquel, D. 2015. "How to Fit Italy into Greek Myth?" In Farney, G. and Bradley, G. eds. *The Peoples of Ancient Italy.* Berlin; Boston: De Gruyter. 11–25.

Castiglioni, M. 2008. "The Cult of Diomedes in the Adriatic: Complementary Contributions from Literary Sources and Archaeology." In Carvalho, J. ed. *Bridging the Gap: Source, Methodology, and Approaches to Religion in History.* Pisa: Edizioni Plus. 9–28.

Champion, C.B. "Timaios (566)." In Worthington, I. ed. *Jacoby Online. Brill's New Jacoby.* Part III. Leiden; Boston: Brill.

Coppola, A. 1988. "Siracusa e il Diomede adriatico." *Prometheus* 14: 221–226.

Delattre, C. 2018. "Paradoxographic Discourse on Sources and Fountains: Deconstructing Paradoxes." In Gerolemou, M. ed. *Recognizing Miracles in Antiquity and Beyond.* Berlin; Boston: de Gruyter. 205–223.

del Corno, D. 1993. "L'immagine di Sibari nella tradizione classica." In Stazio, A. and Cec- coli, S. eds. *Sibari e la Sibaritide. Atti del trentaduesimo Convegno di studi sulla Magna Grecia, Taranto-Sibari, 7–12 Ottobre 1992.* Taranto: Istituto per la storia e l'archeologia della Magna Grecia. 9–18.

de Lisle, C. 2021. *Agathocles of Syracuse. Sicilian Tyrant and Hellenistic King.* Oxford: Oxford University Press.

Doroszewska, J. 2013. "'. . . And she became a man': Sexual Metamorphosis in Phlegon of Tralles' *Mirabilia.*" *Prace Filologiczne. Literaturoznawstwo* 3: 223–241.

Doroszewska, J. 2016. *The Monstrous World. Corporeal Discourses in Phlegon of Tralles' Mirabilia.* Frankfurt am Main: Peter Lang.

Dougherty, C. 1993. *The Poetics of Greek Colonization.* Oxford: Oxford University Press.

Dowdall, L.D. 1985. "On Marvellous Things Heard." In Barnes, J., ed. *Complete Works of Aristotle, Vol. II. The Revised Oxford Translation.* Princeton: Princeton University Press. 1272–1298.

Duffy, W. 2013. "The Necklace of Eriphyle and Pausanias' Approach to the Homeric Epics." *Classical World* 107: 35–47.

Edlund, I. 1987. "The Sacred Geography of Southern Italy in Lycophron's *Alexandra.*" *Opuscula Romana* 16: 43–49.

Eleftheriou, D. 2016. "'MYΘ': The Marginal Annotation on Antigonos' Collection of Extraordinary Stories and Its Meaning." *Auctor* 1: 32–40.

Eleftheriou, D. 2018. *Pseudo-Antigonos di Carystos: Collection d'Histoires Curieuses*. 2 vol. Doctoral thesis. Paris: University of Paris.

Erskine, A. 2005. "Unity and Identity: Shaping the Past in the Greek Mediterranean." In Gruen, E.S., ed. *Cultural Borrowings and Ethnic Appropriations in Antiquity*. Stuttgart: Franz Steiner. 121–136.

Flashar, H. 1981² [1972]. *Aristoteles. Werke in deutscher Übersetzung. Mirabilia*. Berlin: Akademie-Verlag.

Fortenbaugh, W., Huby, P., Sharples, R. and Gutas, D., eds. 1992. *Theophrastus of Eresus. Sources for His Life, Writings, Thought, and Influence*. Vol. 2. Leiden; New York; Köln: Brill.

Fowler, R. 2013. *Early Greek Mythography. Vol. II. Commentary*. Oxford: Oxford University Press.

Frontisi Ducroux, F. 1975. *Dédale. Mythologie de l'artisan en Grèce ancienne*. Paris: Maspéro.

Galvagno, E. 2004. "La Sardegna vista dalla Sicilia: Diodoro Siculo." In Zucca, R., ed. *Logos peri tēs Sardous: le fonti classiche e la Sardegna. Atti del convegno di studi, Lanusei, 29 dicembre 1998*. Rome: Carocci. 27–38.

Geffcken, J. 1892. *Timaios' Geographie des Westens*. Berlin: Weidmannsche Buchhandlung.

Genovese, G. 2009. *Nostoi. Tradizioni eroiche e modelli mitici nel meridione d'Italia*. Rome: L'Erma di Bretschneider.

Gerolemou, M. ed. 2018. *Recognizing Miracles in Antiquity and Beyond*. Berlin; Boston: de Gruyter.

Geus, K. 2016. "Paradoxography and Geography in Antiquity. Some Thoughts About the Paradoxographus Vaticanus." In González Ponce, F.J., Gómez Espelosín, F.J. and Chávez Reino, A. eds. *La letra y la carta: descripción verbal y rapresentación gráfica en los diseños terrestres grecolatinos. Estudios en honor de Pietro Janni*. Sevilla: Universidad de Sevilla. 243–257.

Giacomelli, C. 2023. *Pseudo-Aristotele, De mirabilibus auscultationibus: Edizione critica, traduzione e commento filologico*. Rome: Accademia Nazionale dei Lincei.

Giangiulio, M. 1991. "Filottete tra Sibari e Crotone: Osservazioni sulla tradizione letteraria." In de La Genière, J. ed. *Epeios et Philoctète en Italie: Données archéologiques et traditions légendaires*. Naples: Centre Jean Bérard. 37–54.

Giangiulio, M. 2006. "'Come colosso sulla spiaggia': Diomede in Daunia in Licofrone e prima di Licofrone appunti per una stratigrafia della tradizione." *Hesperìa* 21: 49–66.

Giannini, A. 1966. *Paradoxographorum Graecorum reliquiae*. Milan: Istituto editoriale italiano.

Graf, F. 1982. "Culti e credenze religiose della Magna Graecia." In *Megale Hellas, nome e immagine: Atti del ventunesimo convegno di studi sulla Magna Graecia, Taranto, 2–5 ottobre 1981*. Taranto: Istituto per la storia e l'archeologia della Magna Graecia. 157–185.

Greene, R. 2022. "1680. Paradoxographus Florentinus." In Schorn, S. ed. *Felix Jacoby. Die Fragmente der Griechischen Historiker Continued. Part IV. Biography and Antiquarian Literature. E. Paradoxography and Antiquities. Fasc. 2. Paradoxographers of Imperial Times and Undated Authors [Nos. 1667–1693]*. Leiden; Boston: Brill. 631–782.

Guichard, L.A. 2006. "Posidipo y los prodigios: Una interpretacion de P. Mil. Vogl. VIII 309." *Studi di Egittologia e di Papirologia* 3: 121–133.

Habicht, C. 1985. *Pausanias' Guide to Greece*. Berkeley: University of California Press.

Hall, J.M. 2000. "The East within the Cultural Identity of the Cities of Magna Graecia." In *Magna Grecia e Oriente mediterraneo prima dell'età ellenistica. Atti del trentanovesimo Convegno di studi sulla Magna Grecia. Taranto, 1–5 ottobre 1999*. Taranto: Istituto per la storia e l'archeologia di Magna Grecia. 389–401.

Hansen, W. 1996. *Phelgon of Tralles' Book of Marvels*. Exeter: University of Exeter Press.

Harder, M. 2012. *Callimachus. Aetia*. 2 vols. Oxford: Oxford University Press.

Hardiman, R., trans. 2021. "Antigonus. *Compilation of Marvellous Accounts*." <https://sites.google.com/site/paradoxography/texts/antigonus>

Harris, E. 2018. "Herodotus and the Social Context of Memory in Ancient Greece: The Individual Historian and His Community." In Archibald, Z. and Haywood, J. eds. *The Power of the Individual and Community in Ancient Athens and Beyond: Studies in Honour of John K. Davies*. Swansea: Classical Press of Wales. 79–113.

Hornblower, S. 2013. *Herodotus. Histories Book V*. Cambridge: Cambridge University Press.

Hornblower, S. 2015. *Lykophron:* Alexandra. *Greek Text, Translation, Commentary, and Introduction*. Oxford; New York: Oxford University Press.

Hornblower, S. 2018. *Lykophron's Alexandra, Rome, and the Hellenistic World*. Oxford: Oxford University Press.

Hunter, R. 2015. *Apollonius of Rhodes. Argonautica. Book IV*. Cambridge: Cambridge University Press.

Hunzinger, C. 1995. "La notion de θῶμα chez Hérodote." *Ktèma* 20: 47–70.

Jackson, S. 1990. "Myrsilus of Methymna and the Dreadful Smell of the Lemnian Women." *Illinois Classical Studies* 15: 77–83.

Jacob, C. 1983. "De l'art de compiler: à la fabrication du merveilleux." *Lalies* 2: 121–140.

Jones, W. 1919 and 1935. *Pausanias. Description of Greece*. Vols. 1 and 4. Loeb Classical Library. Cambridge, MA; London: Harvard University Press.

Krevans, N. 2005. "The Editor's Toolbox: Strategies for Selection and Presentation in the Milan Epigram Papyrus." In Gutzwiller, K. ed. *The New Posidippus: A Hellenistic Poetry Book*. Oxford: Oxford University Press. 81–96.

Lacroix, L. 1974. "Héraclès, héros voyageur et civilisateur." *Bulletin de la Classe de Lettres de l'Académie Royale de Belgique* 60: 34–59.

Lamboley, J.-L. 2006. "Myth as an Instrument for the Study of Greek and Indigenous Identities II: Myths in Western Greek Colonies." In Carvahol, J., ed. *Religion, Ritual, and Mythology. Aspects of Identity Formation in Europe*. Pisa: Edizioni Plus. 143–150.

Langerwerf, L. 2018. "'Many are the Wonders in Greece': Pausanias the Wandering Philosopher." In Gerolemou, M. ed. *Recognizing Miracles in Antiquity and Beyond*. Berlin; Boston: de Gruyter. 305–326.

Lepore, E. 1980. "Diomede." In *L'epos greco, Atti del XIX Convegno di Studi sulla Magna Grecia. Taranto 7–12 ottobre 1979*. Taranto: Istituto per la storia e l'archeologia di Magna Grecia. 113–132.

Lightfoot, J. 2021. *Wonder and the Marvellous from Homer to the Hellenistic World*. Cambridge: Cambridge University Press.

Malkin, I. 1994. *Myth and Territory in the Spartan Mediterranean*. Cambridge: Cambridge University Press.

Malkin, I. 1998. *The Returns of Odysseus. Colonization and Ethnicity*. Berkeley; Los Angeles; London: University of California Press.

Mari, M. 2009. "Cassandra e le altre: riti di donne nell'*Alessandra* di Licofrone." In Cusset, C. and Prioux, É. eds. *Lycophron: éclats d'obscurité, Actes du colloque international de Lyon et Saint-Étienne, 18–20 janvier 2007*. Saint-Étienne: PUSE. 405–440.

Mastino, A. 2004. "I miti classici e l'isola felice." In Zucca, R., ed. *Logos peri tēs Sardous: le fonti classiche e la Sardegna. Atti del convegno di studi, Lanusei, 29 dicembre 1998.* Rome: Carocci. 11–26.

Mayhew, R. 2019. *Aristotle's Lost Homeric Problems.* Oxford: Oxford University Press.

Mayor, A. 2018. *Gods and Robots. Myths, Machines, and Ancient Dreams of Technology.* Princeton: Princeton University Press.

Mele, A. 1984. "Crotone e la sua storia." In *Crotone: atti del ventitreesimo Convegno di Studi sulla Magna Grecia. Taranto, 7–10 ottobre 1983.* Taranto: Istituto per la Storia e l'Archeologia della Magna Grecia. 9–87.

Miles, M. 2014. "Burnt Temples in the Landscape of the Past." In Pieper, C. and Ker, J. eds. *Valuing the Past in the Greco-Roman Past.* Leiden; Boston: Brill. 111–145.

Morris, S. 1992. *Daidalos and the Origins of Greek Art.* Princeton: Princeton University Press.

Moscati Castelnuovo, L. 1989. *Siris: Tradizione storiografica e momenti della storia di una città della Magna Grecia.* Galatina: Congedo.

Musso, O. 1985. [Antigonus Carystius.] *Rerum mirabilium collectio.* Naples: Bibliopolis.

Musti, D. 1988. *Strabone e la Magna Grecia.* Padua: Editoriale Programma.

Musti, D. 1999. *I Telchini e le Sirene.* Pisa; Roma: Editoriali e Poligrafici.

Nafissi, M. 1999. "From Sparta to Taras: *Nomina, ktiseis,* and Relationships between Colony and Mother City." In Hodkinson, S. and Powell, A. eds. *Sparta. New Perspectives.* London: Duckworth. 245–272.

Nenci, G. 1998. "L'introduction de l'alphabet en Grèce selon Hérodote." *Revue des Études Anciennes* 100: 579–589.

Oldfather, C.H. 1935. *Diodorus of Sicily. With an English Translation.* Vol. 2. London: William Heinemann.

Pais, E. 1905. "The Temple of the Sirens in the Sorrentine Peninsula." *American Journal of Archaeology* 9: 1–6.

Pajón Leyra, I. 2011a. *Entre ciencia y maravilla. El género literario de la paradoxografía griega.* Zaragoza: Prensas de la Universidad de Zaragoza.

Pajón Leyra, I. 2011b. "Extraordinary Orpheus. The Image of Orpheus and Orphism in the Texts of the Paradoxographers (OF 1065, 787, 790, 793, and 794)." In Herrero de Jáuregui, M., Jiménez San Cristóbel, A., Martínez, E., Álvarez, M. and Tovar, S. eds. *Tracing Orpheus. Studies of Orphic Fragments in Honour of Alberto Bernabé.* Sozomena 10. Berlin; Boston: de Gruyter. 331–337.

Pajón Leyra, I. 2022a. "1682. Anonymous, Collection of Horror Stories Including Ethnography and Mythography (*P. Oxy.* II 218)." In Schorn, S. ed. *Felix Jacoby. Die Fragmente der Griechischen Historiker Continued. Part IV. Biography and Antiquarian Literature. E. Paradoxography and Antiquities. Fasc. 2. Paradoxographers of Imperial Times and Undated Authors [Nos. 1667–1693].* Leiden; Boston: Brill. 829–865.

Pajón Leyra, I. 2022b. "Mythography and Paradoxography." In Smith, R.S. and Trzaskoma, S., eds. *The Oxford Handbook of Greek and Roman Mythography.* Oxford: Oxford University Press. 396–408.

Papalexandrou, N. 2008. "Boiotian Tripods: The Tenacity of a Panhellenic Symbol in a Regional Context." *Hesperia* 77: 251–282.

Pearson, L. 1975. "Myth and archaiologia in Italy and Sicily: Timaeus and his predecessors." *YCIS* 24: 171–195.

Pearson, L. 1987. *The Greek Historians of the West: Timaeus and His Predecessors.* Atlanta: Scholars Press.

Pianu, G. 2004. "Il mito di Aristeo in Sardegna." In Zucca, R., ed. *Logos peri tēs Sardous: le fonti classiche e la Sardegna. Atti del convegno di studi, Lanusei, 29 dicembre 1998.* Rome: Carocci. 96–98.

de Polignac, F. 1997. "Mythes et modèles culturels de la colonisation grecque archaïque." In *Mito e Storia in Magna Grecia, Atti del trentaseiesimo convegno di stui sulla Magna Grecia. Taranto 4–7 ottobre 1996.* Taranto: Istituto per la storia e l'archeologia della Magna Grecia. 167–187.

Porter, J. 2001. "Ideals and Ruins: Pausanias, Longinus, and the Second Sophistic." In Alcock, S., Cherry, J. and Elsner, J. eds. *Pausanias. Travel and Memory in Roman Greece.* Oxford: Oxford University Press. 63–92.

Powell, B. 1991. *Homer and the Origin of the Greek Alphabet.* Cambridge: Cambridge University Press.

Priestley, J. 2014. *Herodotus and Hellenistic Culture. Literary Studies in the Reception of the Histories.* Oxford: Oxford University Press.

Pugliara, M. 2003. *Il mirabile e l'artificio. Creature animate e semoventi nel mito e nella tecnica degli antichi.* Roma: L'Erma di Bretschneider.

Raubitschek, A. 1991. *The School of Hellas: Essays on Greek History, Archaeology, and Literature.* New York: Oxford University Press.

Rosenkranz, B. 1966. "Die Struktur der ps.- isokrateischen Demonicea." *Emerita* 34: 95–129.

Sanna, S. 2004. "La figura di Aristeo in Sardegna." In Zucca, R., ed. *Logos peri tēs Sardous: le fonti classiche e la Sardegna. Atti del convegno di studi, Lanusei, 29 dicembre 1998.* Rome: Carocci. 99–111.

Schachter, A. 1981. *Cults of Boiotia. Volume I. Acheloos to Hera.* London: University of London, Institute of Classical Studies.

Scheer, T. 2018. "Women and *nostoi*." In Hornblower, S. and Biffis, G. eds. *The Returning Hero. Nostoi and Traditions of Mediterranean Settlement.* Oxford: Oxford University Press. 123–146.

Schepens, G. and Delcroix, K. 1996. "Ancient Paradoxography. Origin, Evolution, Production and Reception." In Pecere, O. and Stramaglia, A. eds. *La letteratura di consumo nel mondo greco-latino. Atti del convegno internazionale, Cassino, 14–17 settembre 1994.* Cassino: Università degli Studi di Cassino. 373–460.

Schorn, S. ed. 2022. *Felix Jacoby. Die Fragmente der Griechischen Historiker Continued. Part IV. Biography and Antiquarian Literature. E. Paradoxography and Antiquities. Fasc. 2. Paradoxographers of Imperial Times and Undated Authors [Nos. 1667–1693].* Leiden; Boston: Brill.

Shannon-Henderson, K. 2022. "1667. Phlegon of Tralleis." In Schorn, S. ed. *Felix Jacoby. Die Fragmente der Griechischen Historiker Continued. Part IV. Biography and Antiquarian Literature. E. Paradoxography and Antiquities. Fasc. 2. Paradoxographers of Imperial Times and Undated Authors [Nos. 1667–1693].* Leiden; Boston: Brill. 9–338.

Sharples, R. 1995. *Theophrastus of Eresus. Sources for his Life, Writings, Thought and Influence. Commentary Volume 5. Sources on Biology (Human Physiology, Living Creatures, Botany: Texts).* Leiden; New York; Köln: Brill. 328–435.

Sørensen, S. 2022. "1679. Paradoxographus Florentinus." In Schorn, S. ed. *Felix Jacoby. Die Fragmente der Griechischen Historiker Continued. Part IV. Biography and Antiquarian Literature. E. Paradoxography and Antiquities. Fasc. 2. Paradoxographers of Imperial Times and Undated Authors [Nos. 1667–1693].* Leiden; Boston: Brill. 577–630.

Spittler, J. "1672. Apollonios." In Schorn, S. ed. *Felix Jacoby. Die Fragmente der Griechischen Historiker Continued. Part IV. Biography and Antiquarian Literature. E. Paradoxography and Antiquities. Fasc. 2. Paradoxographers of Imperial Times and Undated Authors [Nos. 1667–1693].* Leiden; Boston: Brill. 387–520.

Stern, J. 1996. *Palaephatus. Peri Apiston. On Unbelievable Tales.* Wauconda: Bochazy-Carducci.

Stern, J. 2008. "Paradoxographus Vaticanus." In Heilen, S. Kirstein, R. Smith, R., Trzaskoma, R., van der Wal, R. and Vorwerk, M. eds. *In pursuit of Wissenschaft: Festschrift für William M. Calder III zum 75. Geburtstag. Spudasmata, Bd. 119.* Hildesheim; New York: Georg Olms Verlag. 437–466.

Vanotti, G. 2007. *Aristotle. Racconti meravigliosi.* Milan: Bompani.

Vian, F. 1981. *Apollonios de Rhodes: Argonautiques. Tome III. Chant IV.* Paris: Les Belles Lettres.

Vian, F. 1987. "Poésie et géographie: les retours des Argonautes." *Comptes rendus des séances de l'Académie des Inscriptions et Belles-Lettres* 131: 249–262.

Webster, G. 2015. *The Archaeology of Nuragic Sardinia.* Sheffeild; Bristol: Equinox.

West, S. 1985. "Herodotus' ephigraphical interests." *Classical Quarterly* 35: 278–305.

Zachos, G. 2013. "Epeios in Greece and Italy. Two Different Traditions in One Person." *Athenaeum* 101: 5–23.

Zucca, R., ed. 2004. *Logos peri tēs Sardous: le fonti classiche e la Sardegna. Atti del convegno di studi, Lanusei, 29 dicembre 1998.* Rome: Carocci.

6   Homer and Homeric exegesis in pseudo-Aristotle's *De mirabilibus auscultationibus* 115

Charles Delattre

Entry 115[1] in the *De mirabilibus auscultationibus* of pseudo-Aristotle is a rather special text if we take into account the whole anthology. Certainly, its subject is not remarkable in itself. It is devoted to a description of the straits of Messina and the flows from the Italian coast to Sicily and back. The paragraph thus fits the section (78–151) that deals with *paradoxa* that can be observed at different points in the Mediterranean world and that go back, for some of them, to historians such as Timaeus, Lycus of Rhegium, and Theopompus.

More precisely, if we admit the changes introduced in the order of the entries by Ciro Giacomelli, the text is directly related to *Mir.* 112–114. *Mir.* 112 refers to a Sicilian "little lake having a circumference about the size of a shield" (λιμνίον τι ἔχον ὅσον ἀσπίδος τὸ περίμετρον) that grows larger and expands when people go into it. *Mir.* 113–114 are also devoted to Sicily and mention, "in the empire of the Carthaginians" (ἐν τῇ ἐπικρατείᾳ τῶν Καρχηδονίων), the mountain Ouranion, which "gives a most delightful air to travellers" (ἡδίστην τινὰ τοῖς ὁδοιποροῦσι προσβάλλειν τὴν ἀναπνοήν), a "spring of oil" (κρήνην ἐλαίου), and "a natural rock, of vast size" (πέτραν αὐτοφυᾶ, μεγάλην τῷ μεγέθει). Unlike the mountain Ouranion, whose charm lies only in the flowers that cover it and which offers itself as a pleasant stay to travelers of all kinds, the spring and the rock behave differently according to the circumstances. The spring in *Mir.* 113–114 is sensitive to the presence of individuals, and even quadrupeds: more oil gushes from it when somebody pure comes close. As for the rock, it has the particularity of "sending up a flame of fire" (φλόγα ἀναπέμπειν πυρός) in summer and a "spring of water" (κρουνὸν ὕδατος) in winter.

Mir. 112 refers to a Polycritus, author of *Sicelica* (*FGrHist* 559), for the mention of the spring.[2] For their part, *Mir.* 113–114 have no determinate source, the author referring only to a general "they say" (φασι). Following Karl Müllenhof,

---

1 According to the ordering in Giacomelli's edition (2023), *Mir.* 130 in Bekker's. Translations of the *Mirabilia* are from Hett (1936), slightly modified with the help of Robert Mayhew's draft translation for a new Loeb edition of pseudo-Aristotle's *Minor Works*. I would like to thank the conference participants, and especially Robert Mayhew, for their comments and suggestions.

2 *Suppl. Hell.* 696 suggests, not without hesitation, identifying the Polycritus mentioned by *Mir.* and listed as *FGrHist* 559, with the Polycritus mentioned by pseudo-Antig. Car. *Mir.* 135, and listed as *FGrHist* 128.

DOI: 10.4324/9781003437819-7

Hellmut Flashar defends the idea that *Mir*. 112–114 go back indirectly to Polycritus, and were transmitted through Lycus of Rhegium.[3] Finally, section 115 begins with a general mention of "quite a large number of authors" (ἄλλοι μὲν πλείους γεγράφασι), but then refers to a "he" (οὗτος) which is likely, in the new set of chapters defined by Ciro Giacomelli, to be identical to Polycritus.[4]

In any case, the length of *Mir*. 115 is rather unusual (257 words) if we compare it to the other sections in *Mir*. The paragraph may be a prose transcription of a versified passage from Polycritus, thus retaining certain characteristics of its source. If this is the case, it means that the author of the transcription considered these features worth preserving for the specific audience of the anthology. If the length of the passage is not due to the details given by Polycritus, but comes from additions made by the author of the entry, the passage is all the more remarkable for what it can tell us about the compositional and writing techniques used here. The natural phenomenon that occupies the straits of Messina, a gigantic swell movement that causes an impressive ebb and flow, is sufficiently surprising to qualify as τερατῶδες, "monstrous." But it is the section itself that becomes like a τέρας, a "monster," in the economy of the *Mir*.

This textual and natural prodigy is related by some commentators to the reality of the currents in the strait of Messina, as does Gabriella Vanotti among others: "I fenomeni meravigliosi qui riportati sono forse da identificare con quelli che ancor oggi si verificano in prossimità dello stretto e sono denominati 'fata Morgana'."[5] Thus, there is no longer any distance between the text and the nature that surrounds us. The description of the *Mir*. is the result of an autopsy, of a personal experience by the source's author (Polycritus?), which is after all possible. But to convince ourselves of this, it is also up to us to make the journey to the straits to verify the veracity of the reported details. We are invited to repeat the experience of Polycritus, and to put the text on top of the landscape of the straits of Messina, in such a way as to remove any rhetorical and aesthetic dimension from the text, as if it were only a vehicle for information that immediately translates the nature of the place. On this interpretation, the text is a perfect representation of a natural fact, the observation of which can be both direct and repeated. If Vanotti's commentary seems reasonable, it is perhaps because it is based on a methodology that is directly familiar to us.[6]

---

3 Flashar (1972) 128, regarding *Mir*. 113: "Zur Beurteilung der Mir. genügt jedenfalls, daß die unmittelbare Quelle mit großer Wahrscheinlichkeit Lykos ist." Flashar also rejects Geffcken's proposal to identify Timaeus as a source, as it is "completely made up out of thin air" ("völlig aus der Luft gegriffen"). See also the commentary in Flashar (1972) 129 at *Mir*. 114, where he adopts Lycus without reservation as an intermediary source.

4 See Giacomelli (2023) regarding *Mir*. 115. While refusing to modify the order of the chapters, Flashar (1972) 136 proposes to find in *Mir*. 115 (identified as *Mir*. 130 in his edition) the same auctorial chain as in *Mir*. 112–114: the "highly stylized" ("hochstilisierte") paragraph goes back "most probably to Lycus of Rhegium, by analogy with *Mir*. 112," and beyond to Polycritus ("hier ist in Analogie zu 112 Lykos von Rhegium die wahrscheinlichste Vermutung").

5 Vanotti (2007) 202.

6 On this empiricism, which differs from the ancient autopsy and on which the idea of science and modernity is built, see Latour (1993) 17–18, who follows Shapin and Schaffer (1985) about Boyle

I would like to propose a radically different reading, which places this passage of the *Mir.* in a very different context: a constellation of what I argue are related texts. It is not the relationship to real space, to the geography of the straits, that I consider important, but the place this paragraph occupies among the various proposals made by certain scholars in the ancient world.

1. Ancient interpretations of Charybdis

Although the name Charybdis is not present in the *Mir.*, it is the existence of this monster in the *Odyssey* and a series of interpretations that associate it with the straits of Messina that explain some of the features in this paragraph.

Charybdis is mentioned three times in the *Odyssey*, in the stories told by Odysseus before the king of the Phaeacians and his court (12.101–107, then 234–244 and 430–446): first, Circe warns Odysseus at the beginning of Book 12 against the dangers he will encounter on his way. She lists the Sirens, the passage between Charybdis and Scylla, and the island of the Sun where that god's oxen graze. Odysseus then develops each episode, in the order in which Circe indicated them. Finally, the shipwrecked Odysseus is sent back to Charybdis, and owes his life only to the wild fig tree, growing on the monster's rock, to which he manages to cling.

The identification of Charybdis with the currents of the straits of Messina is made in the first scholion on the passage[7]:

θαλάσσης δὲ βάθος ἀναποτικόν ἐστι μεταξὺ τοῦ Ἀδρίου καὶ τοῦ Τυρρηνικοῦ πελάγους.

It is an all-consuming sea chasm between the Adriatic and Tyrrhenian Seas.

This equivalence between the abyss that swallows up and spits out Odysseus in the *Odyssey* and the straits of Messina can be found in the Latin commentary tradition, and from there in the encyclopedic tradition, for example in Pliny the Elder, in the first details he gives about the island[8]:

in eo freto est scopulus Scylla, item Charybdis mare verticosum, amb<a>e clarae saevitia.

in these straits is the rock of Scylla, as also Charybdis, a whirlpool of the sea, both of them noted for their perils.

In a number of texts, the straits of Messina are defined as the result of a cataclysmic event, which led to the island of Sicily becoming severed from Italian territory.

---

and Hobbes. Incidentally, Latour uses the Greek term *doxa* to qualify the mechanism for winning approval regarding the definition of a natural phenomenon.

7   *Schol.* BQV Hom. *Od.* 12.104 Dindorf.

8   Plin. *HN* 3.86 (ed. Mayhoff; transl. Bostock and Riley).

This is indicated by Pliny, for example, in the sentence that precedes his mention of Charybdis and Scylla:

> *quondam Bruttio agro cohaerens, mox interfuso mari avulsa, <XV> in longitudinem freto, in latitudinem autem M D p. iuxta Columnam Regiam. Ab hoc dehiscendi argumento Rhegium Graeci nomen dedere oppido in margine Italiae sito.*

In former times it was a continuation of the territory of Bruttium, but, in consequence of the overflowing of the sea, became severed from it; thus forming a strait of 15 miles in length, and a mile and a half in width in the vicinity of the Pillar of Rhegium. It was from this circumstance of the land being severed asunder that the Greeks gave the name of Rhegium to the town situated on the Italian shore.

The name Rhegium, associated with the verb ῥήγνυμι, "to break,"[9] is also associated in Diodorus with a cataclysm that determined the formation of present-day Sicily[10]:

> Φασὶ γὰρ οἱ παλαιοὶ μυθογράφοι τὴν Σικελίαν τὸ πρὸ τοῦ χερρόνησον οὖσαν ὕστερον γενέσθαι νῆσον διὰ τοιαύτας αἰτίας. Τὸν ἰσθμὸν κατὰ τὸ στενώτατον ὑπὸ δυοῖν πλευρῶν θαλάττῃ προσκλυζόμενον ἀναρραγῆναι, καὶ τὸν τόπον ἀπὸ τούτου Ῥήγιον ὀνομασθῆναι, καὶ τὴν ὕστερον πολλοῖς ἔτεσι κτισθεῖσαν πόλιν τυχεῖν τῆς ὁμωνύμου προσηγορίας.

The ancient commentators, that is, say that Sicily was originally a peninsula, and that afterward it became an island, the cause being somewhat as follows. The isthmus at its narrowest point was subjected to the dash of the waves of the sea on its two sides and so a break was made, and for this reason the spot was named Rhegium (*break*), and the city which was founded many years later received the same appellation as the place.

In a longer version, this is also what we find in Strabo. The geographer first details the origins and foundation of Rhegium, then gives a general definition of the city and the region, and finally focuses on the etymology of the name Rhegium[11]:

> ὠνομάσθη δὲ Ῥήγιον εἴθ᾽, ὥς φησιν Αἰσχύλος, διὰ τὸ συμβὰν πάθος τῇ χώρᾳ ταύτῃ · ἀπορραγῆναι γὰρ ἀπὸ τῆς ἠπείρου τὴν Σικελίαν ὑπὸ σεισμῶν ἄλλοι τε κἀκεῖνος εἴρηκεν "ἀφ᾽ οὗ δὴ Ῥήγιον κικλήσκεται."

It was named Rhegium, either, as Aeschylus says, because of the calamity that had befallen this region, for, as both he and others state, Sicily was once

---

9 The use of the verb ῥήγνυται, even without mention of Rhegium, to gloss the Homeric τρὶς δ᾽ ἀναροιβδεῖ is perhaps a remnant of this interpretative tradition (*Schol.* HQ Hom. *Od.* 12.105 Dindorf).

10 Diod. Sic. 4.85.3–4 (transl. Oldfather, with modification for μυθογράφοι and for ἀναρραγῆναι).

11 Strabo 6.1.6 (translations of Strabo are by H.L. Jones, *LCL*).

"rent" from the continent by earthquakes, and so "from this fact," he adds, "it is called Rhegium."[12]

This same cataclysm is the subject of a small paragraph in *Paradoxographus Vaticanus*[13]:

Ἀκύλιος ὁ Ῥωμαῖος ἱστορικός φησι τὴν Σικελίαν πρὸ τοῦ κατακλυσμοῦ μὴ νῆσον εἶναι ὡς σήμερον, ἀλλ᾽ ἤπειρον γενέσθαι συνημμένην τῇ ὕστερον Ἰταλίᾳ· ἐκ δὲ τῆς ἐπικλύσεως τῶν ῥευμάτων τῶν ῥιζῶν ἀποσπασθεῖσαν τοῦ Ἀπεννίνου, κατὰ τὸ Σκύλλαιον ῥαγείσης τῆς ἠπείρου, νῆσον ἀποκαταστῆναι καὶ διὰ τοῦτο Ῥήγιον ἀποκληθῆναι τὸ πλευρὸν τῆς Ἰταλίας ἐκεῖνο.

According to the Roman historian Acilius, Sicily before the cataclysm was not an island as it is today, but was a continental piece attached to what was later Italy; but its foundations were flooded by the overflow of the rivers and it was torn away from the Apennines; the continent was torn apart at the level of the Skyllaion and an island was born; and for this reason that side of Italy was called Rhegium.

These different texts are consistent and build an argument that plays on both the toponymy (Rhegium) and the implicit definition of the straits as a fracture between the island and the continent. The result is a narrative about an event firmly situated in the past, even if the chronology is imprecise and characterized by extreme violence.

The causes of the fracture diverge according to the sources: for Pliny, a catastrophic flood (*interfuso mari*) is at the origin of the straits, and this is also the case for Diodorus (προσκλυζόμενον) and for the *Paradoxographus Vaticanus* (τῆς ἐπικλύσεως τῶν ῥευμάτων). The ebb and flow unleashed by the monster Charybdis in the *Odyssey* are transformed and become the mythical equivalent of a geographical upheaval, as the cataclysmic process still involves the power of running water: under the force of the waves, what was then a peninsula broke away, was swept away by the current and became the autonomous island of Sicily. It is therefore not simply a matter of a flood transforming a landscape and changing its definition from a cartographic point of view. The island of Sicily did not come into being because it was isolated by a new sea channel from the mainland, it was physically and violently displaced.

In the explanation given by Strabo, the break is also a real event, but its causes differ, as it is triggered by earthquakes (ὑπὸ σεισμῶν).[14] The geographer comments afterward on the possibility that this cataclysm was caused by the same underground fires, accompanied by pressurized air, as those that triggered

---

12 Aesch. F 402 Radt.
13 39 Giannini, and Acilius *FGrHist/BNJ* 813 F 3, with a brief commentary in Sørensen (2022) 621. For further references, see the long list by Radt (1985) 440–441.
14 See also Dion. Hal. *Ant. Rom.* 19.2.2, if we accept the anonymous emendation διέστησε σεισμός in *Literarisches Zentralblatt für Deutschland* 1870: 893, mentioned by Radt (1985) 440

earthquakes in the region, even though these have become rare, he says.[15] This particular line of reasoning echoes a theory developed by Posidonius on telluric movements,[16] but it is not impossible that Strabo's argument, which stresses the importance of fire in the process, has also to do with the Homeric comparison that parallels the bubbling of water vomited by Charybdis with a liquid on fire[17]:

ἦ τοι ὅτ᾽ ἐξεμέσειε, λέβης ὣς ἐν πυρὶ πολλῷ
πᾶσ᾽ ἀναμορμύρεσκε κυκωμένη·

When she spewed it back out, like a cauldron over a blazing fire,
    she was all seething turbulence.

A passage from Philo of Alexandria obviously relates to the first group of texts[18]:

<ἢ> τὴν περὶ τὸν ἱερώτατον Σικελικὸν πορθμὸν ᾀδομένην ἱστορίαν ἀγνοεῖτε; <ἣν μὲν γὰρ> τὸ παλαιὸν ἠπείρῳ Ἰταλίᾳ Σικελία συνάπτουσα, μεγάλων δὲ τῶν παρ᾽ ἑκάτερα πελαγῶν βιαίοις πνεύμασιν ἐξ ἐναντίας ἐπιδραμόντων, ἡ μεθόριος ἐπεκλύσθη καὶ ἀνερράγη, παρ᾽ ἣν καὶ πόλις ἐπώνυμος τοῦ πάθους Ῥήγιον κτισθεῖσα ὠνομάσθη. Καὶ τοὐναντίον οὐ προσεδόκησεν ἄν τις ἀπέβη· συνεζεύχθη μὲν γὰρ τὰ τέως διεστῶτα πελάγη κατὰ τὴν σύρρυσιν ἑνωθέντα, ἡ δὲ ἡνωμένη γῆ τῷ μεθορίῳ πορθμῷ διεζεύχθη, παρ᾽ ὃν ἤπειρος οὖσα Σικελία νῆσος ἐβιάσθη γενέσθαι.

Do you not know the celebrated story of the sacred Sicilian straits? In old days Sicily joined on to the mainland of Italy but under the assault on either side of great seas driven by violent winds from opposite directions, the land between them was inundated and broken up and at its side a city was founded, whose name of Rhegium records what happened to it. The result was the opposite of what one would have expected. The seas which were hitherto divided joined together through their confluence, while the land once united was divided by the intervening straits, by which Sicily, which had been mainland, was forced to become an island.

There is unambiguous mention of the original cataclysm remembered by Rhegium, but also of a catastrophic flood (ἐπεκλύσθη), augmented by violent winds. The fracture, which is directly associated with the flood, presumably represents an actual event here, given that Philo's overall argument is about land appearing suddenly or disappearing, as part of a more general reflection on the permanence of the world. The whole of Philo's passage draws on arguments developed

---

15 In contrast, the modern era has seen several destructive earthquakes in the region, in particular in 1783 and 1908.

16 See F 12 Edelstein and Kidd (Diog. Laert. 7.154). However, only the subterranean winds, not the aerial ones, are mentioned in this fragment.

17 Hom. *Od.* 12.237–238. Translations of Homer here and elsewhere are by P. Green (University of California Press).

18 Philo, *De aeternitate mundi* 139 (transl. Colson).

by Theophrastus, but there is scholarly disagreement about which of them precisely go back to the Peripatetic tradition.[19] In particular, it should be noted here that the elements concerning Sicilian volcanism, the Lipari Islands and Mount Etna, which could be traced back to Theophrastus,[20] are found in another section of the *Mir.* (37b–38b and 40), and therefore need not be associated with *Mir.* 112–115, even though Strabo, for example, brings together physical investigation of the straits of Sicily and regional volcanism.

Mir. 115 stands out in this set by the absence of any reference to the cataclysmic event associated with the name Rhegium. Clearly, the process of analysis and commentary associated with the straits of Messina does not rely in the *Mir.* on the same arguments as in the rest of the physical and geographical exegetical tradition common to the Hellenistic and Imperial periods. There is no mention of the past, no reflection on how the places were formed, but the description of a recurrent and observable phenomenon.

2. Mir. *and the waves*

It is necessary to return to the way *Mir.* 115 is structured in order to underline its originality and uniqueness in relation to the rest of the corpus. The phenomenon described appears to be consubstantial with the straits of Sicily, and is the direct consequence of the way in which the places are shaped. A first phenomenon is envisaged: the formation of a large wave, due to the narrowness of the strait, from a "surge of water" (κλύδων) that rushes in from the Tyrrhenian Sea, following a north–south current. These waves hit both Sicily and Italy, at the level of Rhegium. At this point, the waves sometimes collide and sometimes separate. In a second phase, the wave ebbs, in the midst of whirlpools that give off a multitude of chromatic hues. The phenomenon of ebb and flow, which is in itself impressive, is thus reinforced by a second phenomenon that occurs during the first phase, and which mirrors it: to the great wave breaking in the first phase corresponds the "crashing" of the waves against each other (συμπίπτειν), and to the general ebb corresponds the moment when these same waves separate, leaving a glimpse of a "horrible abyss" (βαθεῖαν καὶ φρικώδη).

These waves are clearly explained in *Mir.* by the accumulation of the flow "being carried from a large mass of water which becomes enclosed in a narrow space" (φερόμενον ἐκ μεγάλου πελάγους εἰς στενὸν συγκλείεσθαι), in the case of the primary phenomenon, with the use of the related noun "enclosure" (συγκλεισμός) in turn explaining the secondary phenomenon.[21] As Strabo points out during a lengthy discussion of land formation, the straits are traversed by currents that obey specific laws.[22] In a direct attack on Eratosthenes, Strabo contra-

19 Thphr. F 184 FHS&G, with the commentary by Sharples (1998) 132 and 135–136 on possible sources of Theophrastus. See also Arist. *Pr.* 23.5.932a.2–20 and below.
20 Thphr. F 195–196 FHS&G.
21 I adopt here Giacomelli's text, συγκλεισμός, against Hett's συγκλυσμός, to be found in *Marc. gr.* IV 58.
22 1.3.7: οἱ δὲ πορθμοὶ ῥευματίζονται κατ' ἄλλον τρόπον ("the current through the straits is accounted for by another principle"). The debate in this passage focuses on the role of rivers in the formation of seas and on alluviation.

dicts the analyses of the Alexandrian geographer on the very example of the straits of Sicily. Eratosthenes seemed to base his reasoning on differences in surface level between different parts of the sea (1.3.11). Strabo prefers the analyses of Posidonius and Athenodorus, which he unfortunately does not develop, on the grounds that their discussion "goes deeper into natural science than comports with the purpose of the present work."[23] We are thus reduced to assuming that both scholars relied on landscape features to explain the phenomena of ebb and flow, just as the author of the section does in the *Mir.* The use of the term φυσικώτερον by Strabo is a clue, and the rest of Strabo's text, which emphasizes how particular each strait (Sicily, Chalcis, Byzantium) is in fact, seems to confirm this. There is no general principle to explain the flows in the straits, but specific rules that can only come from their very nature.

Although Strabo does not mention it, this idea is already clearly established by Aristotle, at the beginning of his book II of the *Meteorologica*, where he discusses the origin of the seas (2.1.2.354a.6–12, transl. Lee):

ῥέουσα δ' ἡ θάλαττα φαίνεται κατά τε τὰς στενότητας, εἴ που διὰ τὴν περιέχουσαν γῆν εἰς μικρὸν ἐκ μεγάλου συνάγεται πελάγους, διὰ τὸ ταλαντεύεσθαι δεῦρο κἀκεῖσε πολλάκις. τοῦτο δ' ἐν μὲν πολλῷ πλήθει θαλάττης ἄδηλον· ἡ δὲ διὰ τὴν στενότητα τῆς γῆς ὀλίγον ἐπέχει τόπον, ἀναγκαῖον τὴν ἐν τῷ πελάγει μικρὰν ταλάντωσιν ἐκεῖ φαίνεσθαι μεγάλην.

The sea obviously flows in narrow places where a large expanse of water is contracted by the surrounding land into a small space: but this because the sea ebbs and flows frequently. In a large expanse this motion is unnoticeable; but where the expanse is small because the shore constrict it the ebb and flow which in the open sea seemed small now seems strong.

It should be noted that the name Charybdis comes up a little later in the text, in a discussion of the salinity of seawater (2.2.3.356b6–17), but not in connection with the straits of Messina: Aristotle rejects a fable of Aesop, according to which Charybdis made mountains appear first, then islands, by twice swallowing a vast portion of seawater. In Aristotle's argument, the discussion of sea flows and currents is clearly distinguished from the Homeric figure of Charybdis, at least in the *Meteorologica*.

A more complex passage from pseudo-Aristotle's *Problemata* enriches the description of the straits of Messina. The general discussion is about the fact that "sometimes ships running on the sea in good weather are swallowed up and disappear, such that no wreckage floats up" (23.5.931b39–932a2: ἐνίοτε πλοῖα θέοντα ἐν τῇ θαλάττῃ εὐδίας καταπίνεται καὶ ἀφανῆ γίνεται, ὥστε μηδὲ ναυάγιον ἀναπλεῖν). Aristotle uses the particular example of Messina and introduces, besides the narrowness of the strait, two important factors. First, the winds favor and even create

---

23 1.3.12: ἐχόντων καὶ αὐτῶν φυσικώτερον λόγον ἢ κατὰ τὴν νῦν ὑπόθεσιν. Strabo's text serves as a source for Posidonius, *FGrHist* 87 F 82b.20–21 and Athenodorus, *FGrHist* 746 F 6b 20–21.

the currents, and second, abysmal underwater cavities are likely to swallow up the waves (23.5.932a5–8)[24]:

τὰ δὲ περὶ Μεσήνην ἐν τῷ πορθμῷ πάσχει μὲν τοῦτο διὰ τὸν ῥοῦν (γίνονται γὰρ αἱ δῖναι ὑπὸ τούτου) καὶ καταπίνεται εἰς βυθὸν διὰ ταῦτά τε, καὶ ὅτι θάλαττά τε βαθεῖα καὶ γῆ ὕπαντρος μέχρι πόρρω.

Now ships in the Straits of Messina experience this because of the current (since the whirlpools are produced by this), and they are swallowed up into the abyss both owing to this and also because the sea is deep and the earth is cavernous beneath it to a great distance.

The winds cause currents to meet head-on in the narrow space of a channel, such as the strait of Messina, resulting in violent eddies (23.5.932a10–18):

Ὁ δὲ ῥοῦς γίνεται, ὅταν παυσαμένου τοῦ προτέρου ἀντιπνεύσῃ ἐπὶ τῆς θαλάττης ῥεούσης τῷ προτέρῳ πνεύματι, μάλιστα δὲ ὅταν νότος ἀντιπνεύσῃ. Ἀντιπνέοντα γὰρ ἀλλήλοις τὰ ῥεύματα παρεκθλίβεται ὥσπερ ἐν τοῖς ποταμοῖς, καὶ γίνονται αἱ δῖναι. Φέρεται δὲ ἑλιττομένη ἡ ἀρχὴ τῆς κινήσεως ἄνωθεν ἰσχυρὰ οὖσα. ἐπεὶ οὖν εἰς πλάγιον οὐκ ἔστιν ὁρμᾶν (ἀντωθεῖται γὰρ ὑπ' ἀλλήλων), ἀνάγκη εἰς βάθος ὠθεῖσθαι, ὥστε καὶ ὃ ἂν ληφθῇ ὑπὸ τῆς δίνης, ἀνάγκη συγκαταφέρεσθαι.

Now a current occurs when, the previous wind having stopped, wind blows in the opposite direction over a sea that is flowing under the influence of the previous wind, and especially when it is a south wind blowing in the opposite direction. For the currents blowing at each other in opposite directions thrust each other aside, just as they do in rivers, and whirlpools are formed. And the original motion, being strong, is carried turning round and round from above. Therefore, since it is not possible for the currents to rush sideways (for they push against each other), they must be pushed down into the depths, and so whatever is caught by the whirlpool must be carried down with it.

These whirlpools explain the shipwrecks, the presence of subterranean abysses being the reason why the debris of wrecked ships only exceptionally comes to the surface (23.5.932a.9–10):

φέρουσιν οὖν εἰς ταῦτα ἀποβιαζόμεναι αἱ δῖναι · διὸ οὐκ ἀναπλεῖ αὐτόθι τὰ ναυάγια.

The whirlpools, forcing the ships away (*from the surface*), carry them into these caverns; this is why wreckage does not float up at the spot (*where the ship disappeared*).

---

24 See also Arist. *Pr.* 23.17.933b5–10, which questions the relationship between the violence of the waves and the depth of the sea bed. I have made use here and elsewhere of Robert Mayhew's translation of pseudo-Aristotle's *Problemata* in his Loeb edition.

Mir. 115 shares with the commentaries on the straits of Messina a similar definition of the general topography of Sicily and the Italian side: all of them stress the narrowness of the straits and the strength of the currents that develop there, especially Philo.[25] But these same commentaries integrate these elements to defend an etiology of the name Rhegium, forging the idea of a catastrophic event that occurred in the distant past, while the *Mir.* are part of a reflection that is not specific to Sicily, in discussions on the nature of the sea and of the currents, for which the strait constitutes one example among others. Whether these features come directly from Polycritus' *Sicelica* or are an addition by the compiler who wrote the entry, they are derived from an interpretive tradition that goes back at least to Aristotle. This specific tradition makes *phusis* a world to be explored in the present and an expository principle, not an opportunity to create an origins story to explain how the present world was formed.

3. Mir. *and* thauma

Mir. 115 is, moreover, a particularly controlled textual unit, from the point of view of its construction, as we shall see, and of its objectives: it is, to use an expression of Greta Hawes, the result of a "recognizable interpretative and narrative strategy."[26]

A first clue, still insufficient in itself, lies in the choice of the term, τερατῶδες, which qualifies the whole phenomenon. At first sight, the role of this adjective is trivial: like θαυμαστός and, to a lesser extent, θαυμάσιος, θαυμάζω and θαῦμα,[27] and ἄπιστος, which is much rarer than θαυμαστός,[28] the word τερατώδης[29] qualifies in the *Mir.*, as in other paradoxographical texts,[30] the natural phenomenon that is described. It thus attracts the reader's attention, arouses his interest, and, above all, gives a status to what is depicted: the author validates the existence of the phenomenon, by the mere fact that he describes and qualifies it.[31]

The distribution of the different terms within the *Mir.*, however, introduces distinctions between the three groups. Θαῦμα and its family are indeed used throughout the text, on a regular basis, without distinction between the parts of the *Mir.* All these occurrences can therefore be considered as stylistically unmarked lexis,

---

25 Philo, *De aeternitate mundi* 139: μεγάλων δὲ τῶν παρ᾽ ἑκάτερα πελαγῶν βιαίοις πνεύμασιν ἐξ ἐναντίας ἐπιδραμόντων ("under the assault on either side of great seas driven by violent winds from opposite directions").

26 Hawes (2014) 3.

27 30 (832b): θαυμασιώτατον; 40 (833a); θαυμαστόν; 61 (835a): θαυμαστόν; 79 (836a): θαυμαστόν; 84 (836b): θαυμαστήν; 92 (837b): θαῦμα; 102.1 (839a): θαυμαστόν and θαυμάσιον; 126 (841b): ἄν τις θαυμάσειεν; 127 (841b): θαυμαστόν; 135.3 (842b): θαυμάσειεν ἄν τις; 122 (844b): θαυμαστόν (2 occurrences).

28 In addition to *Mir.* 115, ἄπιστον appears in 126 (841b) and in 121 (844a). The noun ἀπιστία is absent from the text.

29 Τερατώδης appears three times, in addition to *Mir.* 115, in 101.1 (838b), in 101.2 (839a), and in 126 (841b). The noun τέρας is absent from the text.

30 The author of the *Mir.* however never uses either παράδοξον, or the expressions παρὰ δόξαν or παρὰ φύσιν, unlike ps.-Antig. Car. or Pausanias.

31 For the example of Pausanias, see Delattre (2018) 205–223.

resonating with the title given to the text in the major manuscripts, Περὶ θαυμασίων ἀκουσμάτων.[32] On the other hand, τερατώδης and ἄπιστον are less frequent, and appear in certain passages where the author draws his or her reader's attention particularly, by concentrating the number of expressions.

This is the case in *Mir.* 121–122, where two occurrences of θαυμαστόν are themselves preceded by ἄπιστον, and in *Mir.* 101.1–102.1, where τερατώδης appears twice, in 101.1 and 101.2, and precedes *Mir.* 102.1, where the author uses both θαυμαστόν and θαυμάσιον. *Mir.* 126–127, on the other hand, combine the three semantic families: a combination of τερατώδης and ἄπιστον completes ἄν τις θαυμάσειεν and θαυμαστόν. The presence of τερατώδης and ἄπιστον at the same time in our section 115 signals it in turn as one of the most noteworthy passages in the *Mir.*

One might have thought that the use of τερατῶδες in *Mir.* 115 had the role of reviving the semantics of the adjective, by qualifying a natural phenomenon (the ebb and flow) which would carry with it the memory of the "monstrous prodigy" (τέρας) that is Charybdis. But the few uses of τερατώδης in the *Mir.* do not allow us to go in this direction. Certainly, in *Mir.* 101, the adjective is applied to an acoustic phenomenon, the sound of drums and cymbals, laughter, clamors, and castanets, coming from a tomb on the island Lipara: the sounds are reminiscent of a Dionysian procession that has become a ghost parade. The phenomenon narrated in *Mir.* 101.2 also has to do with the island Lipara and concerns a tomb as well: a drunken individual fell into a cave where he remained for three days. He was thought to be dead when he was found on the fourth day, but woke up in the middle of his own burial. Here the author expresses his skepticism, which partly motivates the use of the adjective τερατωδέστερον. In both paragraphs, then, the adjective τερατώδης is associated with death and its mysteries, and with phenomena which, like Charybdis, are not necessarily a matter of nature, but of tales: see the use of μυθολογοῦσι ("legend has it") in *Mir.* 101.1, and of λέγουσι ("they say") in *Mir.* 101.2, whose ending emphasizes that all this is μυθωδέστερον ("more like legend"). However, the phenomenon described in *Mir.* 126 does not allow us to generalize this sense of τερατώδης that we have just identified: the text only mentions tame falcons coming to the aid of children to hunt small birds. Even if the order of nature seems temporarily overturned, since the falcons go so far as to abandon the prey they catch to the children on their own, the text does not describe anything equivalent to a ghostly apparition, an unexpected resurrection or a legendary monster. The phenomena qualified by τερατώδης are too diverse for the adjective to define a uniform category into which Charybdis would fit.

So it is clearly not the possible allusion to Charybdis that τερατώδης would provide, but its association with ἄπιστον that reinforces the importance of the passage in the eyes of the reader, and induces the latter to pay more attention to this entry than to others. But this association can then act like a trigger, and push the reader to reinterpret the first sentence of the entry: when the author of the compilation

---

32 Even if manuscripts H, P, R, and G have παραδόξων instead of θαυμασίων. P, R, and G are copies of a now lost *x*, and *x* and H make up the branch γ of the *stemma codicum* (see the *stemma riassuntivo* in Giacomelli [2021] 269).

writes that "about the Sicilan strait, a good number of people have written" (περὶ δὲ τοῦ πορθμοῦ τῆς Σικελίας καὶ ἄλλοι μὲν πλείους γεγράφασι), is he referring only to the strictly geographical tradition? Should we not include Homer, who precisely plays such an important role in the very establishment of a geographical discourse, for example, in Strabo? This means that τερατῶδες in *Mir.* 115 does indeed refer to Charybdis, but not by a direct and transparent allusion to the monster: it is the insistence with which the author of the text qualifies the phenomenon described that prompts the reader to plunge into his memory and mobilize references that a quick reading would have allowed him to ignore.

4. Mir. *and* Odyssey

One might think that the figure of Charybdis plays only a distant role in *Mir.* 115. After all, her name does not appear, and the whole description is in line with what could be the account of a direct observation of the straits of Sicily. No literary references or mythological memories need to be summoned for the entry to achieve its aims: to turn the straits into an example of "wonder" (θαῦμα), to arouse the curiosity and reflection of the audience. If Charybdis is present, it is perhaps only as the origin of a tradition that draws attention to the straits of Messina. In that case, *Mir.* 115 would be only what it appears to be, a careful description of a natural phenomenon whose violence attracts attention.

A close reading of the three passages linked to Charybdis from the *Odyssey*, in juxtaposition with the description of the straits of Sicily in *Mir.* 115, might suggest, however, that there is a closer connection. More precisely, it seems to me that *Mir.* 115 can be read as an expanded paraphrase of and commentary on the relevant verses in the *Odyssey*, that is, it is a long gloss.

The first mention of Charybdis, in Circe's words reported by Odysseus himself,[33] associates this monster with Scylla, and underlines the narrowness of the passage between the two: they are "within easy bowshot" (12.102: καί κεν διοϊστεύσειας), each on a rock or promontory (σκόπελος), one high and the other lower. Unlike Scylla, which directly devours passing sailors, Charybdis is invisible, her body is not described: only the effects of her actions are perceptible, a gigantic ebb and flow that occurs three times a day. The equivalence between what could be a natural phenomenon and the monster is underlined by the etymological play that makes Charybdis (Χάρυβδις) the one that "sucks in" (ἀναρρυβδεῖ), the word being repeated three times in three verses in slightly different forms.[34]

The scholia that focus on ebb and flow in this passage mostly question the frequency and duration of the phenomenon,[35] or gloss the verb ἀναρυβδεῖν with the nouns ἄμπωτις, "engulfment of a liquid," and ἀναρρόφησις, "engulfment of

---

33 Hom. *Od.* 12.101–107.

34 Hom. *Od.* 12.104: ἀναρρυβδεῖ (the variant ἀναρροιβδεῖ is also attested); 105: ἀναρυβδεῖ; 106: ῥυβδήσειεν. The use of κεν ῥύσαιτο ("could save you") in v. 107 could be a final etymological play on the part of the poet.

35 See in particular *schol.* Q Hom. *Od.* 12.105 Dindorf [τρὶς μὲν γάρ τ᾽ ἀνίησιν].

a mush," and emphasize the "mortal danger" of the place (τόπος . . . ἐπικίνδυνος, θανάτου μεστός), which leads to the sinking of many ships where the flow "breaks" (ῥήγνυται), which is why "this flow is also called breakage" (διὸ τὸ ὕδωρ καλεῖται ῥαχία).[36] *Mir.* 115 is composed from a different perspective: the two great movements of ebb and flow described in the text correspond to only one of the three daily episodes through which Charybdis manifests itself, and amplify to the extreme the alternation indicated by verses 12.104–106:

τῷ δ᾽ ὑπὸ δῖα Χάρυβδις ἀναρρυβδεῖ μέλαν ὕδωρ.
Τρὶς μὲν γάρ τ᾽ ἀνίησιν ἐπ᾽ ἤματι, τρὶς δ᾽ ἀναρυβδεῖ,
δεινόν ·

Beneath this divine Charybdis sucks in the dark water –
thrice daily she spews it out upward, thrice sucks it all back
with fearsome effect.

It is perhaps also the third mention of Charybdis, where a single ebb and flow is mentioned, that motivates the reduction of the general description in *Mir.* to a single ebb and flow (*Od.*, 12.430–438):

ἦλθον ἐπὶ Σκύλλης σκόπελον δεινήν τε Χάρυβδιν.
Ἡ μὲν ἀνερρύβδησε θαλάσσης ἁλμυρὸν ὕδωρ·
αὐτὰρ ἐγὼ ποτὶ μακρὸν ἐρινεὸν ὑψόσ᾽ ἀερθείς,
τῷ προσφὺς ἐχόμην ὡς νυκτερίς· . . .
νωλεμέως δ᾽ ἐχόμην, ὄφρ᾽ ἐξεμέσειεν ὀπίσσω
ἱστὸν καὶ τρόπιν αὖτις·

I came to the headland of Skylle and to dread Charybdis, who now
sucked in the salt seawater; but I reached up
as high as I could, caught hold of the tall fig tree,
and clung to it like a bat. . . . So there
I clung on persistently, waiting for her to spew back
the mast and keel again.

The description of the first wave in *Mir.* 115 seems to transcribe some features of the second mention of Charybdis in the *Odyssey*. The text of *Mir.* emphasizes the height of the wave, its foam, and the general resemblance of the phenomenon with those produced by storms:

κῦμα μετέωρον αἴρειν σὺν πολλῷ βρόμῳ ἐπὶ πάνυ πολὺν τόπον τῆς ἄνω φορᾶς,
ὥστε τοῖς μακρὰν ἀπέχουσι σύνοπτον εἶναι τὸν μετεωρισμόν, οὐχ ὅμοιον

---

36 *Schol.* HQ Hom. *Od.* 12.105 Dindorf. We thus find here the verb ῥήγνυμι, not associated with Rhegium, as in Diodorus or Strabo, but with the very idea of "shipwreck" (ναυάγιον), the verb ῥήγνυμι then appearing itself as a gloss of ἄγνυμι, whose radical ἀγ- forms the second term of ναυ-άγιον.

φαινόμενον θαλάσσης ἀναφορᾷ, λευκὸν δὲ καὶ ἀφρῶδες, παραπλήσιον δὲ τοῖς συρμοῖς τοῖς γινομένοις ἐν τοῖς ἀνυπερβλήτοις χειμῶσι.

[A] wave is carried high in the air with a loud noise over a wide space upwards, so that, when hurled high in the air, it can be seen by those who are a long way off, not like the rising of the sea but white and foamy, and like the surges occurring in violent storms.

It is these characteristics precisely that the epic poet emphasizes, in a slightly different order (*Od.* 12.238–239):

πᾶσ᾽ ἀναμορμύρεσκε κυκωμένη· ὑψόσε δ᾽ ἄχνη
ἄκροισι σκοπέλοισιν ἐπ᾽ ἀμφοτέροισιν ἔπιπτεν.

[S]he was all seething turbulence, and the flying spray
was flung up so high that it rained down on the tops of both headlands.

In particular, the mention of "foam" (ἀφρῶδες) in *Mir.* could be a gloss on the hapax ἀναμορμύρεσκε, ἄφρος being associated with the verb μορμύρω in Homer.[37] And it is the phenomenon of reflux that this same passage in Homer may have inspired. Specifically, the author of *Mir.* 115 emphatically describes the moment when the waves, after colliding on the headlands, separate:

Ἐπειδὰν δὲ προσπεσὸν τὸ κῦμα πρὸς ὁποτερονοῦν τῶν τόπων καὶ μετεωρισθὲν ἕως τῶν ἄκρων πάλιν εἰς τὴν ὑπορρέουσαν θάλασσαν κατενεχθῇ, τότε δὴ πάλιν σὺν πολλῷ μὲν βρυχηθμῷ μεγάλαις δὲ καὶ ταχείαις δίναις τὴν θάλασσαν ἀναζεῖν καὶ μετεωρίζεσθαι κυκωμένην ἐκ βυθῶν, παντοδαπὰς δὲ χρόας μεταλλάσσειν· ποτὲ μὲν γὰρ ζοφεράν, ποτὲ δὲ κυανῆν, πολλάκις δὲ πορφυρίζουσαν διαφαίνεσθαι.

But when the wave, falling on either of the spots, and flung as high as the promontories, dashes back again into the sea flowing below, then again with a vast roar and with huge swift eddies, the sea boils up and is hurled high, seething from the depths and changing to every kind of colour: for sometimes it appears dark and sometimes blue, and then again purple.

The "boiling" metaphor (ἀναζεῖν) recalls the "cauldron set on fire" (λέβης ὣς ἐν πυρὶ πολλῷ) of verse 237, and the κυκωμένην ("seething") could be a direct borrowing from v. 238 (πᾶσ᾽ ἀναμορμύρεσκε κυκωμένη, "she was all seething turbulence"). As for the colors listed by the author of the entry, "dark and sometimes blue, and then again purple" (ζοφεράν, ποτὲ δὲ κυανῆν, πολλάκις δὲ πορφυρίζουσαν), they can be interpreted as an amplification of

---

37 Hom. *Il.* 5.599: a river; 18.403: the Ocean; 21.325: the river Scamander. See also Leonidas of Tarentum, *Anth. Pal.* 16.182; Quint. Smyrn. 14.578; Ael. *NA* 14.26; Philostr. *Imag.* 10; Clem. Al. *Paed.* 1.6.40.1. Hsch. μ 1676, *s. v.* μορμύρων, glosses the term with ἀφροὺς ἀποβάλλων ("rejecting foam").

the "κυανέη" that is revealed at the bottom of the abyss when the reflux occurs (*Od.* 12.237–243):

ἤ τοι ὅτ᾽ ἐξεμέσειε, λέβης ὣς ἐν πυρὶ πολλῷ
πᾶσ᾽ ἀναμορμύρεσκε κυκωμένη· ὑψόσε δ᾽ ἄχνη
ἄκροισι σκοπέλοισιν ἐπ᾽ ἀμφοτέροισιν ἔπιπτεν.
Ἀλλ᾽ ὅτ᾽ ἀναβρόξειε θαλάσσης ἁλμυρὸν ὕδωρ,
πᾶσ᾽ ἔντοσθε φάνεσκε κυκωμένη, ἀμφὶ δὲ πέτρη
δεινὸν βεβρύχει, ὑπένερθε δὲ γαῖα φάνεσκε
ψάμμῳ κυανέη·

[W]hen she spewed it back out, like a cauldron over a blazing fire,
she was all seething turbulence, and the flying spray
was flung up so high that it rained down on the tops
of both headlands; when she sucked the salt seawater back,
she exposed her wild inner turmoil, while all around
the rock roared: beneath it the bottom was revealed,
black with sand.

One of the remarkable features of *Mir.* 115 is that it also punctuates the description of each phase of the ebb and flow, including in the secondary episodes, with remarks that make the phenomenon a spectacle, and which can be grouped together in pairs. In the first stage, the rising up (of water) can be seen by those who are a long way off."[38] But in the second stage, this spectacle is rhetorically canceled: the wave-clash is "impossible to describe, and unbearable to look at."[39] The impossibility for a spectator to witness the flow to the end is redoubled by the impossibility for the author to describe it accurately, and therefore by the impossibility for the audience to imagine it exactly. Paradoxically, the author puts forward his inability to produce a vivid (ἐναργές) discourse so that the audience can all the better picture the monstrous (τερατῶδες) character of the phenomenon.

As for the third stage, it becomes a spectacle that imposes itself on the spectators without them being able to avoid it: the scene is "so deep and terrifying to those who are forced to see it that many cannot control themselves, but grow dizzy and fall down from fear."[40] But it is precisely flight that becomes the solution for those who witness the general ebb: "No creeping beast can bear either to hear or to see it, but all flee to the foot of the mountain."[41] Here again, the two

---

38 τοῖς μακρὰν ἀπέχουσι σύνοπτον εἶναι τὸν μετεωρισμόν.
39 ἄπιστον μὲν διηγεῖσθαι, ἀνυπομόνητον δὲ τῇ ὄψει θεάσασθαι.
40 οὕτω βαθεῖαν καὶ φρικώδη τὴν ἄποψιν ποιεῖν τοῖς ἐξ ἀνάγκης θεωμένοις, ὥστε πολλοὺς μὲν μὴ κρατεῖν ἑαυτῶν, ἀλλὰ πίπτειν σκοτουμένους ὑπὸ δέους. It should be noted that the theme of vision makes a discreet appearance here, as a reprise of the first pair of remarks.
41 οὔτε (οὐδὲ Hett) ἀκούειν οὐδὲν ἑρπετὸν οὔθ᾽ ὁρᾶν ὑπομένειν, φεύγειν δὲ πάντα πρὸς τὰς ὑποκειμένας ὑπωρείας.

complementary remarks contradict and cancel each other out. The four remarks thus constitute a coherent system, organized around the pairs seeing/not seeing, forced to see/fleeing not to see. This system makes sense in paradoxographical writing, as a spectacular staging of what *thauma* is, a vision that imposes itself and that can both stupefy, freeze in place, and drive people to flee.[42] These fictitious spectators are thus, in a way, the ideal audience, or rather the audience whose feelings the author of the entry wishes to exacerbate to the extreme. But they are also those mentioned in the *Odyssey*, namely, Odysseus and his men (*Od.* 12.243–244):

τοὺς δὲ χλωρὸν δέος ᾕρει.
Ἡμεῖς μὲν πρὸς τὴν ἴδομεν δείσαντες ὄλεθρον·

Pale fear now possessed my men,
we gazed at Charybdis in terror, fearing destruction.

Again, *Mir.* 115 amplifies a narrative technique set out in just a few lines in the *Odyssey*. The spectacle is, moreover, total, even more so than in the epic, given that the spectator's *phantasia* is mobilized not only by images, but by sounds: the text stresses at several points the auditory dimension of the phenomenon, drawing attention, for example, to the "massive whooshing sound" (πολλῷ ῥοίζῳ) made by the waves as they break from the Tyrrhenian Sea against the promontories, or to the "massive roar" (σὺν πολλῷ βρόμῳ) of the wave as it rises in the air from the upward rush. The last part of the description accumulates the movements of the swell, the colors and the noise (σὺν πολλῷ μὲν βρυχηθμῷ, "with a vast roar"): not to endure this spectacle is not to be able "to listen . . . or to watch it" (οὔτε ἀκούειν . . . οὔθ' ὁρᾶν). *Thauma* is a multisensory experience in which the audience finds itself immersed as if it were in the middle of the strait of Messina, swept along by the crashing waves.

5. Mir. *115: a naturalistic rewriting of the* Odyssey?

The attention paid by the author of *Mir.* 115 to the diversity of perceptible sensations (movements, noise, colors) shows the care he takes to compose a description that strikes the audience. In contrast to many other sections in the *Mir.*, the text does not simply list in a few lines the original or remarkable character of a natural phenomenon. All the rhetorical techniques of *ekphrasis* are mobilized here, so as to produce, according to Theon's famous definition, "a descriptive speech which brings the thing shown vividly before the eyes."[43]

This *ekphrasis* takes up a number of characteristics of Charybdis, as described in the three relevant passages of the *Odyssey*. Several operations are mobilized: the

---

42 Hunzinger (2005); D'Angour (2011) 148–150.
43 Theon *Prog.* 118: λόγος περιηγηματικὸς ἐναργῶς ὑπ' ὄψιν ἄγων τὸ δηλούμενον; see Webb (2009) 51, as well as Webb (1999), Elsner (2002) and Webb (2015).

author of the entry in particular transposes verses into prose, and he keeps certain lexical elements, or proposes equivalents in most cases. The result is a paraphrase that is also an explanatory commentary. The text is also at the crossroads of several exegetical traditions: it is part of a paradoxographical anthology, as the introductory sentence emphasizes and as the mentions of possible spectators more subtly suggest; it incorporates a number of elements specific to the geographical tradition, in particular that found in Strabo, the prodromal stages of which are to be found in Aristotle's *Meteorologica* and pseudo-Aristotle's *Problemata*. It can thus be defined as a *naturalistic* (φυσικόν) interpretation of the *Odyssey*. We can also parallel some of its assumptions with those governing, for example, Diodorus' and Strabo's texts on the straits of Messina, which may go back to Theophrastus and even Aristotle. And we must contrast it with the allegorical interpretations that make Charybdis a symbol of "debauched and insatiable spending at banquet,"[44] just as Scylla may be a figure of the harlot.[45]

Within the *Mir.* itself, one can contrast this section devoted to the straits of Messina, which is part of a tradition of commentary on book 12 of the *Odyssey*, with the one that integrates a discussion of the Symplegades, in *Mir.* 105.1 (839b-840a).[46] In the latter case, there is no *ekphrasis*, no general definition, but a sequence of remarks: the section's author starts with the definition of the Ister basin, continues with a discussion of the course of the Argonauts, incorporates the Wandering Rocks (Πλαγκταί) and the island Aethaleia, and ends with a quote from the *Odyssey* (12.67–68) about the same Wandering Rocks, which are otherwise identified with the Clashing Rocks (Συμπλήγαδες). It is not the relationship to the *Odyssey* that distinguishes the two *Mir.* entries, but rather the nature of the statement, the mode of enunciation, and the nature of the intellectual operations involved. The implied unified commentary of *Mir.* 115 is radically different from the mentions of the Argonauts' expedition for which Homer is summoned at the end of *Mir.* 105 as a quick witness, the general purpose of this passage seeming to be the highlighting not of the Ister, which nevertheless opens the section, but of the surprising character of the Wandering Rocks and of the island Aethaleia.

Should the characteristics of *Mir.* 115 be traced back to Polycritus? It is difficult to say, given the limited evidence we have on this author. In the final chapter of his *Aristotle's Lost Homeric Problems*, Robert Mayhew examines the evidence for an

---

44 Heraclit. *Probl. Hom.* 70.10: Καὶ Χάρυβδὶς μὲν ἡ δάπανος ἀσωτία καὶ περὶ πότους ἄπληστος εὐλόγως ὠνόμασται. See also Clem. Al. *Protr.* 12.1.3, with brief commentary by Hunter (2018) 218.

45 Scylla is a prostitute in another Heraclitus' *De incredibilibus* 2: ἦν δὲ αὕτη νησιῶτις καλὴ ἑταίρα καὶ εἶχε παρασίτους λαιμούς τε καὶ κυνώδεις: "she was an island woman who was a courtesan of great beauty; her parasites swallowed everything like dogs." For an interpretation of Scylla as a pirate ship, see Palaephatus 20:Ἦν καὶ ναῦς τριήρης ταχεῖα τότε, ᾗ ὄνομα Σκύλλα, καὶ κατεπεγέγραπτο ἐπὶ τῆς πρῴρας. Αὕτη δ᾽ ἡ τριήρης τὰ λοιπὰ τῶν πλοίων συλλαμβάνουσα πολλάκις εἰργάζετο βρῶμα: "at that time there was a fast trier, whose name was Scylla – it was written on its bow. This trier frequently seized other ships and made them its food."

46 On *Mir.* 105.1 and its possible connection to Theopompus, see Flashar (1972) 122–124, as well as Pietro Zaccaria, p. 86–88 and Stefan Schorn, n. 3 p. 32 in this volume.

interpretation by Aristotle of the Sirens and the Isle of the Sun, emphasizing how this interpretation is based on naturalistic commentary, "defending Homer (where possible) according to rational principles of literary criticism, but without relying on allegorical interpretation." Given the place of Charybdis, between the Sirens and the Isle of the Sun in book 12, could it be that *Mir.* 115 should somehow be traced back not to Polycritus, but to Aristotle, or to Aristotle via Polycritus? There is no argument outside the text of the *Mir.* to say so at present, and it would there-fore be foolhardy to claim so. But it is tempting to find in the Peripatetic *Homeric Problems* a complement to the observations that unite *Mir.* 115 to the *Meteoro-logica* of the same Aristotle.

Works cited

D'Angour, A. 2011. *The Greeks and the New. Novelty in Ancient Greek Imagination and Experience.* Cambridge; New York: Cambridge University Press.

Delattre, C. 2018. "Paradoxographical Discourse on Sources and Fountains." In Gerolemou, M. ed. *Recognizing Miracles in Antiquity and Beyond.* Berlin; New York: de Gruyter. 205–223.

Elsner, J. 2002. "The Genres of Ekphrasis." *Ramus* 31: 1–18.

Flashar, H. 1972. *Aristoteles, Mirabilia.* Berlin: Akademie Verlag.

Giacomelli, C. 2021. *Ps.-Aristotele, De mirabilibus auscultationibus. Indagini sulla storia della tradizione e ricezione del testo.* Berlin; New York: De Gruyter.

Giacomelli, C. 2023. *Pseudo-Aristotele, De mirabilibus auscultationibus: Edizione critica, traduzione e commento filologico.* Rome: Accademia Nazionale dei Lincei.

Hawes, G. 2014. *Rationalizing Myth in Antiquity.* Oxford; New York: Oxford University Press.

Hett, W.S. 1936. *Aristotle. Minor Works.* Cambridge, MA: Harvard University Press.

Hunter, R. 2018. *The Measure of Homer. The Ancient Reception of the Iliad and the* Odys-sey. Cambridge: Cambridge University Press.

Hunzinger, C. 2005. "La perception du merveilleux: *thaumazô* et *théèomai.*" In Villard, L. ed. *Études sur la vision dans l'Antiquité classique.* Rouen: Publications des Universités de Rouen et du Havre. 29–38.

Latour, B. 1993. *We Have Never Been Modern.* Cambridge: Harvard University Press.

Mayhew, R. 2019. *Aristotle's Lost Homeric Problems: Textual Studies.* Oxford: Oxford Uni-versity Press.

Radt, S. 1985. *Tragicorum Graecorum Fragmenta. Aeschylus.* Vol. 3. Göttingen: Vanden-hoeck & Ruprecht.

Shapin, S. and Schaffer, S. 1985. *Leviathan and the Air-Pump. Hobbes, Boyle, and the Ex-perimental Life.* Princeton: Princeton University Press.

Sharples, R.W. 1998. *Theophrastus of Eresus. Sources for His Life, Writings, Thought and Influence. Commentary. Volume 3.1. Sources on Physics (texts 137–223), with Contribu-tions on the Arabic Material by Gutas, D.* Leiden; Boston: Brill.

Sørensen, S. 2022. "1679. Paradoxographus Vaticanus." In Schorn, S. ed. *Felix Jacoby. Die Fragmente Der Griechischen Historiker Continued. Part IV. Biography and Antiquarian Literature. E. Paradoxography and Antiquities. Fascicle 2. Paradoxographers of the Im-perial Period and Undated Authors [Nos. 1667–1693].* Leiden; Boston: Brill. 579–632.

Vanotti, G. 2007. *Aristotele. Racconti meravigliosi.* Milan: Bompiani.

Webb, R. 1999. "Ekphrasis Ancient and Modern: The Invention of a Genre." *Word and Image* 15: 7–18.

Webb, R. 2009. *Ekphrasis, Imagination and Persuasion in Ancient Rhetorical Theory and Practice*. Farnham: Ashgate.

Webb, R. 2015. "Sight and Insight. Theorizing Vision, Emotion and Imagination in Ancient Rhetoric." In Squire, M. ed. *Sight and the Ancient Senses*. London; New York: Routledge. 205–219.

7 Suspicious toponyms in the *De mirabilibus auscultationibus*

Textual problems, "forgeries," and methodological issues*

Ciro Giacomelli

Most paradoxographical works seem to "have little or nothing to offer to students of ancient historiography."[1] Derivative as they are, such collections of excerpts give us a unique even if profoundly distorted glimpse of their sources, only partially mitigated by the comparative study of parallel passages, an exercise which can sometimes be misleading and should be attempted with great caution.[2] In assessing the more historiographical sections of the pseudo-Aristotelean *Mir.*, one of the few points of reference offered to the historian is the – very broadly speaking – geographical or local framing of the θαύματα, the sequence of which allows us to trace the underlying sources of the text.[3]

Since the beginning of the modern era, *Mir.* presented its readers with many textual puzzles, the most notable of which involve a significant number of toponyms. Over the course of at least four centuries, many scholars have attempted to correct the transmitted text of *Mir.* conjecturally, heavily relying on a rather facile use of the parallel passages (Aelian, Stephanus of Byzantium, and even the *Mirabilia* ascribed to Antigonus of Carystus) while rarely paying attention to the overall structure of the text and to the dynamics of its manuscript transmission.

In this chapter, I will try to offer some insights into the early stages of the production of the paradoxographical collection focusing on the almost invariable format of each θαῦμα. Once the dubious toponyms are fully understood in their

---

* All quotations from *Mir.* Are given according to the critical text of my edition. The English translation is Dowdall's, with occasional changes for my argument's sake, often based on Robert Mayhew's translation (in preparation). Other translations from Greek and Latin are mine, unless specified otherwise. For the *sigla codicum*, I shall refer to Giacomelli (2021) xvi–xvii and to my 2023 edition. A short overview of the transmission of the text is in Giacomelli (forthcoming). I would like to thank Robert Mayhew, Stefan Schorn, and Pietro Zaccaria for their precious remarks on this text.

1 Schepens and Delcroix (1996) 378.

2 The contribution of paradoxographical texts – and particularly of *Mir.* – to our understanding of lost historiographical works is great, as shown by the chapters collected in this volume. Nevertheless, the way ancient compilers of such collections selected their material is intrinsically biased toward the θαυμάσιον and therefore distorted and "distorting," to quote Cyrill Mango's definition of Byzantine literature: see Mango (1975).

3 See, for example, Pajón Leyra (in this volume) and Flashar (1972) 40–50.

DOI: 10.4324/9781003437819-8

context, is it possible to try to amend them, carefully distinguishing what could be the result of a corruption and what should be left untouched.

1. Paradoxography and "geographical singularization"

Modern scholars classify several miscellaneous texts under the umbrella term paradoxography, thus giving a false impression of unity and substantial similarity to a diverse group of works. Looking at the body of texts published by Anton Westermann under the label of ΠΑΡΑΔΟΞΟΓΡΑΦΟΙ, *Paradoxographers*, we find works that have very little in common. It is, however, possible to group together a few texts that are similar in structure: the Aristotelian *Mirabilia*, the Ἱστοριῶν παραδόξων συναγωγή attributed to Antigonus of Carystus, the *Mirabilia* attributed to Apollonius and three later anonymous collections, namely, the *Paradoxographi Florentinus*, *Palatinus*, and *Vaticanus*, and the lost Περὶ παραδόξων by Damascius of Damascus (described briefly by Photius, *Bibl.* 130). All these texts lack an introduction – or any editorial grace – and consist in collections of excerpts arranged according to variable criteria.[4]

The chronology of these works is far from being firmly established due to their derivative nature: while it is sometimes possible to date the source of single *mirabilia* or of a series of anecdotes, it is almost impossible to determine the time at which they were collected in the form in which we read them today. Antigonus' collection, for instance, used to be considered the oldest of its kind and the only (almost) completely preserved piece of Hellenistic paradoxography, supposedly dating from the 3rd century BC. However, in the last decades, many doubts have been cast on its chronology, leaving modern scholarship without what used to be the firmest and most extensive point of reference for the early stages of the genre.[5]

The most striking feature these collections of excerpts share with the pseudo-Aristotelian *Mir.* is the almost invariable structure of each chapter – or, more precisely, of each θαῦμα, since the boundaries of a single chapter are not always easily discerned – where the event described is almost inevitably circumscribed to a single specific location. Such a technique, very common in paradoxographical texts of any period, has been called "geographical isolation or singularization of the phenomenon described."[6]

---

4 The text of Apollonius, the three anonymous paradoxographers, and the testimonies on Damascius can now be read in the recent editions published in Schorn (2022): *FGrHist* 1672 (Apollonius: J. Spittler); 1679 (*Paradoxographus Vaticanus*: S.L. Sørensen); 1680 (*Paradoxographus Florentinus*: R. Greene); 1681 (*Paradoxographus Palatinus*: S.L. Sørensen), and 1668 (Damascius: S. Schorn). Photius' stylistic remark on such kind of collections arranged in short chapters is quite telling: Κεφαλαιώδης δὲ αὐτῷ ἐν τούτοις ὁ λόγος, καὶ οὔτε ἄκομψος οὔτε τὸ σαφὲς ὑπερορῶν, ὡς ἐν διηγήμασι τοιούτοις. On the terminology used by Photius, and specifically on the adjective κεφαλαιώδης, recurring in the description of other collections of *paradoxa* similar in their structure, see Schorn (2022) 350, commentary to *FGrHist* 1668 T 1.

5 Cf., for example, Schepens and Delcroix (1996) 376–377. The most recent studies on Antigonus are Dorandi (2017) and the annotated edition compiled by Eleftheriou (2018).

6 Schepens and Delcroix (1996) 392. See also Geus (2016) 244.

Mir. offers geographical singularization of the θαύματα in almost every chapter of the older kernel of *mirabilia* (chapters 1–151, in Bekker's edition) as well as the much later *appendix* (chapters 152–178) – probably compiled when such collections of novelties were already more rigidly canonized – which is even more consistent in this respect.[7] Only chapters 11–14, 21, 60, 64–67, 76 (77), 77(76), 147, 165,[8] 177, and 178 do not offer any geographical hint and, unsurprisingly, they are mostly concerned with zoological matters.

Geographical singularization is one of the few "original" stylistic features of *Mir.*: the compiler of the συναγωγή, who clearly had access to several written sources, only had to copy out what he considered worthy of inclusion in his collection of marvels; but, in order to do so, he first had to subtract from his excerpts any rational explanation, and then search for a plausible geographical location, either deriving it from the texts he was reading or, more interestingly, giving an arbitrarily restrictive interpretation of the work he was excerpting.[9] In doing so, our compiler often made significant mistakes, wrongly interpreting his source text or omitting crucial details we can only infer from parallel texts. This kind of innovation goes back to the earliest phases of the production of the treatise, and a modern editor of *Mir.* should therefore be extremely careful in dealing with them, trying to distinguish the different layers of corruption and editing the text accordingly.[10]

Giving a full account of the corrupted – or supposedly corrupted – toponyms in *Mir.* exceeds the scope of this chapter, since that would require a book-length treatment and it would mainly elaborate on the briefer critical commentary that will be published alongside my new critical edition of the text. Instead, I will limit myself to an illustration of some examples – which will serve as a methodological guide through the critical apparatus of my edition – and provide an insight into the compiler's workshop.

2. Suspicious toponyms

When dealing with toponyms in *Mir.*, it is always worth consulting parallel texts and later *testimonia*; but when it comes to altering the text transmitted by the manuscripts, some care is always advisable. In this first part of the chapter, I will deal with some problematic toponyms which contradict the evidence offered by other ancient authors but should not, in my view, be amended conjecturally. In all of these cases, the most cautious course of action for a textual critic is to present the evidence in the clearest possible way, resisting the urge to contaminate it in an attempt at harmonizing the text of *Mir.* with other sources.

---

7 See Giacomelli (2021) 21–44.

8 This chapter depends on the same source as *Mir.* 164, namely Nicander, *Theriaka* vv. 145–165, with the ancient scholium: see Jacques (2002) 94–96 and Giacomelli (2021) 32–33.

9 Cf. Schepens and Delcroix (1996) 392.

10 On such methodological issues, see most recently Dorandi (2013) 46–47, dealing with the text of Diogenes' *Lives*, where it is often arduous "to determine . . . which errors can be traced back to Diogenes (or to his sources) and which, on the contrary, should be attributed to accidents during the transmission of the text. The risk of correcting Diogenes himself is constantly present."

2.1 *Λυδίαν or Λυκίαν?* (Mir. 39)

The survey moves from chapter 39, a short narration of a certain fire that can be seen burning in Anatolia for an entire week:

> Λέγεται δὲ καὶ περὶ **Λυδίαν** ἀναζέσαι πῦρ πάμπληθες, καὶ καίεσθαι ἐφ' ἡμέρας ἑπτά.
>
> It is said that in Lydia a vast amount of fire blazes up and continues burning for seven days.

The δέ connects this chapter syntactically to a series of *mirabilia* dealing with fires (chapters 33–41), analyzed attentively by Myrto Garani, and it is within this frame that the passage should be read.[11]

All manuscripts agree on the reading Λυδίαν, but the conjecture Λυκίαν – espoused by Robert Mayhew and Garani – was originally put forward by Gotthold Ephraim Lessing in 1773 and, with seeming independence, by Johannes Beckmann in 1786, the first scholar to publish a full commentary on *Mir*. Beckmann observes[12]:

> Non statim occurrit alius scriptor, qui montis in Lydia flagrantis mentionem fecerit, quo in tractu horribiles ac frequentes terrae motus fuisse vel e Strabonis lib. I . . . constat. . . . Num forte legendum est Λυκίαν? De Lyciae montibus ignivomis testimonia dabimus ad cap. 139.
>
> I cannot think of another author mentioning a burning mountain in Lydia, where terrible and frequent earthquakes take place, as one can tell from Strabo's book I. Should we read Λυκίαν instead? We will give references about Lycian mountains that emit fire, in the commentary on chapter 139 [= 135(127)]

The conjecture seems almost banal given that Lycia is well known for its ever-burning fires, whereas, as Beckmann rightly pointed out, there seems to be no other classical author mentioning burning mountains in Lydia.[13] Even the connection with chapter 135(127) seems to support such a correction and, as pointed out by

---

11 Garani (forthcoming).

12 See Beckmann (1786) 82 and Praechter (1905) 386 (with reference to Lessing's edition of the poem).

13 On Lycian fires, see Ctesias, as reported by Photius *Bibl.* 72.46a34–37 (= *FGrHist* 688 F 45 § 44): Καὶ ὅτι πῦρ ἐστιν ἐγγὺς Φασήλιδος ἐν Λυκίᾳ ἀθάνατον, καὶ ὅτι ἀεὶ καίεται ἐπὶ πέτρας καὶ νύκτα καὶ ἡμέραν, καὶ ὕδατι μὲν οὐ σβέννυται, ἀλλὰ ἀναφλέγει, φορυτῷ δὲ σβέννυται. ≈ Plin. *NH* 2.236 (= *FGrHist* 688 F 45εβ): *flagrat in Phaselitis mons Chimaera, et quidem inmortali diebus ac noctibus flamma; ignem eius accendi aqua, extingui vero terra aut fimo Cnidius Ctesias tradit.* See also Ps.-Antigonus, *Mir.* 166 (= Callimachus F 407 XXXVIII Pf. = *FGrHist* 688 F 45εα): Περὶ δὲ πυρὸς Κτησίαν φησὶν ἱστορεῖν, ὅτι περὶ τὴν τῶν Φασηλιτῶν χώραν ἐπὶ τοῦ τῆς Χιμαίρας ὄρους ἐστὶν τὸ καλούμενον ἀθάνατον πῦρ· τοῦτο δέ, ἐὰν μέν τις ὕδωρ ἐμβάλῃ, καίεσθαι βέλτιον, ἐὰν δὲ φορυτὸν ἐπιβαλὼν πήξῃ τις, σβέννυσθαι.

Garani, a parallel sequence of *miracula ignum* in Pliny the Elder "strengthens this suggestion."[14]

I will not discuss the Plinian parallel, which, in my opinion, is only fitting for chapter 135(127) – as Garani herself acknowledged – and I will not argue that the text of the manuscripts should be defended at all costs by reading Lydia. Nevertheless, a closer intertextual framing of this chapter suggests that one should take a more cautious approach.

There is no obvious ancient parallel for chapter 39, but we know of at least two later authors who are supposedly quoting this passage of *Mir.* in their own works: Leo Choirosphactes (9th–10th century), in a poem on hot springs which, until 1925, was attributed to Paulus Silentiarius,[15] and the Byzantine historiographer George the Monk (mid-9th century).[16]

In a longer passage on hot springs and water-related phenomena, George the Monk makes a generic reference to fires that burn beneath the earth and that remind us of Hell and the eternal punishment of the sinners.[17]

ὅτι δὲ πῦρ ἔστιν ὑποκάτω τῆς γῆς, πειθέτω σε τὸ ἐν Σικελίᾳ καὶ ἐν **Λυκίᾳ** προφανῶς ἀναδιδόμενον καὶ μέντοι καὶ ἐν ἄλλοις διαφόροις τόποις ὁμοίως παραδεικνύμενον πῦρ εἰς τὴν φοβερὰν γέενναν προδήλως πάντας κατακαῖον, ὅσοι τὰ τοῦ πυρὸς ἔργα πεπράχασιν.

You should be convinced that there is fire underneath the surface of the earth by the [fire] which is sent up in Sicily and Lycia for everyone to see, as well as that which is shown in other different places, which clearly burns in Geena all those who practice the works of fire.

The German scholar Karl Praechter was the first to connect this short passage with the text of *Mir.* 39. Praechter claimed that this quotation seems to support Lessing's correction Λυκίαν and argued that George the Monk was reading *Mir.*, or a closely related text, as demonstrated by a similar passage in the poem attributed to Paulus Silentiarius – a text which, according to Praechter, preceded or even inspired George's narrative. George's quotation would, in other words, provide evidence of a now-lost manuscript reading Λυκίαν instead of Λυδίαν, the text transmitted by all the existing manuscripts of *Mir.*, thus confirming a modern conjecture.[18]

The ties between *Mir.* and George's *Chronicle* are, however, far from being firmly established, and the mention of hot springs and waters in George's *text* is in no way comparable to the poem by Choirosphactes, who composed his verses almost 50 years later.[19]

---

14 Garani (forthcoming). The sequence of *mirabilia ignum* in Pliny closely follows the one we find in *Mir.*, but the anecdote described in chapter 39 has no actual parallel in the *Naturalis Historia*.
15 See Mercati (1923–1925) and Giacomelli (2021) 328, both with previous literature.
16 Cf. Giacomelli (2021) 338–339.
17 Greek text from de Boor (1904) 440.
18 Praechter (1905) 386–387.
19 Praechter (1905) 387: "Daß auch hier die Beziehungen zu Ps.-Aristoteles obwalten, ergiebt sich daraus, daß der Grund der Erörterung genau der gleiche ist wie bei dem sog. Paulos Silentiarios.

A much closer parallel to George's passage can instead be found in the *Martyrium Pionii*, a 3rd–4th-century hagiographic text which, in the last speech of Pionius of Smyrna, reports some interesting details about water and fire *mirabilia* with a view to showing to the pagans how terrible the wrath of God can be.

The similarity between the structure of Pionius' argument and George's text is striking; but it is also worth noticing that, according to the Martyr, everyone in the early 3rd century knew of the "land of the Lydian Decapolis," "scorched by fire," "an example of men's impiety even to this day."[20]

ὑμεῖς ὁρᾶτε καὶ διηγεῖσθε Λυδίας γῆν Δεκαπόλεως κεκαυμένην πυρὶ καὶ προκειμένην εἰς δεῦρο ὑπόδειγμα ἀσεβῶν, Αἴτνης καὶ Σικελίας καὶ προσέτι Λυκίας καὶ τῶν νήσων ῥοιγδούμενον πῦρ. εἰ καὶ ταῦτα πόρρω ἀπέχει ἀφ᾽ ὑμῶν, κατανοήσατε τοῦ θερμοῦ ὕδατος τὴν χρῆσιν, λέγω δὴ τοῦ ἀναβλύζοντος ἐκ γῆς, καὶ νοήσατε πόθεν ἀνάπτεται ἢ πόθεν πυροῦται εἰ μὴ ἐκβαῖνον ἐν ὑπογαίῳ πυρί. λέγετε δὲ καὶ ἐκπυρώσεις μερικὰς καὶ ἐξυδατώσεις, ὡς ὑμεῖς ἐπὶ Δευκαλίωνος ἢ ὡς ἡμεῖς ἐπὶ Νῶε. μερικὰ γίνεται ἵνα ἐκ τῶν ἐπὶ μέρους τὰ καθόλου γνωσθῇ.

You yourselves see and testify how the land of the Lydian Decapolis is scorched by fire and remains as an example of men's impiety even to this day; you know the volcanic fire of Etna and Sicily and even Lycia and the islands. And even though this has been kept away from you, consider your familiarity with hot water, I mean the sort which gushes up from the earth: how else could it be enkindled and heated unless it emerged from an underground fire? Consider, too, the partial conflagrations and floods, such as you know of, for example, in the case of Deucalion, and we know in the case of Noah. They are partial and occur in this way so that we may comprehend the nature of the whole from the part.

Since it is impossible to envisage a dependence of Pionius on the pseudo-Aristotelian *Mirabilia*, it is worth looking at the Lydian reference included in his speech. As it turns out, a region in northeastern Lydia, characterized by the remains of volcanism, was indeed called κεκαυμένην, "scorched/burnt-up" already by Strabo, who, quoting Xanthus the Lydian, refers to the burnt-up country divided between Lydians and Mysians.[21] It is not surprising that, following a suggestion put forward by Friederich Creuzer in 1806, *Mir.* 39 was included among the fragments of Xanthus' Λυδιακά (Strabo's source) in Müller's *Fragmenta historicorum Graecorum*.[22]

Auch hier handelt es sich um die Erklärung der warmen Quellen. Die Erklärung ist die auch von Paulos vertretene."

20 For the text and the English translation, I quote from Musurillo (1972) 143–144. See also Bastiaens (1987) 162–163 and the commentary at 459.

21 Strab. 13.4.11.628C = 3.652.14 Radt = *FGrHist* 765 F 13b. On the region, usually called "katakekaumene," see Daubner (2013).

22 Müller (1853) 37, see also Creuzer (1806) 165, who refutes Beckmann's conjecture. On the (im)possible dependence of *Mir.* on Xanthus' Λυδιακά, see Flashar (1972) 86 and Vanotti (2007) 153. Stefan Schorn kindly provided me with a passage from Jacoby's unpublished commentary on Xantus' fragments, which goes in a similar direction: "Der Exkurs rechtfertigt nicht Creuzers Zuweisung . . .

While the slight similarity between *Mir.* and the *Chronicle* of George the Monk should be regarded as a mere coincidence, it is quite clear that the Byzantine scholar and politician Leo Choirosphactes – who quotes extensively from *Mir.* in many sections of his poem on thermal baths – was referring to *Mir.* 39 while paraphrasing the sequence of *mirabilia* related to burning fires (vv. 40–56 = 43–56 Gallavotti)[23]:

οἶδεν φέρειν τοιαῦτα
Τιτανία Μηδίας
καὶ Περσικὴ Ψιττάκη
καὶ **Λυδία** πλουτοῦσα
μεταλλόχρυσον γαῖαν.
Ἡρακλέων στηλῶν δὲ
πόρρω πέφυκε πλεῖστα,
ἐν δ' αὖ γε Πιθηκούσαις
καὶ Λιπάρῃ τῇ νήσῳ
ἄφεγγές ἐστιν ἆσθμα,
ὃ νυκτί περ παμφαῖνον
πέμπει λίθους θειώδεις
πολυψόφους βροντώδεις
ὄψει τὸ πᾶν δηλοῦντας.

Such things [*viz.* fires] can produce Titania of Media, and Psittace of Persia, and the rich land of Lydia, full of golden metal, but the greatest part of these [marvelous fires] happens out of the Pillars of Hercules. In Pithecusa and in the island of Lipari there is a flame that does not shine but is clearly visible at night and ejects sulphureous rocks, which produce a thunderous noise and make everything clear to see.

Leo's poem follows the sequence of fire-related *mirabilia* in *Mir.* 35–36 and 39 (Τιτανία Μηδίας καὶ Περσικὴ Ψιττάκη καὶ Λυδία πλουτοῦσα μεταλλόχρυσον γαῖαν), 37 (Ἡρακλέων στηλῶν δὲ πόρρω πέφυκε πλεῖστα), and 38 (ἐν δ' αὖ γε Πιθηκούσαις καὶ Λιπάρῃ τῇ νήσῳ ἄφεγγές ἐστιν ἆσθμα), and there can be no doubt that, in the early 10th century, Choirosphactes was reading Λυδίαν in his source text, as we still do in our manuscripts.

This later quotation of *Mir.* does not necessarily support the text as transmitted by the manuscripts: Lydia could well be an error for Lycia, but it must have happened before the first quarter of the 10th century. The compiler of *Mir.* could in fact have been referring "to the eternal flames of Mount Chimaera, in Lycia, which can still be visited today," as Mayhew puts it, and it is easy to explain the corruption of Λυκίαν in Λυδίαν palaeographically as a simple misreading of a letter (be

---

von Ps.-Aristot. Θαυμ. ἀκ. 39 . . .. An sich ist natürlich nicht ausgeschlossen, dass die Thaumasiographen (direkt oder indirekt) auch Xanthos benutzten. Aber § 39 gehört sicher nicht dazu."

23 I quote from Gallavotti (1990) 86–87. On Leo's dependence on *Mir.*, see Giacomelli (2021) 328–331.

it in majuscule or minuscule script). But can we really rule out the possibility that the transmitted text might be correct? This short overview of the Byzantine reception of this chapter makes the seemingly easy conjecture Λυκίαν far less obvious. Hence, following all modern editors of *Mir.*, I preferred to print the text as transmitted in the manuscripts, preserving this almost isolated mention of Lydian fires and leaving the final choice to historians.[24]

2.2 Lydia or Media? (Mir. 20(19))

A problematic mention of the toponym Lydia can also be found in the first part of chapter 20(19), which deals with the production of honey. Only, in this case, we can count on an ancient parallel: Aelian's *De natura animalium*, apparently relying on the same source as *Mir.*, commonly identified with Theophrastus' Περὶ μελίτων (*On Honey*).[25]

> Φασὶ δὲ καὶ ἐν **Λυδίᾳ** ἀπὸ τῶν δένδρων τὸ μέλι συλλέγεσθαι πολύ, καὶ ποιεῖν ἐξ αὐτοῦ τοὺς ἐνοικοῦντας ἄνευ κηροῦ τροχίσκους, καὶ ἀποτέμνοντας χρῆσθαι διὰ τρίψεως σφοδροτέρας.

> They say that in **Lydia** also the honey is gathered from trees in abundance, and that the inhabitants form out of it small wheels[26] without wax, and cutting off portions by very violent rubbing make use of it.

The text should be compared with Aelian, *NA* 5.42[27]:

> Ἐν **Μηδίᾳ** δὲ ἀποστάζειν τῶν δένδρων ἀκούω μέλι, ὡς ὁ Εὐριπίδης ταῖς Βάκχαις ἐν τῷ Κιθαιρῶνί φησιν ἐκ τῶν κλάδων γλυκείας σταγόνας ἀπορρεῖν.

> I learn that in **Media** honey drips from the trees, just as Euripides in the *Bacchae* [714] says that on Cithaeron sweet drops flow from the boughs.

The Median setting of the anecdote seems to be confirmed by Strabo (2.1.14.72), who describes of a kind of honey dripping from trees common in that region: the location given in *Mir.* is obviously inconsistent with these texts and, in all probability, objectively wrong as well.

Modern scholars may wonder whether Λυδίᾳ should be considered a corruption of Μηδίᾳ. Since it is impossible to assess at which stage the name was altered, amending it could wipe out important evidence for establishing the relationship between *Mir.* and Aelian. Assuming, as most scholars do, that the anecdote on honey in *Mir.* and in Aelian came independently from the lost Theophrastean

---

24 Beckmann's conjecture is considered "unadvisable" by Flashar (1972) 85; even Giannini (1965), usually prone to correct the text extensively in light of parallels, left Λυδίαν untouched.

25 See Oikonomopoulou (forthcoming).

26 I adopt here Mayhew's rendering of τροχίσκους; Dowdall translates "balls."

27 Text and translation are taken from Scholfield (1959) 338–339; the most recent critical edition by García Valdés *et al.* (2009), to which I refer for the chapter subdivision of each book, is not satisfactory; see Wilson (2010).

treatise Περὶ μελίτων and other lost sources on honey, the variation Λυδίᾳ/Μηδίᾳ could be explained as an innovation (or a simple mistake) introduced by the compiler of *Mir.* or, alternatively, as an old error stemming from the archetype of the manuscript transmission of the text, since it is documented by all surviving witnesses: harmonizing the three passages (Strabo, Aelian and *Mir.*) mechanically by correcting the pseudo-Aristotelean text would be rather dangerous, since, as Oikonopoulou points out, "The two references seem, overall, distinct in focus."[28]

If, on the other hand, we postulate that Aelian is here quoting *directly* from *Mir.*, as Scholfield implies, the first option should be ruled out and Aelian's text would become an important and particularly old witness of the secondary transmission of *Mir.*, allowing us to correct the text of the Byzantine manuscripts.[29]

The situation, however, is quite complicated: Aelian's *De natura animalium* offers several passages comparable to the wonders described in *Mir.* (I list 54 in my *apparatus locorum*), but only two of them (*NA* 5.14 and 9.20) explicitly reference Aristotle in relation to anecdotes which are substantially identical to what we read in *Mir.* 25 and 135(127), and could actually be direct quotations from the paradoxographical collection. But this certainly does not seem to be the case for the honey-related *mirabilia* in chapters 16–22, which do not include details present in Aelian's account.[30]

The philologist is confronted once more with a textual problem that goes beyond the letter of the text itself, and I decided that the safest, if not most satisfactory, course of action would be to present the readers with the evidence arising from the investigation of the manuscript tradition, while acknowledging Aelian's alternative reading in the critical apparatus.

2.3 Amphipolis: Mir. 126(118)

Chapter 126(118) describes a peculiar fowling technique which can be witnessed in Thrace. In this case, the geographical singularization of the episode is crucial to the understanding of the sources of the chapter. A close parallel for the anecdote, which is enriched with "marvelous" details in *Mir.*, can be found in Aristotle's *Historia animalium* 8(9).36.620a33–620b4, on which other accounts of the same episode seem to rely: Plin. *HN* 10.23; Ael. *NA* 2.42; Ps.-Antigonus' *Mirabilia* ch. 28; and Philo *De animalibus* 37 (preserved only in the Armenian translation). Thphr. *Od.* 2.4 deals with another anecdote seemingly taking place in the same location.

According to the account we read in *Mir.*, this special style of falconry can be witnessed in Amphipolis, in Thrace – Περὶ . . . τὴν Θρᾴκην τὴν ὑπὲρ Ἀμφίπολιν – while the *Historia Animalium* places it in an otherwise unknown Thracian location "once" called Κεδρειπολιός – the manuscripts offer slightly different readings: κέδρει πόλιος, κεδρειπόλιος, κεδροπόλει and κέδρει πόλει – a name that caught the

28 Oikonomopoulou (forthcoming).
29 Scholfield (1959) 304.
30 See Section 4.2.

attention of Wilhelm Dittenberger, who interpreted it as the only literary mention of an ancient Thracian dynasty called Ketriporis.[31]

Two of the parallel texts reporting the anecdote seem to support the reading of the *Historia animalium*: Theophrastus reads Κεδροπόλιος, Antigonus has Κεδριπολι [!], while Aelian and Philo limit themselves to relating that the hunting practice takes place in Thrace.[32]

The mention of Amphipolis in *Mir.* could, therefore, easily be construed as a corruption of a difficult toponym, and one could even be tempted to correct the text of *Mir.* based on the two parallels – only, Pliny (*HN* 10.23), who mentions the same episode, had not offered us a location that is so close to the one described in *Mir.* that it almost sounds like a direct translation: *in Thraciae parte super Amphipolim.*[33]

According to some commentators, Ἀμφίπολιν could be the modern name of the mysterious Κεδροπόλιος, and the compiler of *Mir.* would just have decided to update his source text or to offer a more precise location of the event.[34] Oddone Longo more convincingly suggests that Amphipolis was a better-known Thracian city, and that this would be the best candidate for the altered toponym in *Mir.* (but this takes us beyond the scope of my discussion).[35] Defining the exact relationship among these texts is, in fact, far from easy: most scholars argue that *Mir.* is here copying the text of *HA*; but that the crucial innovation (ὑπὲρ Ἀμφίπολιν) is missing in *HA*, yet witnessed by Pliny implies that those two texts derive from the same source, possibly a different arrangement of the material that we read in *HA* 8(9), where the toponym Κεδρειπολιός was changed in the more familiar Amphipolis.[36]

Once it is established that *Mir.* and Pliny go back independently to a common source, which was not identical with Aristotle's *HA* as we read it, the question of the exact location of the anecdote becomes less stringent for the textual critic: the compiler of *Mir.* surely read ὑπὲρ Ἀμφίπολιν in his lost source – be that correct or not – and amending such a reading based on other parallels would be unsound from a methodological point of view.[37]

---

31 Dittenberger (1879). On the name, see also Schnieders (2019) 876–877 (with reference to previous literature). Dittenberger's hypothesis is very attractive, but in order to support his reconstruction, the German scholar had to correct all the passages reporting the name; cf. also, Dittmeyer (1907) 389–390 (app. *Ad loc.*) and Louis (1964) 103 n. 1. Louis prefers the reading Κεδρειπόλει following a conjecture by Bekker, based on its proximity to text of family **α**, in Louis' words "la plus proche du texte des meilleurs manuscrits." Balme (2002) 428, more conservative in his approach, prints Κεδρειπολιός following βLʳrc.

32 For a translation of Philo's account in his *De animalibus*, surviving only in Armenian, see Terian (1981) 84–85.

33 No variants are recorded in the manuscripts: see Mayhoff (1875) 164.

34 Cf. Flashar (1972) 130.

35 Longo (1986–1987) 43.

36 Flashar (1972) 130 writes of *HA* "bzw. Theophrast, περὶ ἤθους καὶ φρονήσεως ζῴων" as "Quelle." Vanotti (2007) 198 only states that the source "potrebbe essere Teofrasto," while a dependence on Timaeus should be ruled out.

37 On this chapter, see the detailed analysis offered by Zaccaria (in this volume).

2.4 Κάπαν: Mir. 95

The toponyms examined so far are unanimously attested in the manuscript tradition of *Mir.* and they become problematic only when compared with the evidence coming from other sources. Conversely, the manuscripts of *Mir.* are divided on the toponym which I shall examine now. It involves the name of a river, situated in Campania, near Cumae (or Paestum), the waters of which can petrify the objects that are cast into it (*Mir.* 95.2):

> Εἶναι δὲ λέγουσιν ἐν ἐκείνοις τοῖς τόποις περὶ τὴν Κύμην ποταμόν τινα †Κάπαν† ὀνομαζόμενον, εἰς ὅν φασι τὸν πλείω χρόνον τὸ ἐμβληθὲν πρῶτον περιφύεσθαι καὶ τέλος ἀπολιθοῦσθαι.

They state moreover that in those parts about Cumae there is a certain river called †Kapan†, and they say that whatever is thrown into this is after a considerable time coated over, and finally turns into stone.

The manuscripts of *Mir.* present us with several variant readings: κάπαν **GP**, σκετὸν **G^γρ R**, or σκεπὸν **P^γρ**, κετὸν **β**, κακέτταν **B** (or possibly κακέ[π]ταν, since the text of this manuscript is damaged here).

The history of the manuscript transmission of *Mir.* could provide a basis for the analysis of these discrepancies. The text of manuscripts **GPR** was contaminated in the margins with the reading of **β** (the consensus of manuscripts **TF**); κετόν, with the addition of an initial σ. The scribe of manuscript **R**, which was derived from the same model as **GP**, only copied out the γράφεται variant, inserting it in the main text, as he did in several other cases.[38] The puzzling reading of **B** (κακέτταν) can be partially corrected with the aid of the mediaeval Latin translation of Bartholomew of Messina (φ, second half of the 13th century) – who had access to a text close to **B** yet independent from this manuscript – which reads *Cetum*, a mere transliteration of the text we read in **β**. The rather confused picture we get from the critical apparatus can be reduced to two options, equally viable from a purely stemmatic point of view: Κετόν and Κάπαν. None of these hydronyms is attested elsewhere, while the ancient parallels seem to refer to the rivers Sele or Salsum, both located in the same area.[39]

Bekker, following the *editio princeps* of *Mir.*, printed the text of **β**, and the reading Κετόν was also kept by all subsequent editors. More recently, Silvia Miscellaneo (1996) has argued that the reading Κάπαν should be interpreted as the ancient name of the river Capodifiume, which flows near Paestum. Even if the evidence presented by Miscellaneo might not be strong enough to support her reconstruction, I think she is right in challenging the superiority of the so-far dominant text, and so I chose to print the reading of **γ** between *obeloi*, thus informing the readers that the text transmitted in the manuscripts should be regarded as corrupted.

---

38 See Giacomelli (2021) 258–261.

39 See, for example, Strabo 5.4.13; Plin. *HN* 2.226; Sen. *QNat.* 3.20.3 and 31.29; Sil. *Pun.* 8.579–580. On this point, see also the brief remarks in Flashar (1972) 115.

3. Redactional imperfections

Suspicious toponyms can sometimes be more than mere textual brain teasers: in some cases, they offer us unexpected insights on the process of collecting excerpts from which *Mir.* originated. In this section, I will illustrate the paradoxographer's *modus operandi* with two examples.

3.1 Deer in Achaea: Mir. 5

Mir. 5 presents us with an anecdote concerning a deer in Achaea, but the location of the θαῦμα is not as obvious as it may seem at first reading.

Φασί τινας ἐν Ἀχαΐᾳ τῶν ἐλάφων, ὅταν ἀποβάλωσι τὰ κέρατα, εἰς τούτους τοὺς τόπους ἔρχεσθαι ὥστε μὴ ῥᾳδίως εὑρεθῆναι· τοῦτο δὲ ποιεῖν διὰ τὸ μὴ ἔχειν ᾧ ἀμυνοῦνται, καὶ διὰ τὸ πονεῖν τοὺς τόπους ὅθεν τὰ κέρατα ἀπέβαλον·. Πολλαῖς δὲ καὶ κισσὸν ἐπιπεφυκότα ἐν τῷ τῶν κεράτων τόπῳ ὁρᾶσθαι.

Men say that some of the stags in Achaea, when they have shed their horns, proceed to places of such a kind that they cannot be easily found; and that they act in this way because they have no means of defense, and also because the parts from which they have shed their horns give them pain; and it is stated that, in the case of many of these animals, ivy is seen growing in the place of the horns.

The closest parallel to this passage can be found in Aristotle's *HA* 8(9).611a25–29 and 611b8–20. This last section offers us the rare adjective ἀχαΐνης, on which the location ἐν Ἀχαΐᾳ seems to be based.

Ἤδη δ᾽ εἴληπται ἀχαΐνης ἔλαφος ἐπὶ τῶν κεράτων ἔχων κιττὸν πολὺν πεφυκότα χλωρόν, ὡς ἁπαλῶν ὄντων τῶν κεράτων ἐμφύντα ὥσπερ ἐν ξύλῳ χλωρῷ.

An Achaeine stag has been caught with a quantity of green ivy grown over its horns, it having grown apparently, as on fresh green wood, when the horns were young and tender (trans. D'Arcy Thompson).

As previously observed, defining the connection between *Mir.* and *HA* 8(9) is far from easy; and, in this case, it again seems quite clear that *Mir.* relies on a lost source in which the zoological material included in *HA* 8(9) was presented in a slightly modified form.[40]

---

40 On the connection between *HA* 8(9) and the lost zoological works of Theophrastus and the Peripatetic school, see Schnieders (2019) 200–215, with references to earlier bibliography, and Mayhew (forthcoming 2), who analyzes the connection between *Mir.* and *HA* coming to the conclusion "that the parallels between *HA* 8(9) and the *Mirabilia* are explained by the fact that (something like) the notebooks posited by Balme and Gotthelf [*i.e.* lost notebooks in which the zoological material later included in *Mir.* and *HA* was first written] were not by Aristotle, but by Theophrastus, and that these can be identified with known (if lost) works of his."

In his recent commentary on this passage, Stefan Schnieders rightly points out that the interpretation of the adjective ἀχαΐνης presents some difficulties. According to at least four later sources – an ancient scholium to Apollonius Rhodius (4.175), a gloss in Hesychius' dictionary (α 8811), a scholium to Ps.-Antigonus' *Mirabilia* 70, and a passage in Eustathius' commentary on the *Iliad* – ἀχαΐνης refers to the age of the deer and not to a specific geographical origin: ἀχαῖναι ποιά τις ἐλάφων ἡλικία ("ἀχαῖναι, a certain age of the deer").[41] The same adjective is also used in *HA* 2.506a23 (τῶν δ' ἐλάφων αἱ ἀχαῖναι καλούμεναι), where again it could be regarded as a locative or as indicating the age of the deer.[42]

The evidence supporting a geographical interpretation of the adjective, preferred by Schnieders in his commentary, is however extremely scarce and the single passage unambiguously referring to it can be found only in this chapter of *Mir.* I, therefore, believe that, in this case, we are facing an obvious error perpetrated by the compiler of *Mir.* (or his lost source), who did not understand this rare adjective, specifically referring to a biological detail, and who preferred to singularize geographically a rather nonmarvelous cervine detail. The local framing of each wonder, as stated in the introductory remarks, functions here only as an intra-textual reference and should be understood accordingly.[43]

3.2 *A miraculous Olive tree:* Mir. 51

Chapter 51 – which enjoyed a rich secondary tradition since the middle-Byzantine era – is a particularly complicated case of layered anecdote that presents us with several textual and redactional problems. I will present it in the form I printed in my critical edition, with only a selection of the variant readings. I will not elaborate on the many difficulties concerning the first part (beginning with the description of the Olive tree, which is corrupted in the manuscript tradition) and I will rather focus on the second section of the anecdote and, more precisely, on the mention of the Attic river Ilissus.

|12| *1.* Ἐν τῷ Πανθείῳ ἐστὶν ἐλαία, καλεῖται δὲ καλλιστέφα|13|νος· ταύτης πάντα τὰ φύλλα ταῖς λοιπαῖς ἐλαίαις ἐναντία |14| πέφυκεν· ἔξω γὰρ ἀλλ' οὐκ ἐντὸς ἔχει τὰ λευκά. ἀφίησί |15| τε τοὺς πτόρθους ὥσπερ ἡ μύρτος εἰς τοὺς στεφάνους συμ|16|μέτρους. ἀπὸ ταύτης φυτὸν λαβὼν ὁ Ἡρακλῆς ἐφύτευσεν |17| Ὀλυμπίασιν, ἀφ' ἧς οἱ στέφανοι τοῖς ἀθληταῖς δίδονται.

|18| *2.* Ἔστι δὲ αὕτη παρὰ τὸν Ἰλισσὸν ποταμόν, σταδίους ἓξ |19| τοῦ ποταμοῦ ἀπέχουσα· περιῳκοδόμηται δέ, καὶ ζη|20|μία μεγάλη τῷ θιγόντι

41 Cf. Schnieders (2019) 727. The Antigonean scholium, not mentioned by Schnieders, can be read in Musso (1985) 44.
42 See Louis (1964) 170 n. 7; Kitchell (2014) 46.
43 *Pace* Keller (1909) 277–279, and subsequent literature.

ταύτης ἐστίν. ἀπὸ ταύτης δὲ τὸ |21| φυτὸν λαβόντες, Ἠλεῖοι ἐφύτευσαν ἐν Ὀλυμπίᾳ, καὶ τοὺς |22| στεφάνους ἀπ' αὐτῆς ἔδωκαν.

14 ἔξω γὰρ ἀλλ' οὐκ] ἔξω γὰρ οὔκ, ἀλλ' Kuster: ἀλλ' om. **OAld.** | λευκά Σ in Aristoph. et Sud. (cf. Bodaeus a Stapel: χλωρά **βx**: χλορά **B • 18** αὕτη] ἄλλη Gohlke | Ἰλισσὸν] Ἐλίσσης aut Ἐλίσσοντος Hemsterhuis | ἐξ Hemsterhuis (def. Ziegler); cf. *sex* Beccaria: ἑξήκοντα ω (ἑξ > ξ): ὀκτὼ Giann. ex Σ in Theocritum

1. In the Pantheon there is an olive-tree, which is called that 'of the beautiful crowns.' But all its leaves are contrary in appearance to those of other olive-trees; for it has the pale-green outside, instead of inside, and it sends forth branches, like those of the myrtle, suitable for crowns. From this kind of tree Heracles took a shoot, and planted it at Olympia, and from it are taken the crowns which are given to the competitors.

2. This tree is near the river Ilissus, six stadia distant from the river. It is surrounded by a wall, and a severe penalty is imposed on anyone who touches it. From this the Eleians took the shoot, and planted it in Olympia, and from it they took the crowns which they bestowed.

Since at least the 12th century, when the Byzantine John Tzetzes quoted this passage secondhand, taking it from a scholium to Aristophanes' *Plutus*, the text was perceived as subdivided into two distinct sections, the second openly contradicting the first, to the point that Tzetzes was convinced that the latter part of the anecdote relied on a different source.[44]

The chapter begins by mentioning the olive tree planted in Olympia by Heracles and then abruptly goes on to describe *another* tree of the same kind, planted near the Ilissus River, near Athens, in Attica, from which the Eleians supposedly took a shoot and planted it in Olympia. The two sections of the chapter clearly go back to two different traditions concerning the Olympian olive tree: the first connected with Heracles, the second – possibly a regional version of the myth – stating that the Eleians planted the miraculous tree in Olympia, taking a shoot from a preexisting Athenian one, which was kept with great care.[45]

Unable to recognize the different layers of the story, modern scholars have tried to amend the text by modifying the name of the river Ilissus and trying to harmonize the two sections forcibly.[46] For instance, the two options put forward by the Dutch scholar Tiberius Hemsterhuis are mere attempts at restoring the

44 On Tzetzes and *Mir.* see Giacomelli (2021) 344–348.
45 Such a detail may well refer to the sacred olive trees believed to be originating from the one donated to the city by the goddess Athena, the oil of which was handed as a prize to the winners of Panhellenic competitions: see Arist. *Ath. Pol.* 60.2–3 (on the sacred Olive trees, see also Ar. *Nub.* 1005 with schol. *ad loc.* and Lys. 7.7). On this subject, see also Kyle (1996).
46 A good summary of the many conjectures proposed in the past centuries was published by Jacobi (1930) 85–88.

name of a river flowing near Olympia, and should be regarded with great caution since they rely on extremely uncertain evidence. Some have even located
the Pantheon described here in Athens, thus finding a *terminus post quem* for
Mir. in the first mention of an Athenian Pantheon built by the Roman emperor
Hadrian.[47]

Gabriella Vanotti was the first modern scholar who recognized the stratification
of the chapter, but her explanation of such a contradictory account is not tenable:
according to her reconstruction, some later (Byzantine?) scholiast contaminated
the original text of *Mir.* with a different version of the myth. Such an interpretation
goes back to Vanotti's dubious understanding of the quotation of John Tzetzes and
on her approach to the transmission of the text of *Mir.*, influenced by old theories
summarized by Konrat Ziegler in the entry *Paradoxographie* published in Pauly-
Wissowa, where the manuscript transmission of *Mir.* is presented as divided into
different "redactions" of the text.[48]

We can now say with certainty that the text we read in the manuscripts, as far as
the second part of the chapter is concerned, was read in the exact form we read it
today as early as the 10th century – almost 200 years before Tzetzes – by the compilers of the *Suda* and by the Aristophanic scholiasts, who, as we know quite well,
had access to the same sources.[49]

My explanation of the layered text of chapter 51 is therefore of a completely
different nature: the obvious discrepancy between the two parts of the anecdote
should be regarded as one of the few instances in which we can glimpse the work
of the compiler, who juxtaposed two different traditions concerning the Olympian
olive tree – the second coming from an Attic or even Athenian source – and never
polished the resulting text by eliminating this major inconsistency, which is still
preserved in the chapter.

From a purely methodological point of view, athetizing the latter part of
the anecdote or correcting the transmitted toponyms would only muddle the
evidence.

4. Scribal errors and later conjectures

The philological puzzles so far described put the textual critic in a paradoxical
situation: the transmitted text is doubtful, but any editorial intervention would
result in a methodologically flawed falsification of the evidence. There are many
other such cases in *Mir.*, but often the kind of textual corruption we face can be
dealt with in a more orthodox fashion, correcting it conjecturally and by a wise
use of parallel texts and later *testimonia*. Most of the work, it should be noted,
has already been done by notable scholars of the past, from the Florentine Pier
Vettori (1499–1585) to the most recent edition of *Mir.* published by Alessandro
Giannini.

---

47 Cf. Ziegler (1949) 722; see also Flashar (1972) 90 and Vanotti (2007) 155–156.
48 Vanotti (1981). See also Giacomelli (2021) 345–347.
49 For a discussion of these passages, see Giacomelli (2021) 340–342.

4.1 Aethalia: Mir. 93

Chapter 93 deals with an island in Tyrrhenia, where copper and – later – iron were extracted. Although the source of this chapter remains a matter of debate,[50] the geographical location is clear: the place described here is Elba, not far from the Tuscan coast, exploited by the inhabitants of Populonia. The transmitted toponyms are evidently corrupted: the island is called Θάλεια, while the city of Populonia is named Ποπάνιον (Πωπάνιον in β). Both names are otherwise unattested and therefore doubtful, at least from an orthographical point of view.

|26| Ἐν δὲ τῇ Τυρρηνίᾳ λέγεταί τις νῆσος Αἰθάλεια ὀνο|27|μαζομένη, ἐν ᾗ ἐκ τοῦ αὐτοῦ μετάλλου πρότερον μὲν χαλ|28|κὸς ὠρύσσετο, ἐξ οὗ φασὶ πάντα κεχαλκευμένα παρ᾽ |29| αὐτοῖς εἶναι, ἔπειτα μηκέτι εὑρίσκεσθαι, χρόνου δὲ διελ|30|θόντος πολλοῦ φανῆναι ἐκ τοῦ αὐτοῦ μετάλλου σίδηρον, ᾧ |31| νῦν ἔτι χρῶνται <οἱ> Τυρρηνοὶ τὸ καλούμενον Ποπλώνιον οἰ|32|κοῦντες.

26 Αἰθάλεια Victorius (1582) 389 (cf. Steph. Byz. s.v. Αἰθάλη [A 120 Billerbeck]: θάλεια ω **31** Ποπλώνιον Victorius (1582) 388–389: ποπάνιον **Bx**: πωπάνιον β

In Tyrrhenia there is said to be a certain island named Aethalia, in which out of a certain mine in former days copper was dug, from which they say that all the copper vessels among them have been wrought; that afterwards it could no longer be found: but, when a long interval of time had elapsed, from the same mine iron was produced, which the Tyrrhenians, who inhabit the town called Poplonium, use to the present day.

This chapter is not included among the many passages of *Mir.* quoted by Stephanus of Byzantium in his *Ethnika*, a late-antique geographical dictionary arranged alphabetically. In this work, under the name Αἰθάλη, we read a detailed treatment of the toponym: the island, Stephanus reports, was called Αἰθάλη by Hecataeus,[51] but Αἰθάλειαν by Philistus[52] and the later grammarians Herodianus and Orus.[53] As for the name of the city, Populonia, the closest parallels are in Diodorus Siculus (possibly relying on the same source of *Mir.*) and Strabo, both offering the reading Ποπλώνιον.[54] It seems that, in this case, the corruption of the two names should be regarded as a mere scribal mistake, already witnessed in the archetype of the manuscript tradition, rather than an innovation due to the compiler of *Mir.*

---

50 See Schorn (in this volume), with reference to previous literature. The source of the chapter has been identified with Timaeus since Geffcken (1892) 149–150, but Flashar (1972) 113 was not convinced and preferred to name Posidonius; Vanotti (2007) 174–175 is more cautious.

51 *FGrHist* 1 F 59.

52 *FGrHist* 556 F 2.

53 A 120 Billerbeck: Αἰθάλη· νῆσος Τυρσηνῶν, Ἑκαταῖος Εὐρώπῃ. ἔοικε δὲ κεκλῆσθαι διὰ τὸ σίδηρον ἔχειν τὸν ἐν αἰθάλῃ τὴν ἐργασίαν ἔχοντα. Φίλιστος δὲ ἐν ε´ Σικελικῶν Αἰθάλειαν αὐτὴν καλεῖ, καὶ Ἡρωδιανὸς καὶ Ὧρος.

54 Diod. Sic. 5.13.1 (included in *FgrHist* 566 F 164) and Strabo 5.2.6.

Both instances are easy to explain palaeographically, particularly the corruption of Ποπλώνιον to Ποπάνιον, which could have been caused by an erroneous reading of the majuscule script. It should be noted that these remarkable errors strengthen the reconstruction of the overall *stemma codicum*: given that all manuscripts transmit the same wrong reading, we can safely assume that they all stem from a single archetype, produced between late-antiquity and the middle-Byzantine age.[55]

Pier Vettori was the first scholar to put forward both corrections; he first annotated them in the margins of his Aldine edition of *Mir.*, now preserved in Munich, and then published them in a revised edition of his *Variarum lectionum libri*, printed in Florence in 1582.[56]

Vettori did not limit himself to chapter 93 and he went on to correct chapter 94, dealing with another Tyrrenian city, called Οἰναρέα, the name of which, according to Vettori, should be corrected in Οὐολατέρρα, *Volterra*, since the description offered by the compiler of *Mir.* fits well with the one found in Strabo (5.2.6). However, Vettori did not know that a toponym close to Οἰναρέα, Οἶνα, is also preserved by Stephanus of Byzantium, who quotes directly from *Mir.*, attesting that a reading similar (if not identical) to the one found in the medieval manuscripts was already circulating in late antiquity.[57] Stephanus' testimony does not imply that the reading transmitted by the manuscript should be regarded as correct: Οἰναρέα could indeed still be an ancient mistake for Volterra and a modern editor of the text might choose to amend it accordingly. Such a correction, however, is less obvious than it would seem, and Jacques Heurgon has more recently defended the reading Οἰναρέα, which, according to his reconstruction, should not be regarded as a reference to Volterra, as Vettori thought, but to the nearby *Volsinii*, Bolsena, in northern Latium, thus turning a purely textual problem into a historiographical one.[58]

4.2 Cyprus or Gyarus? Mir. 25

|22|Ἐν Γυάρῳ τῇ νήσῳ λέγεται τοὺς μῦς τὸν σίδηρον |23| ἐσθίειν.

22 Γυάρῳ Cagnatus (1587) 153 et Holstenius (1684) 88 (vd. et Anon. Bas.): κύπρῳ βγ

It is said that in the island of Gyaros the mice eat iron.

---

55 Cf. Giacomelli (2021) 54–55.

56 On Vettori's philological work on the text of *Mir.*, see also Giacomelli (2021) 264–267.

57 O 22 Billerbeck, ὡς Ἀριστοτέλης Περὶ θαυμασίων ἀκουσμάτων. On the importance of Stephanus' quotations from *Mir.*, see Giacomelli (2021) 317–328.

58 Heurgon 1969. See also Vanotti (2007) 175–176. It goes without saying that the reference to Volsinii, destroyed by the Romans in 265 BC, would not be "un (*sic*) utile indicazione cronologica *ante quem* per la composizione dell'opera," as stated by Vanotti. The compiler of *Mir.* was obviously reporting the anecdote secondhand and any *ante quem* chronological detail that can be inferred from the text of a single chapter should be limited to the source of *Mir.*

The chapter at hand describes mice eating iron, an anecdote that can be read in several parallel texts, one of which, Aelian's *NA* 5.14, is a fragment taken from Aristotle that could even come from *Mir.* itself[59]:

Ἐν τῇ Γυάρῳ [Holstenius, Hercher: Πάρῳ codd.] τῇ νήσῳ Ἀριστοτέλης λέγει μῦς εἶναι καὶ μέντοι καὶ τὴν γῆν σιτεῖσθαι τὴν σιδηρῖτιν.
    In the island of Gyarus Aristotle says that there are rats and that they actually eat iron ore (text and trans. Scholfield).

The name of the island in which the marvelous event can be witnessed is transmitted as Πάρῳ by the manuscripts of Aelian and as Κύπρῳ, Cyprus, in those of *Mir.* The earliest evidence for the latter can be found in the *Excerpta Constantiniana de animalibus* (2.369), compiled in the 10th century from a manuscript close to branch γ, and in a short, 11th century quote by Michael Psellus, who had access to a manuscript of *Mir.* (*Ep.* 126.53–54 Papaioannou).[60]
    Another early parallel, Pliny (8.222), depends on Theophrastus' *On Swarms of Creatures*, as attested by Photius' summary (F 359A, ll. 53–54 FHS&G). Pliny situates the event *in Gyara insula*, in the small island of Gyaro, in the northern Cyclades, not far from Andros and Tinos, and such testimony fits well with the account in Stephanus of Byzantium, who reports an authentic Antigonean fragment,[61] and pseudo-Antigonus' *Mirabilia*.[62]
    The first to point out the obvious discrepancy between *Mir.* and the other parallel texts was the Veronese doctor Marsilio Cagnati (1545–1612), who published his conjecture in 1587, almost a century before Lucas Holste did.[63] It should be observed that, in the second part of the 16th century, an anonymous German scholar had already proposed the same solution in an addendum to a never published Latin translation of the text now preserved in a Basel manuscript: "puto . . . pro ἐν Κύπρῳ legendum ἐν Γυάρῳ."[64]
    The reading ἐν Κύπρῳ, attested by all extant manuscripts and by the 10th century *Excerpta de animalibus*, seems an obvious corruption of the rarer toponym Γυάρῳ, an error that could easily be due to both paleographic and cultural circumstances. Correcting it seems safe enough: indeed, if Aelian is here quoting directly from *Mir.*, the reading Πάρῳ we find in the manuscripts of *NA* would support this reconstruction, since such an error could be easily explained as a misreading of the majuscule script ΓΥΑΡΩ > ΠΑΡΩ, resulting, again, in a better-known toponym. Such an error could be a mistake made by Aelian himself, or an easy blunder of a careless scribe.

---

59 F 270.13 Gigon = 317 Rose. I am thankful to Pietro Zaccaria for a fruitful exchange on this point.
60 See Giacomelli (2021) 334 and 342–344 (on Psellus).
61 Γ 114 Billerbeck = F 51A Dorandi. See also Dorandi (1999) xxiii–xxiv.
62 18a1 Musso = F 51B Dorandi. On this passage, see also Dorandi (2017) 63.
63 Cagnatus (1587) 153.
64 Giacomelli (2021) 176 n. 4.

4.3 Tius: Mir. 73

|15| Ἐν Ἡρακλείᾳ δὲ τῇ ἐν τῷ Πόντῳ καὶ **ἐν τῇ Τίῳ** |16| γίνεσθαί φασιν ὀρυκτοὺς ἰχθῦς, τούτους δὲ μάλιστα κατὰ τὰ |17| ποτάμια καὶ τὰ ἔνυδρα χωρία.

15 τῇ Τίῳ Giann. ex Ath., cf. Rose περὶ Τίον: ῥηγίῳ ω

At Heraclea in Pontus and **in Tius**, they say there are fish obtained by digging, especially in places near rivers, and such as are well watered.

This chapter is one of a series of excerpts connected to Theophrastus' *On Fish* (*Mir.* 71–73), a short treatise known in antiquity under different names.[65] As we shall argue, despite the close proximity between the two texts, we should not assume *Mir.* to be a simple copy of the Theophrastean treatise handed down by the Byzantine manuscripts under the tile Περὶ ἰχθύων, insofar as it preserves seemingly authentic pieces of additional information that are nowhere to be found in the latter.

Such a reconstruction can benefit from a closer look at the toponym καὶ ἐν Ῥηγίῳ transmitted by the manuscripts. The anecdote about the "excavated fish" described in this chapter takes place in Pontus and a mention of Rhegium, in southern Italy, seems clearly out of place. A modern editor of the text could be tempted to think that the compiler of *Mir.* made up the name of the city, since the Theophrastean *On Fish* (7.58–64 Sharples) only says that this kind of fish can be seen περὶ Ἡράκλειαν καὶ ἄλλοθι τῶν ἐν τῷ Πόντῳ (around Heraclean and somewhere else in the Pontus).[66]

A more complete quote of the same Theophrastean anecdote found in Athenaeus (8.331c) points, however, to a different solution.[67] Athenaeus specifies that the "excavated fish," described by Theophrastus (ἱστοροῦντος περὶ αὐτῶν Θεοφράστου) can be found ἐν Ἡρακλείᾳ . . . καὶ περὶ Τίον, mentioning a location in Pontus, not far from Heraclea, which fits perfectly with the geographical frame and seems to preserve an authentic passage of Theophrastus' original work.

Valentin Rose was the first to correct the odd reading Ῥηγίῳ to περὶ Τίον, restoring the name of the settlement mentioned by Athenaeus,[68] while Giannini

---

65 See Sharples (1992) 347–348. The text, transmitted directly by the MS Vaticanus gr. 1302 and its descendants, is also known as Fragment 171 Wimmer. A detailed study of the intertextual relation between *Mir.* and *Pisc.* can be read in Mayhew (forthcoming 1).

66 See also Sen. *QNat.* 3.16.5: *Inde, ut Theophrastus affirmat, pisces quibusdam locis eruuntur*, and, most importantly, Plin. *HN* 9.175–176: *Piscium genera etiamnum a Theophrasto mira produntur. . . . Circa Heracleam et Cromnam et multifariam in Ponto unum genus esse quod extremas fluminum aquas sectetur cavernasque sibi faciat in terra atque in his vivat, etiam reciprocis amnibus siccato litore; effodi ergo, motu demum corporum vivere eos adprobante.*

67 On the quotations from the Περὶ ἰχθύων in Athenaeus, see briefly Sharples (1992) 348–349.

68 Rose (1863) 358

only perfectioned Rose's conjecture to τῇ Τίῳ, a reading which offers a paleo-graphical explanation of the corruption in Ρηγίῳ.[69] Again, the origin of this rather easy mistake is to be found in both cultural and mechanical circum-stances: a simple misreading of the majuscule script (ΤΗΤΙΩ > ΡΗΓΙΩ) led an ancient copyist to think of the more famous Italian city of Rhegium instead of the name of the lesser-known Pontic settlement of Tium, preserved by Athenaeus.

In this case, I believe we have to conclude that this was a scribal error and not an innovation introduced by the compiler of *Mir.* Restoring the name of Tium, based on the parallel quote by Athenaeus, also offers an important piece of evi-dence concerning the actual relationship among the parallel texts, and puts into question the commonly held view according to which this set of chapters is a series of excerpts taken from the extant Theophrastean treatise *On Fish*: if the toponym Tium is to be restored in *Mir.*, we have to infer that the compiler of *Mir.* and Athenaeus quoted (directly or indirectly) from a Theophrastean text which only partially corresponds to the treatise transmitted under the name of Theophrastus by the medieval manuscripts.[70]

5. Conclusions

The suspicious toponyms analyzed here – only a fraction of the problematic pas-sages of *Mir.* – can and should be used as methodological warnings to anyone interested in the text and its interpretation. In the first two sections of the chapter, I have taken into account toponyms that seem corrupt but should actually be left in the state in which they were transmitted by the manuscripts, either because they could have been drawn from the compiler's source or because there are no convinc-ing alternatives. In Section 3, which should be regarded as a subsection of the set of examples preceding it, I dealt with a peculiar feature of *Mir.* and of any excerpt collection of the same kind: redactional imperfection and made-up locations. Some chapters betray a complex and contradictory layering of the sources while oth-ers, as shown by *Mir.* 5, present the reader with a setting which is the result of

---

69 Giannini (1965) 252.
70 Flashar (1972) 102 argues that "[k]onkret ist Mir. 71 aus De pisc. in sicc. deg. 2 . . . exzerpiert," while noticing that "[e]s fehlt jedoch bei Theophrast die Ortsangabe" (the same is true for chap-ter 73). Flashar mentions as direct Quelle the treatise *On Fish* also for chapters 72–73. For a similar approach, see Sharples (1992) 349 and n. 10, who, states that the few innovations in *Mir.* "could easily have been added by the compiler," specifically referring to the reading Ῥηγίῳ but at the same time failing to mention the name of the river in chapter 71. Cf. also Vanotti (2007) 162, who is more cautious in dealing with the treatise *On Fish* transmitted by the manuscripts. It is very unlikely, to say the least, that the compiler of *Mir.* could independently conjecture the name of a Greek settle-ment on the southern coast of the Black Sea later mentioned, in the same Theophrastean context, by Athenaeus. It is worth noticing that Robert Mayhew (forthcoming 1) comes to the same conclusion for *Mir.* 71–74, stating that "we should not simply assume that De piscibus must have been the work from which the compiler of the *Mirabilia* extracted material, which became our *Mir.* 71–74, and that that is all there is to it – i.e., that there is no plausible alternative relationship between *Mir.* 71–74 and *De piscibus* worth considering."

two co-factors: the need to frame each marvelous anecdote geographically, and the compiler's lack of understanding of his source. Again, correcting the text here would falsify it, obliterating precious evidence that could be helpful in understanding the structure of the work.

Finally, I have described instances in which correcting the text transmitted by the manuscripts seems the best approach, even if it should be attempted with great caution, always bearing in mind the possible implications of each conjecture.

To summarize, any reader of *Mir.* should keep in mind that the numerous errors present in the work may be due to (1) the sources employed by the compiler, which are now almost all completely lost and cannot be directly controlled; (2) the compiler's own carelessness and limited understanding of the source; (3) the inevitable degradation of the text in the process of its manuscript transmission, including contamination and ancient conjectural attempts (as it is apparent in branch γ).[71] An editor can act on the text only in this last case, but it should be noted that distinguishing these three layers of corruption is never easy: in fact, it is often impossible.

What should a historian expect from a critical edition of *Mir.*? I think a good answer to this question was outlined more than a century ago by Karl Müllenhoff, as reported by Johannes Geffcken: such texts need to be dissected and analyzed and the presentation of the evidence should be as clear and useful as possible. Müllenhoff's model for such enterprises was the magisterial edition of Solinus, published first in 1864 by Theodor Mommsen.[72] Not only should the textual critics provide all the evidence that can be gathered from the study of the manuscript tradition, but it is among their duties to provide a clear overview of the structure of each chapter, pointing out parallel passages and possible sources. In accomplishing such a task, there is, however, very little space for conjectural virtuosity, especially when dealing with names and places: the state of the transmitted text is such that any intervention, even the most unassuming and apparently obvious, could result in a reduced understanding of the intricate web of intertextual connections underlying each chapter – one of the features that make *Mir.* so interesting to the eye of contemporary scholars.

A modern reader of *Mir.* – and of other paradoxographical collections of its kind – does not require a text "edited in the conventional manner" and "established by the methods appropriate to literary texts." Quoting Edward John Kenney, what such user actually needs and should expect from a critical edition of it "is not the text of this or that editor, which is merely liable to engender a sense of false security, but a do-it-yourself kit."[73]

---

71 Giacomelli (2021) 48–49.
72 Geffcken (1892) 83.
73 Quotes are from Kenney (1969) 183, reviewing Gelsomino's Teubner edition of Vibius Sequester.

Works cited

Balme, D.M. 2002. *Aristotle, Historia Animalium. Vol. I: Books I-X. Text*. Cambridge: Cambridge University Press.

Bastiaens, A.A.R. 1987. *Atti e passioni dei Martiri*. Introduzione di Bastiaens, A.A.R. Testo critico e commento a cura di Bastiaens, A.A.R., Hilhorst, A., Kortekaas, G.A.A., Orbán, A.P., van Assendelft, M.M. Traduzioni di Chiarini, G., Kortekaas, G.G.A., Lanata, G., Ronchey, S. Milan: Arnoldo Mondadori.

Beckmann, J. 1786. Ἀριστοτέλους περὶ θαυμασίων ἀκουσμάτων. *Aristotelis liber de mirabilibus auscultationibus*. Göttingen: Vandenhoek.

de Boor, C. 1904. *Georgii monachi chronicon*. Vol. I. Leipzig: Teubner.

Cagnatus, M. 1587. *Variarum observationum libri quatuor*. Rome: apud Bernardinum Donangelum.

Creuzer, F. 1806. *Historicorum Graecorum antiquissimorum fragmenta*. Heidelberg: In officina Mohrii et Zimmeri Academica.

Daubner, F. 2013. "Katakekaumene." In Bagnall, R.S., Brodersen, K., Champion, C.B., Erskine, A. and Huebner, S.R. eds. *The Encyclopedia of Ancient History*. Hoboken, NJ: Wiley-Blackwell. 3705–3706.

Dittenberger, W. 1879. "Ketriporis von Thrakien." *Hermes* 14: 289–303.

Dittmeyer, L. 1907. *Aristoteles, De animalibus historia*. Leipzig: Teubner.

Dorandi, T. 1999. *Antigone de Caryste. Fragments*. Paris: Les Belles Lettres.

Dorandi, T. 2013. *Diogenes Laertius. Lives of Eminent Philosophers*. Cambridge: Cambridge University Press.

Dorandi, T. 2017. "La ricezione del sapere zoologico di Aristotele nella tradizione paradossografica." In Sassi, M.M. ed. *La zoologia di Aristotele e la sua ricezione dall'età ellenistica e romana alle culture medievali*. Pisa: Pisa University Press. 59–80.

Eleftheriou, D. 2018. *Pseudo-Antigonos de Carystos: collection d'histoires curieuses*. Doctoral thesis. Université Paris Nanterre.

Flashar, H. 1972. *Aristoteles, Mirabilia*. Berlin: Akademie-Verlag.

Gallavotti, C. 1990. "Planudea (X). 37. L'anacreontica *de thermis* di Leone Magistro." *Bollettino dei Classici* ser. 3 11: 78–103.

Garani, M. forthcoming. "*Miracula ignium*: Theophrastus, Mir. 33–41, Pliny HN 2.236–238." In Hellmann, O., Mayhew, R. and Zucker, A. eds. *The Aristotelian* Mirabilia *and Early Peripatetic Natural Science*. London: Routledge.

García Valdés, M., Llera Fueyo, L.A. and Rodríguez-Noriega Guillén, L. 2009. *Claudius Aelianus De Natura Animalium*. Berlin; New York: W. de Gruyter.

Geffcken, J. 1892. *Timaios' Geographie des Westens*. Berlin: Weidmannsche Buchhandlung.

Geus, K. 2016. "Paradoxography and Geography in Antiquity. Some Thoughts about the *Paradoxographus Vaticanus*." In González Ponce, F.J., Gómez Espelosín, F.J. and Chávez Reino, A. eds. *La letra y la carta: descripción verbal y representación gráfica en los diseños terrestres grecolatinos. Estudios en honor de Pietro Janni*. Sevilla: Universidad de Sevilla. 243–257.

Giacomelli, C. 2021. *Ps.-Aristotele, 'De mirabilibus auscultationibus' Indagini sulla storia della tradizione e ricezione del testo*. Berlin-Boston: W. de Gruyter.

Giacomelli, C. 2023. *Pseudo-Aristotele, De mirabilibus auscultationibus: Edizione critica, traduzione e commento filologico*. Rome: Accademia Nazionale dei Lincei.

Giacomelli, C. forthcoming. "The text of the *De mirabilibus auscultationibus*. Observations on its Structure and Transmission." In Hellmann, O., Mayhew, R. and Zucker, A. eds. *The Aristotelian* Mirabilia *and Early Peripatetic Natural Science*. London: Routledge.

Giannini, A. 1965. *Paradoxographorum Graecorum reliquiae*. Milan: Istituto editoriale italiano.

Heurgon, J. 1969. "Oinarea-Volsinii." In Altheim-Stiehl, R. and Stier, H.E. eds. *Beiträge zur Alten Geschichte und deren Nachleben. Festschrift für Franz Altheim zum 6.10.1968*. Vol. I. Berlin: De Gruyter. 273–279.

Jacobi, F. 1930. Πάντες Θεοί. Doctoral thesis. Halle; Wittenberg: Karras, Krober et Nietschmann.

Jacques, J.M. 2002. *Nicandre, Les thériaques. Fragments iologiques antérieurs à Nicandre*. Paris: Les Belles Lettres.

Keller, O. 1909. *Die Antike Tierwelt. Vol. I. Säugetiere*. Leipzig: W. Engelmann.

Kenney, E.J. 1969. "Quot editores tot Vibii." *The Classical Review* 19: 183–185.

Kitchell, K.F. 2014. *Animals in the Ancient World from A to Z*. London; New York: Routledge.

Kyle, D.G. 1996. "Gifts and Glory: Panathenaic and Other Greek Athletic Prizes." In Neils, J. ed. *Worshiping Athena. Panathenaia and Parthenon*. Madison: University of Wisconsin Press. 106–136.

Longo, O. 1986–1987. "Caccia coi falchi in Tracia?" *Atti e memorie dell'Accademia Patavina di Scienze Lettere ed Arti* 99(3): 39–45.

Louis, P. 1964. *Aristote, Histoire des animaux. Vol. III. Livres VIII-X*. Paris: Les Belles Lettres.

Mango, C. 1975. *Byzantine Literature as a Distorting Mirror. An Inaugural Lecture delivered before the University of Oxford on 21 May 1974*. Oxford: At the Clarendon Press.

Mayhew, R. forthcoming 1. "Pseudo-Aristotle, *De mirabilibus auscultationibus* 71–74 and Theophrastus' *De piscibus*." In Hellmann, O., Mayhew, R. and Zucker, A. eds. *The Aristotelian* Mirabilia *and Early Peripatetic Natural Science*. London: Routledge.

Mayhew, R. forthcoming 2. "*Historia animalium* 8(9).5, *De mirabilibus auscultationibus* 5 & 75, and two of Theophrastus' lost works on animals." In Connell, S. ed. *Philosophical Essays on Aristotle's* Historia Animalium. Leiden; Boston: Brill.

Mayhoff, C. 1875. *C. Plini Secundi Naturalis Historiae libri XXXVII. Vol. II. Libri VII-XV*. Leipzig: Teubner.

Mercati, G. 1923–1925. "Intorno all'autore del carme εἰς τὰ ἐν Πυθίοις θερμά (Leone Magistro Choirosphaktes)." *Rivista degli Studi Orientali* 10: 212–248.

Miscellaneo, S. 1996. "Cuma o Posidonia? (Nota a Ps. Arist. mir. ausc. 95)." *Hesperìa. Studi sulla grecità di Occidente* 7: 111–119.

Müller, C. and Müller, T. 1853. *Fragmenta historicorum Graecorum*. Vol. I. Paris: Didot.

Musso, O. 1985. *Antigonus Carystius, Rerum mirabilium collectio*. Naples: Bibliopolis.

Musurillo, H. 1972. *The Acts of the Christian Martyrs*. Oxford: Clarendon Press.

Oikonomopoulou, K. forthcoming. "*Mir.* 16–22 and Theophrastus' lost *On Honey*." In Hellmann, O., Mayhew, R. and Zucker, A. eds. *The Aristotelian* Mirabilia *and Early Peripatetic Natural Science*. London: Routledge.

Praechter, K. 1905. "Kritisch-exegetisches zu spätantiken Philosophen." *Philologus* 64: 385–390.

Rose, V. 1863. *Aristoteles pseudepigraphus*. Leipzig: Teubner.

Schepens, G. and Delcroix, K. 1996. "Ancient Paradoxography: Origin, Evolution, Production, and Reception." In Pecere, O. and Stramaglia, A. eds. *La letteratura di consumo nel mondo greco-latino: Atti del convegno internazionale, Cassino, 14–17 settembre 1994*. Cassino: Università degli Studi di Cassino. 373–460.

Schnieders, S. 2019. *Aristoteles, Historia Animalium Buch VIII und IX*. Berlin: W. de Gruyter.

Scholfield, A.F. 1959. *Aelian, On the Characteristics of Animals. Vol. I. Books 1-5.* Cambridge, MA: Harvard University Press.

Schorn, S. ed. 2022. *Felix Jacoby. Die Fragmente der Griechischen Historiker Continued. Part IV. Biography and Antiquarian Literature. E. Paradoxography and Antiquities. Fasc. 2. Paradoxographers of Imperial Times and Undated Authors [Nos. 1667–1693].* Leiden; Boston: Brill.

Sharples, R.W. 1992. "Theophrastus: *On Fish.*" In Fortenbaugh, W.W. and Gutas, D. eds. *Theophrastus: His Psychological, Doxographical and Scientific Writings.* New Brunswick; London: Transaction. 347–385.

Terian, A. 1981. *Philonis Alexandrini de animalibus. The Armenian Text with an Introduction, Translation, and Commentary.* Ann Arbor: Scholars Press.

Vanotti, G. 1981. "Appunti sul *De mirabilibus auscultationibus.*" *Giornale filologico ferrarese* 4: 83–88.

Vanotti, G. 2007. *Aristotele, Racconti meravigliosi.* Milan: Bompiani.

Victorius, P. 1582. *Variarum lectionum libri XXXVIII.* Florence: Apud Iunctas.

Wilson, N. 2010. "Review of García Valdés *et al.* (2009)." *Exemplaria Classica* 14: 403–405.

Ziegler, K. 1949. "Paradoxographoi." *Paulys Realencyclopaedie der classischen Altertumswissenschaft* XVIII(3): 1137–1166.

8 Ps.-Plutarch's *On Rivers*, the *Mirabilia*, Stobaeus 4.36, and paradoxographical literature

Søren Lund Sørensen

An essay on ps.-Plutarch's *On Rivers* needs no justification in a volume on ps.-Aristotle's *Mirabilia*, since a substantial portion of reports at the very end of the latter, the so-called appendix, are found in the former work (*Mir.* 158–160; 162–163; 166–167; 171; 173–174; 175). A few introductory remarks on ps.-Plutarch and his work will, however, be useful.

1. ps.-Plutarch *On Rivers*

The treatise *On Rivers* (*De fluviis*) is found in two manuscripts: *Palatinus Heidelbergensis gr. 398* (*P*) (9th century) and *Suppl. grec. 443A* (*B*) (14th century),[1] in which Plutarch of Chaeronea is given as the author.[2] An anonymous note in the margin of *P* questions the authenticity of the text, as it differs far too much from Plutarch in terms of "spirit" (διάνοια) and "style" (φράσις).[3] Scholarship has (almost) unanimously espoused this view, and the text is considered to have been written by someone other than Plutarch.[4] The attribution to the famous Chaeronean is, however, already present in references and quotations from late antiquity.[5] *On Rivers* shares this attribution to Plutarch with the so-called *Minor Parallels*, a treatise transmitted among the manuscripts of Plutarch.[6] In addition, the two texts, which are concerned with different things – mythology and natural phenomena on the one hand,[7] biography and moral qualities on the other – have a number of elements in common: a somewhat poor but still learned prose style[8]; alternative

---

1 Dorda *et al.* (2003) 91–97; Delattre (2011) 12–14. *B* derives from *P*, cf. the collation by Poidomani (2016).
2 Useful introductions to ps.-Plutarch's *On Rivers* may be found in the many entries in *Brill's New Jacoby* and *Die Fragmente der Griechischen Historiker* on the various bogus authors, but see in particular the biographical essays and introductions to Diocles of Rhodes, *BNJ*² 302; Thrasyllus of Mendes, *BNJ* 622; Clitophon of Rhodes, *FGrHist* 1781 and the commentary on Demaratus, *BNJ*² 42 F 1.
3 *In margine P:* ψευδεπίγραφον τοῦτο. πόρρω γὰρ τῆς Πλουτάρχου μεγαλοφυίας ἥ τε διάνοια καὶ ἡ φράσις. Cf. Delattre (2011) 8–9.
4 Dorda *et al.* (2003) 33–44; Delattre (2011) 8–10; Brodersen (2022) 13.
5 Dorda *et al.* (2003) 10–14; Delattre (2011) 9–10; 14–17; Brodersen (2022) 7–10.
6 De Lazzer (2000) 7–38; Boulogne (2002) 240–241; Ibáñez Chacón (2014).
7 For the structure of *On Rivers*, cf. now Delattre (2017).
8 Hercher (1851) 5–9; Dorda *et al.* (2003) 23–30; Delattre (2011) 32–37; Brodersen (2022) 14.

DOI: 10.4324/9781003437819-9

mythologies otherwise unattested; and quotations from works and authors which are not known from other sources and generally held to be fabrications by the author.[9] As for the date, the presence of names such as Juba of Mauretania (ca. 50 BC to AD 23) (*FGrHist/BNJ* 275 F 5) as a *terminus post* and a possible quotation by Clement of Alexandria (around AD 200)[10] as a *terminus ante quem* have led scholars to posit a date for *On Rivers* as well as the *Minor Parallels* at the end of the second or beginning of the 3rd century AD.[11]

On Rivers is not concerned with waters alone but also with mountains, stones, and plants. Each of the 25 chapters, which are of various length, commences with an etiology, explaining the various names of a river by resorting to mythology. This is followed by the description of a stone or plant in the river or near a mountain in the vicinity of the river. Occasionally, a thematic and sympathetic link connects all phenomena with the initial myth.[12]

Scholars have assigned the work to different genres.[13] In particular, *On Rivers* shares several thematic parallels with paradoxographical literature, such as an interest in rivers, plants, stones, and mountains – all of which are connected to more or less known toponyms – as well as etiologies.[14] A notable difference with regard to etiologies is the use of mythology in *On Rivers*. Although mythology plays a role in paradoxographical literature, *On Rivers* engages in actual mythography.[15]

Similar to paradoxographical collections, the author of *On Rivers* took great pride in providing references for his wonderous and miraculous reports – the major difference being the dubious identity of most of these sources.

Ps.-Plutarch thus engages in "pseudo-paradoxography,"[16] but to what end? Delattre has emphasized the value of the term "fictionality" by which we may avoid such labels as "true" and "false." This may be enhanced by Martin Hose's discussion of fictional versus pragmatic structures of communication, in which the former comprises a theoretically endless horizon of meaning and the latter demands a reflection of reality.[17] The alternative mythologies in *On Rivers*, for example, do not constitute new mythologies per se but rather redactions of familiar myths. In

---

9 Hercher (1851) 17–24; Jacoby (1940); Boulogne (2002) 228–231; Dorda *et al.* (2003) 60–66; Cameron (2004) 127–134; Delattre (2011) 24–30; Ibáñez Chacón (2014) 62–64. See also Paola Ceccarelli's biographical essay on Aristides of Miletus, *BNJ* 286, Ken Dowden's essay "Ptolemy Chennos and the citing of 'bogus authors'" in his edition of Antipater of Acanthus, *BNJ*² 56, and the commentary on Dorotheus of Athens, *BNJ* 145 F 2.

10 Clem. Al. *Protr.* 3,42,7 = [Plut.] *Par. min.* 20Aa; 20Ba.

11 Dorda *et al.* (2003) 32–33; Delattre (2011) 10–11; Brodersen (2022) 14–15.

12 Dorda *et al.* (2003) 14–23; Delattre (2011) 30–32.

13 Dorda *et al.* (2003) 44–60; Delattre (2011) 38–42.

14 Cf. Pajón Leyra (2009) 222–223, 474–475; Ibáñez Chacón (2008/2009) 400–401.

15 The role of mythology in paradoxographical literature has previously been underestimated. See, however, Greene in the present volume.

16 Ziegler (1949) 1164; Giannini (1964) 132 n. 206, 140; Delattre (2011) 57. In this "genre" may be included Aelian and Ptolemy Chennos.

17 Delattre (2011) 57–59; Hose (2008) 177; 196. On fictionality in ancient historiography, cf., for example, Syme (1983); Bowersock (1994) *passim*, but especially 1–27; Borghart and De Temmerman (2010); Ibáñez Chacón (2014).

this way, the author engages with a mythological substratum familiar to his readers. This may also be the case for the paradoxographical elements in *On Rivers*.

As the name Plutarch has been attached to both *On Rivers* and the *Minor Parallels* since antiquity, we find it plausible that the author is deliberately embarking on an intertextual dialogue with the "real" Plutarch, and the words "parody" and "mimicry" seem merited. This is surely the case for the *Minor Parallels*, which are juxtaposed with the much more comprehensive *Parallel Lives*. In these, Plutarch assesses the moral qualities of famous Greek and Roman statesmen and refers to his sources throughout. The moral qualities, or lack thereof, is equally the subject of the *Minor Parallels* but also of the treatise *On Rivers* (imitating, inter alia, a well-established genre concerned with hydrological phenomena). The etiologies in *On Rivers* are almost always connected with vices or moral failings (e.g., hubris, rape, incest, bestiality, human sacrifice, and murder of one's kin).[18] In turn, these tragic elements result in the suicide of an eponymous person, often by jumping into the river or, to use a technical term, performing a *katapontismos*.[19] In both treaties, the veracity of these immoral acts is, as mentioned, vouchsafed by the aforementioned authors and works invented ad hoc. The ingenuity of the author with regard to alternative mythology, pseudo-paradoxography, and fictionality is supposedly in stark contrast with the poor style of both works.[20] May we assume that the difference in style between Plutarch and ps.-Plutarch is intentional? If this is the case, ps.-Plutarch could be adding insult to injury by deliberately aiming at a Greek prose stylistically less impressive than that current in the 2nd and 3rd centuries AD – in the heyday of the Second Sophistic. In Flavius Philostratus' biographies of prominent men of the Second Sophistic, we come upon the case of Aspasius of Ravenna (2nd–3rd century AD), who excelled in the Greek prose style of Classical Athens but missed clarity to such a degree that the documents issued by him in the capacity of *ab epistulis Graecis* missed their audience by being completely unintelligible.[21] Seen in this light, ps.-Plutarch stands out as a very original writer, whose counter-sophistic, literary resistance to the society of his day is achieved in different ways[22]: (1) exclusively focusing on the vices of Greeks and Romans, past and present, in terms of atrocious acts[23]; (2) using paradoxography, a well-established but at the same time somewhat disrespected genre, and literature on rivers; (3) ridiculing the learned obsession with educated references by ingeniously providing such to bogus authors and fictitious works of well-known authorities[24]: (4) intentionally resorting to a Greek prose that did not meet the standards for proper

18 On these themes, cf. Hughes (1991); Robson (2002); Ibáñez Chacón (2008/2009); Malheiro Magalhães (2022).
19 Dorda *et al.* (2003) 16–17. See the introduction to Timolaus, *BNJ* 798.
20 On the style of the *Minor Parallels*, cf. De Lazzer (2000) 28–31.
21 Philostr. *VS* 2.33.
22 On literary resistance in the Roman empire, cf. Whitmarsh (2013); Ursin (2019) 266–272; on counter-sophistic rhetoric, cf. Winter (2002) 113–122; Henderson (2011). A similar characteristic may be given for the late first-century AD Ptolemy Chennos, cf. Hose (2008).
23 As opposed to the real Plutarch, who excuses himself when paying too much attention to vices, for example, in *Demetr.* 1.1–7.
24 This is particularly evident in the introduction to the *Minor Parallels*.

style set out by the proponents of the so-called Second Sophistic – persons often belonging to the upper echelons of society in terms of wealth, prestige, and influence; and finally (5) attributing both treatises to Plutarch of Chaeronea, who is often held to have been a forerunner of the Second Sophistic as well as a famous stylist.[25]

During our work editing paradoxographical collections and fragments for one of the volumes of *Die Fragmente der Griechischen Historiker IV*, we have observed how most of the paradoxographical reports transmitted preserve a principle of organization by which they are associated with each other by geography, theme, or prominent words. This is the case for the two anonymous collections: *Paradoxographus Palatinus* (*FGrHist* 1681) and *Paradoxographus Vaticanus* (*FGrHist* 1679). In addition, it appears to have been the case also for the paradoxographical collection of strange customs compiled by Nicolaus of Damascus (*FGrHist* 1659). An organization principle different from what would seem reasonable to the modern reader appears to have been widespread for collections of information in general, and not necessarily paradoxographical, and is resorted to, for instance, in the *Physiologus* and in Ampelius.[26] Accordingly, we assume that a similar principle is also to be recognized in ps.-Aristotle, for which reason the commentary of Irene Pajón Leyra is much awaited.[27]

It is, however, by no means an easy task sorting out the organization principle of an ancient text. In the case of *Mir.* 152–178, the appendix, and especially the 11 reports taken from *On Rivers* (*Mir.* 158–160; 162–163; 166–167; 171; 173–174; 175), we have found no good explanation for the order in which they occur. As for *On Rivers* itself, Charles Delattre has attempted to explain the organizational principle as alphabetical.[28] While this may be the case, there is no need to assume that only one such principle is lurking in ps.-Plutarch. Each chapter consists of several reports supposedly connected only by the toponym and myth around which the entire chapter revolves; but at times individual reports are, however, connected by other themes. Furthermore, such themes recur across several chapters.

Accordingly, our overall impression of ps.-Plutarch is that of a sharp, intelligent, and at times witty author with a particular sense of detail, for which we hope to provide examples.[29]

Recently, Paula Ceccarelli has argued that the 11 reports found in both works have been taken from a common source, rather than having been lifted from *On Rivers* and inserted into *Mirabilia* at a later date.[30] In most of the mutual reports, there are only microscopic if any differences, consisting mainly in the change from direct (ps.-Plutarch) to indirect speech (*Mir.*). In other cases, obscure names of, mainly, stones found in rivers are distorted (e.g., *Mir.* 159). In addition, all mythological passages in ps.-Plutarch are, as will be discussed subsequently in the section on Stobaeus, absent in *Mir.* Two reports do, however,

25 Schmitz (2014).
26 The first study appears to be that of Rosenkranz (1966), dealing with speeches of pseudo-Isocrates.
27 See Pajón Leyra in this volume.
28 Delattre (2017). Along similar lines, Brodersen (2022) 16–17 proposes to see the treatise as a parody of scientific literature and compares it to Lucian's *True Stories*.
29 Unless otherwise indicated, translations are my own.
30 Cf. the commentary on Agatharchides of Samos, *BNJ* 284 F 3.

stand out. The first is *Mir.* 175, the last report for which a parallel may be found in ps.-Plutarch:

> Ἐν Ἀρτέμιδος Ὀρθωσίας βωμῷ ταῦρον ἵστασθαι χρύσειον, ὃς κυνηγῶν εἰσελθόντων φωνὴν ἐπαφίησιν.
>
> On the altar of Artemis Orthosia stands a golden bull, which emits a sound when hunters enter.

In ps.-Plutarch 21, the river Caicus as well as mount Theuthras in Mysia are described at length. On the latter stands a temple for Artemis Orthosia to which belongs an altar with a humanoid sculpture of a boar, which according to myth had taken refuge in the sanctuary when being hunted. The sound emitted is a human voice commanding hunters to take pity (φείδεσθαι). Apart from the words κυνηγῶν εἰσελθόντων φωνήν, the verbal parallels are few. ps.-Plutarch's κάπρον may by scribal mistake have been changed to ταῦρον, but the entire passage in *Mir.* seems to have been conflated to the point of important information being left out, for example, what sort of sound is emitted? Why should hunters take an interest in bulls?[31] Although obvious differences may be highlighted between *Mir.* 175 and ps.-Plutarch 21, the former appears to depend on the latter.

The other chapter to be mentioned here is *Mir.* 173, the only report in the appendix in which a source is given. Here also ps.-Aristotle may despite differences be seen to depend on ps.-Plutarch.

> Ἐν ὄρει Βερεκυνθίῳ γεννᾶσθαι λίθον καλούμενον μάχαιραν, ὃν ἐὰν εὕρῃ τις, τῶν μυστηρίων τῆς Ἑκάτης ἐπιτελουμένων ἐμμανὴς γίνεται, ὡς Εὔδοξός φησιν.
>
> In mount Berecynthius a stone is produced called *machaira*. If one of those celebrating the mysteries of Hekate finds it, he goes mad, as Eudoxus says.

Compare [Plut.] *Fluv.* 10.5:

> γεννᾶται δ' ἐν αὐτῷ λίθος καλούμενος μάχαιρα· ἔστι γὰρ σιδήρῳ παραπλήσιος· ὃν ἐὰν εὕρῃ τις τῶν μυστηρίων ἐπιτελουμένων τῆς θεᾶς, ἐμμανὴς γίνεται, καθὼς ἱστορεῖ Ἀγαθαρχίδης ἐν τοῖς Φρυγιακοῖς.
>
> A stone is produced in it called *machaira*, for it is similar to iron. If one of those celebrating the mysteries of the goddess finds it, he goes mad, just as Agatharchides records in the *Phrygiaca*.

The report is taken from the chapter on the river Marsyas in Phrygia. Near the river lies mount Berecynthius, where the mysteries of Cybele take place. That the stone is called "sword" (μάχαιρα) is a pun either on the flaying of Marsyas by Apollo or the self-mutilation of the followers of Magna Mater. Attention has, however, centered around the attribution of this report to Eudoxus[32] or Agatharchides, respectively,

---

31 See also Flashar (1972) 153; Vanotti (2007) 218; Vespa (2020) 417–420.
32 Eudoxus of Rhodes, *FGrHist/BNJ* 79 F 5; Eudoxus of Cnidus F 338 Lasserre.

and Paula Ceccarelli has diligently digested the scholarly discussions on the identity of Eudoxus and Agatharchides.[33]

Nonetheless, rather than suggesting that ps.-Plutarch altered the name of Eudoxus, be it the Rhodian or the Cnidian, to Agatharchides, we believe Agatharchides is the correct reading. According to Photius, Agatharchides of Cnidus is reported to have composed a Συναγωγὴ θαυμασίων ἀνέμων (*Collection of Remarkable Winds*).[34] The title may be corrupt, but it indicates the existence of a paradoxographical work by Agatharchides of Cnidus. ps.-Plutarch elsewhere invents titles for actual paradoxographers: Archelaus will be mentioned below, but several unknown works by (a) Ctesias are similarly referred to in *On Rivers*.[35] ps.-Plutarch's intention seems to have been to confuse his reader who would identify Agatharchides in this chapter with either Agatharchides of Samos, known only from *On Rivers*, or the famous Cnidian. With the Cnidian in mind, the excerptor will have confused him with the yet more famous Eudoxus of Cnidus, leading to this name being perpetuated in *Mir.* 173.

Having argued that *Mir.* relies on ps.-Plutarch with no intermediate source, we shall move on to the excerpts in the gigantic *florilegium* of Stobaeus, allowing us to draw conclusions as to the use, misuse, and occasional abuse of ps.-Plutarch's work.

2. Stobaeus 4.36 and *On Rivers*

The first observation to be made is that Stobaeus appears to have excerpted a cluster of reports found in the latter part of ps.-Plutarch's work, that is, chapters 20, 21, 24, and 25, with the addition of chapters 6, 8, and 16, the final report also being found in *Mir.* 166. Furthermore, Stobaeus has inserted *Mir.* 18 between reports from chapters 20 and 6 of *On Rivers*. There appears to be no obvious connection between the pseudo-Aristotelian report on Trapezuntian boxwood honey and stones that may alleviate amblyopia and quartan fever, respectively, and [Plut.] *Fluv.* 20 and 6. Granted, the entire 36th chapter of Stobaeus' fourth book is concerned with diseases and remedies, but a guiding principle is not easily observed.

The lengthy chapter begins and ends with quotations from classical Athenian authors, Euripides and Plato. The excerpts from *On Rivers* are preceded by a quotation from Demosthenes with general comments on the healthy and sick body. Now the quotation from Demosthenes includes the word σῶμα, which is found also in the quotation from Diogenes preceding it, and again in the chapter preceding Diogenes, which is a quotation from none other than Socrates. The eighth chapter does not include the word σῶμα, but may be connected to the saying of Socrates by the words νόσος/νοῦσος. In chapter 7 are found several forms of this noun, and in chapters 1–3 and 5 other forms of νόσος are to be observed. Connecting chapters 6 and 7 is more difficult, but the two infinitives λέξαι and λέγειν may be a clue. Whereas chapter 5, as mentioned, picks up on the noun νόσος again, no obvious connection between

---

33   See the commentary on Agatharchides of Samos, *BNJ* 284 F3, and more recently Giacomelli (2021) 37–39.

34   *FGrHist/BNJ* 86 T 2.

35   [Plut.] *Fluv.* 9.2 = *FGrHist* 688 F 74; 21.5 = *FGrHist* 688 F 73; 28.6.

chapter 4 and its immediately surrounding chapters is observable. It should, however, be mentioned that chapter 7 is identical to chapter 5, with the addition of a few
verses. Textual difficulties due to transmission may thus be at play.

An overall use of an organizational principle is observable at least for the initial
part, but we have not been able to establish a link between the 11th chapter from
Demosthenes and the collection of reports from *On Rivers*.

The cluster of pseudo-Plutarchan reports is, however, neatly rounded off by a
quotation from Porphyry's work *On Styx*, in which the willow (ἰτέα) is mentioned,
thereby connecting this chapter to *On Rivers* 25.3, in which a plant (βοτάνη),
καρπύλη, is described.

Regardless of the authenticity of the following header (ἐκ τῆς Τροφίλου),[36] Porphyry's report on the willow being able to ruin semen and sexual inclination may
be connected to the ps.-Aristotelian report on the penis of the marten (*Mir.* 12). The
following chapters from *Mir.* all have animals as their common theme.

After the four reports from *Mir.* follows a quotation from Alcmaeon, which
repeats the noun νόσος, prominent in the first part of Stobaeus' lengthy chapter.
Furthermore, the quotation from Alcmaeon is perhaps connected to the final ps.-
Aristotelian report, which speaks of a diet (τὸ ἐπιφαγεῖν) consisting of locusts (*Mir.*
139), by the word τρόφη.

Chapters 29–31, which include quotations from Diocles of Carystus and Erasistratus, are found in the same order, albeit with textual differences, in the so-called
Placita philosophorum, a doxographical collection transmitted among the works
of Plutarch.[37] In addition, these three quotations are united by the words πλῆθος
and νόσος.

The final quotation, from Plato's *Symposium*, is difficult to connect to these four
chapters, but perhaps chapter 29 provides a hint: Alcmaeon speaks of the need for
a balance between extremes to obtain health. The list of antonyms given is reminiscent of a passage in Plato's *Symposium*, which follows immediately on the passage
on the hiccup, that is, the concluding quotation in Stobaeus' 36th chapter.[38]

---

36 See the defense of Trophilus by Stefan Schorn in the introduction to *FGrHist* 1677 and the discus-
    sion of Gertjan Verhasselt (forthcoming).

37 Aët. *Plac.* 5,30: Ἀλκμαίων τῆς μὲν ὑγείας εἶναι συνεκτικὴν <τὴν> ἰσονομίαν τῶν δυνάμεων, ὑγροῦ
    θερμοῦ ξηροῦ ψυχροῦ πικροῦ γλυκέος καὶ τῶν λοιπῶν· τὴν δ' ἐν αὐτοῖς μοναρχίαν νόσου ποιητικήν·
    φθοροποιὸν γὰρ ἑκατέρου μοναρχία· καὶ νόσων αἰτία ὡς μὲν ὑφ' ἕξεως, ὑπερβολὴ θερμότητος ἢ
    ψυχρότητος· ὡς δ' ἐξ ἧς, διὰ πλῆθος <τροφῆς> ἢ ἔνδειαν· ὡς δ' ἐν οἷς, ἢ αἷμα ἔνδον ἢ ἐγκέφαλος·
    τὴν δὲ ὑγείαν τὴν σύμμετρον τῶν ποιῶν κρᾶσιν. Διοκλῆς πλείστας τῶν νόσων δι' ἀνωμαλίαν τῶν ἐν
    τῷ σώματι στοιχείων καὶ τοῦ καταστήματος.
    Ἐρασίστρατος τὰς νόσους διὰ πλῆθος τροφῆς καὶ <δι'> ἀπεψίας καὶ φθορᾶς, τὴν δ' εὐταξίαν καὶ
    αὐτάρκειαν εἶναι ὑγείαν. ("Alcmaeon [says] that a balance of properties ensures health – wet, warm,
    dry, cold, bitter, sweet and the rest – whereas the sole rule among these creates illness, for sole rule
    [of one] causes destruction of every other [quality] and is the cause of illnesses, e.g. in a condi-
    tion *by which*, e.g a surplus of heat or cold, or *from which*, e.g. through an excess of food or a lack
    thereof, or *in which*, e.g. internal blood or the brain. He says that health is an equal portion of the
    qualities. Diocles says that most illnesses are caused by an anomaly of the elements in the body and
    the weather. Erasistratus says that illnesses are due to excess food, indigestion and deterioration, but
    health is due to moderation in diet and self-suffiency.")

38 Pl. *Smp.* 186d-e: δεῖ γὰρ δὴ τὰ ἔχθιστα ὄντα ἐν τῷ σώματι φίλα οἵον τ' εἶναι ποιεῖν καὶ ἐρᾶν
    ἀλλήλων. ἔστι δὲ ἔχθιστα τὰ ἐναντιώτατα, ψυχρὸν θερμῷ, πικρὸν γλυκεῖ, ξηρὸν ὑγρῷ, πάντα τὰ

This proposed organization principle for the 36th chapter is, admittedly, far from bulletproof, but it does go to show how reports on a common theme may be associated with each other by other themes or prominent words.

Coming back to the reports lifted from *On Rivers*, the only internal connection preserved between the various reports is that of a stone or plant serving as a remedy against a disease or as a protection against demons (δαιμόνια). Thus, Stobaeus chose only reports thematically related to the general subject of diseases and remedies. Stobaeus for no obvious reason, omitted several reports found in *On Rivers* that similarly provide remedies, protection, etc.:

11.2 A stone *pausilupos*, which alleviates pain
14.2 A plant *halinda*, which makes one endure cold
18.5 A plant, *selene*, protecting against vermin
19.2 A plant, *kenchritis*, delivering people from madness
24.2. A stone, *mynda*, protecting against wild beasts

How, then, does Stobaeus present reports found worthy of inclusion in chapter 36? First of all, we may notice that he always includes the beginning of a chapter from *On Rivers*, that is, he provides the toponym around which all reports in a chapter revolve in the ps.-Plutarchan work:

Stob. *Flor.* 36.12: Ἀγάθωνος Σαμίου ἐν δευτέρῳ Περὶ ποταμῶν Ἴναχος ποταμός ἐστι τῆς Ἀργείας χώρας. γεννᾶται δ' ἐν αὐτῷ βοτάνη κύουρα καλουμένη, πηγάνῳ παρόμοιος, ἣν αἱ γυναῖκες, ὅταν ἀκινδύνως ἐκτρῶσαι θέλωσιν, ἐν οἴνῳ βεβρεγμένην τοῖς ὀμφαλοῖς ἐπιτιθέασιν.

In the second book of Agathon of Samos' *On Rivers*: Inachus is a river in the territory of Argos. In it grows a plant called *kyura*, similar to a rue, which women, when they wish to have an abortion without trouble, place on their navels after it has been steeped in wine.

[Plut.] *Fluv.* 18.1–3: (1) Ἴναχος ποταμός ἐστι τῆς Ἀργείας χώρας· ἐκαλεῖτο δὲ τὸ πρότερον Καρμάνωρ. Ἁλιάκμων δὲ τῷ γένει Τιρύνθιος ἐν τῷ Κοκκυγίῳ ποιμαίνων ὄρει καὶ κατ' ἄγνοιαν τῇ Ἥρᾳ συγγινόμενον τὸν Δία θεασάμενος ἐμμανὴς ἐγένετο καὶ μεθ' ὁρμῆς ἐνεχθεὶς ἔβαλεν ἑαυτὸν εἰς ποταμὸν Καρμάνορα, ὃς ἀπ' αὐτοῦ Ἁλιάκμων μετωνομάσθη· προσηγορεύθη δὲ Ἴναχος δι' αἰτίαν τοιαύτην. Ἴναχος, Ὠκεανοῦ παῖς, φθαρείσης τῆς θυγατρὸς αὐτοῦ Ἰοῦς ὑπὸ Διός, τὸν θεὸν βλασφήμοις λοιδορίαις ἐπέπληττεν κατόπιν ἀκολουθῶν· ὁ δὲ ἀναξιοπαθήσας ἔπεμψεν αὐτῷ Τισιφόνην, μίαν τῶν Ἐρινύων· ὑφ' ἧς ἐξοιστρηλατούμενος ἔβαλεν ἑαυτὸν εἰς ποταμὸν Ἁλιάκμονα, ὃς ἀπ' αὐτοῦ Ἴναχος μετωνομάσθη.

τοιαῦτα· τούτοις ἐπιστηθεὶς ἔρωτα ἐμποιῆσαι καὶ ὁμόνοιαν ὁ ἡμέτερος πρόγονος Ἀσκληπιός, ὥς φασιν οἵδε οἱ ποιηταὶ καὶ ἐγὼ πείθομαι, συνέστησεν τὴν ἡμετέραν τέχνην. ("For indeed he has to make the most vicious opponents in the body befriend and desire each other. The most contrary qualities are most hostile – cold vs. warm, bitter vs. sweet, dry vs. wet, and all the others. Skilled in producing love and unanimity, as say these poets (and also I believe), our forefather Asclepius composed this science of ours.")

(2) γεννᾶται δ' ἐν αὐτῷ βοτάνη κύνουρα καλουμένη, πηγάνῳ προσόμοιος· ἣν αἱ γυναῖκες, ὅταν ἀκινδύνως ἐκτρῶσαι θελήσωσιν, ἐν οἴνῳ βεβρεγμένην τοῖς ὀμφαλοῖς ἐπιτιθέασιν. (3) εὑρίσκεται δ' ἐν αὐτῷ καὶ λίθος βηρύλλῳ παρόμοιος· ὃν ἐὰν κρατήσωσιν οἱ ψευδομαρτυρεῖν ἐθέλοντες, μέλας γίνεται. κεῖνται δὲ πολλοὶ ἐν τῷ τεμένει τῆς Προσυμναίας Ἥρας, καθὼς ἱστορεῖ Τιμόθεος ἐν Ἀργολικοῖς· μέμνηται δὲ τούτων καὶ Ἀγάθων ὁ Σάμιος ἐν β' Περὶ ποταμῶν.

(1) Inachus is a river in the territory of Argos. Previously it was called Carmanor. When Haliacmon, by birth a Tirynthian, was herding on mount Congygiona he accidentally observed Zeus copulating with Hera and became mad and was taken by impulse and threw himself into the river Carmanor, which was named Haliacmon after him. It was renamed Inachus for the following reason: Inachus, the son of Oceanus, whose daughter Io had been despoiled by Zeus began to abuse the god with blasphemous reproaches whilst pursuing him. He did, however, become angry and sent Tisiphone, one of the Erinyes, against him. Being plagued by her he threw himself into the river Haliacmon, which was renamed Inachus after him. (2) In it grows a plant called *kyura*, similar to a rue, which women, when they wish to have an abortion without trouble, place on their navels after it has been steeped in wine. (3) There is also found in it a stone similar to a beryl, which, if those wishing to make a false testimony lay hands on it, becomes black. Many (of these stones) lie in the sanctuary of Hera Proshymnaia, just as Timotheus records in *Argolica*. Agathon the Samian also mentions these things in the second book of *On Rivers*.

Stobaeus does, however, omit the entire mythological passage, so characteristic of *On Rivers*. Thereby, the reader is not informed of the etymology of the names borne by the river prior to the name Inachus. Omitting the myth, the link between the two stories of rape and the *kynura* plant that can induce a miscarriage, which may have been desired by the unfortunate Io, is lost. Stobaeus is, however, not interested in mythology, and his reason for excluding the lengthy passage is understandable.

Throughout the 36th chapter, Stobaeus always introduces his quotations by adding the source from which they were taken. This is also the case with the reports lifted from *On Rivers*, and in the present example the report is attributed to Agathon of Samos. The Samian is, as may be inferred from *On Rivers*, only mentioned after the third report, on a stone, where he is found in the company of Timotheus. Throughout *On Rivers*, the initial, mythological passage is never attributed to an authority, whereas Stobaeus needed to find an author. But it is impossible to say why he choose Agathon over Timotheus.

Similarly, in the report on the plant *axalla*, found in the Euphrates, Stobaeus attributes this to Chrysermus of Corinth:

Stob. *Flor.* **4.36.13:** Χρυσέρμου Κορινθίου ἐν τῷ ιγ' Περὶ ποταμῶν

Εὐφράτης ποταμός ἐστιν τῆς Παρθίας. γεννᾶται δ' ἐν αὐτῷ λίθος ἀετίτης καλούμενος, ὃν αἱ μαῖαι ταῖς δυστοκούσαις ἐπὶ τὰς γαστέρας ἐπιτιθέασι, καὶ παραχρῆμα τίκτουσιν ἄτερ ἀλγηδόνος. εὑρίσκεται δ' ἐν αὐτῷ καὶ βοτάνη ἄξαλλα, μεθερμηνευομένη θερμόν· ταύτην οἱ τεταρταΐζοντες, ὅταν ἐπὶ τοῦ στήθους τιθῶσιν, ἀπαλλάσσονται παραχρῆμα.

In the 13th book of Chrysermus the Corinthian's *On Rivers*:

The Euphrates is a river in Parthia. In it is produced a so-called eagle-stone, which midwives place on the stomachs of women suffering in child-birth, and immediately they give birth without pain. There is found in it also a plant, *axalla*, which translates as *thermos*. Those suffering from quartan fever place this one on their breast and are immediately relieved.

[Plut.] *Fluv.* **20.1.3:** (1) Εὐφράτης ποταμός ἐστι τῆς Παρθίας κατὰ Βαβυλῶνα πόλιν . . .

(2) γεννᾶται δ' ἐν αὐτῷ λίθος ἀετίτης καλούμενος· ὃν αἱ μαῖαι ταῖς δυστοκούσαις ἐπὶ τὰς γαστέρας ἐπιτιθέασι καὶ παραχρῆμα τίκτουσιν ἄτερ ἀλγηδόνος.

(3) γεννᾶται δ' ἐν αὐτῷ καὶ βοτάνη ἄξαλλα καλουμένη, μεθερμηνευομένη θερμόν· ταύτην οἱ τεταρταΐζοντες, ὅταν ἐπὶ τοῦ στήθους θῶσιν, ἀπαλλάττονται παραχρῆμα τῆς ἐπισημασίας, καθὼς ἱστορεῖ Χρύσερμος Κορίνθιος ἐν ιγ' Περὶ ποταμῶν.

(1) The Euphrates is a river in Parthia by the city Babylon . . . (2) In it is produced a so-called eagle-stone, which midwives place on the stomachs of women suffering in childbirth, and immediately they give birth without pain. (3) There is found in it also a plant, *axalla*, which translates as *thermos*. Those suffering from quartan fever place this one on their breast and are immediately relieved of the symptom, just as Chrysermus the Corinthian records in book 13 of *On Rivers*.

It is, however, only the third report that is explicitly attributed to Chrysermus. Nonetheless, in all of his quotations from *On Rivers*, Stobaeus subsumes several parts of a chapter under one authority.

Still, exceptions to the observations mentioned above may be noted, for example, Stobaeus at one time preserves the etymology of a river:

Stob. *Flor.* **4.36.16:** Καλλισθένους Συβαρίτου ἐν ιγ' Γαλατικῶν Ἄραρ ποταμός ἐστι τῆς Κελτικῆς, τὴν προσηγορίαν εἰληφὼς παρὰ τὸ ἡρμόσθαι τῷ Ῥοδανῷ· καταφέρεται γὰρ εἰς τοῦτον κατὰ τὴν χώραν τῶν Ἀλλοβρόγων.

In the 13th book of Callisthenes of Sybaris' *Galatica*: Arar is a river in Gaul, which has received its name from the fact that it is joined to the Rhodanus. For it enters this (river) in the territory of the Allobroges.

[Plut.] *Fluv.* **6.1:** Ἄραρ ποταμός ἐστι τῆς Κελτικῆς, τὴν προσηγορίαν εἰληφὼς παρὰ τὸ ἡρμόσθαι τῷ Ῥοδανῷ· καταφέρεται γὰρ εἰς τοῦτον κατὰ τὴν χώραν τῶν Ἀλλοβρόγων.

Arar is a river in Gaul, which has received its name from the fact that is joined to the Rhodanus. For it enters this (river) in the territory of the Allobroges.

And at another he preserves the previous name of a river:

Stob. *Flor*. 4.16.18: Θρασύλλου ἐν τοῖς Αἰγυπτιακοῖς

Νεῖλος ποταμός ἐστι τῆς Αἰγύπτου, ἐκαλεῖτο δὲ τὸ πρότερον Μέλας. γεννᾶται δ᾽ ἐν αὐτῷ λίθος κυάμῳ παρόμοιος, ὃν ἂν κύνες ἴδωσιν, οὐχ ὑλακτοῦσι· ποιεῖ δ᾽ ἄριστα πρὸς τοὺς δαιμονιζομένους· ἅμα γὰρ <τῷ> αὐτὸν προστεθῆναι ταῖς ρισὶν ἐξέρχεται τὸ δαιμόνιον.

In Thrasyllus' *Aegyptiaca*:

The Nile is a river in Egypt, which was previously called Melas. In it is produced a stone similar to a bean. If dogs see it, they do not bark. It is most useful against those possessed by demons. For when it is placed before the nose, the demon exits.

[Plut.] *Fluv*. 16.1: Νεῖλος ποταμός ἐστι τῆς Αἰγύπτου κατὰ πόλιν Ἀλεξάνδρειαν. Ἐκαλεῖτο δὲ τὸ πρότερον Μέλας ἀπὸ Μέλανος τοῦ Ποσειδῶνος.

The Nile is a river in Egypt in the city of Alexandria. It was previously called Melas after Melas, the son of Poseidon.

Two reports excerpted by Stobaeus are also found in *Mir*. The text-critical value of these parallel versions is obvious, but so are the insights concerning intention and use that may be gathered from an examination of these two examples.

Stob. *Flor*. 4.36.17: Ἀρχελάου ἐν α᾽Περὶ ποταμῶν Λυκόρμας ποταμός ἐστι τῆς Αἰτωλίας, μετωνομάσθη δ᾽ Εὔηνος. γεννᾶται δ᾽ ἐν αὐτῷ βοτάνη σάρισα προσαγορευομένη, λόγχη παρόμοιος, ποιοῦσα πρὸς ἀμβλυωπίας ἄριστα.

In book 1 of Archelaus' *On Rivers*: Lycormas is a river in Aetolia and it was renamed Euenus. In it grows a plant called *sarisa*, similar to a spear, which has the best effects against dim-sightedness.

***Mir*. 171:** παρὰ Λυκάρμῳ ποταμῷ γεννᾶσθαι βοτάνην λόγχη παρόμοιον, συντελοῦσαν πρὸς ἀμβλυωπίαν ἄριστα.

By river Lycormas a plant similar to a spear grows, which has the best effects against dim-sightedness.

[Plut.] *Fluv*. 8.1–2. (1) Λυκόρμας ποταμός ἐστιν Αἰτωλίας· μετωνομάσθη δ᾽ Εὔηνος δι᾽ αἰτίαν τοιαύτην. Ἴδας, ὁ Ἀφαρέως παῖς, δι᾽ ἐρωτικὴν ἐπιθυμίαν Μάρπησσαν ἁρπάσας, ἀπήνεγκεν εἰς Πλεύρωνα. κατηχηθεὶς δὲ περὶ τῶν συμβεβηκότων ὁ Εὔηνος, ἐπεδίωκεν τὸν ἐπίβουλον τῆς ἰδίας θυγατρός· γενόμενος δὲ κατὰ Λυκόρμαν καὶ τῆς συλλήψεως ἀπελπίσας, ἑαυτὸν εἰς ποταμὸν ἔβαλεν, ὃς ἀπ᾽ αὐτοῦ Εὔηνος μετωνομάσθη.

(2) γεννᾶται δ᾽ ἐν αὐτῷ βοτάνη, λόγχη παρόμοιος, ποιοῦσα πρὸς ἀμβλυωπίας ἄριστα.

(1) Lycormas is a river in Aetolia, and it was renamed Euenus for the following reason: Idas, the son of Aphareus, had grabbed Marpessa out of sexual desire and carried her to Pleuron. When Euenus had been informed of what had happened he began to pursue the man who had preyed on his daughter. When he had come to Lycormas and had despaired of capturing him, he threw himself into the river, which was renamed Euenus after him.

(2) In it grows a plant called *sarisa*, similar to a spear, which has the best effects against dim-sightedness.

Like Stobaeus, ps.-Aristotle omits the myth and jumps to the report itself. Furthermore, ps.-Aristotle includes only the toponym and its designation as a river as well as the plant and its paradoxographical effect. Finally, ps.-Aristotle does not quote Archelaus as his source.

Interestingly, the name of the authority is missing also from *On Rivers* and is found only in Stobaeus. It is very unusual for ps.-Plutarch not to provide a source or to limit himself to only one for a chapter. At the end of *On Rivers* 8, Dercyllus is given as the authority, but for a different report, on the flower *leukoïon*, which grows on the mountain Myenus in the vicinity of the river Lycormas. We may thus assume that the reference to Archelaus fell out of ps.-Plutarch's text sometime after Stobaeus. Only once does ps.-Aristotle give the authority of one of the reports lifted from *On Rivers*, but this is the exception to the rule, and we shall shortly return to this passage.

In addition, Archelaus is no poor candidate for a paradoxographical author. Several fragments of a collection by the title *Peculiar Natures* (Ἰδιοφυῆ) are attributed to an author by this name, active in Alexandria, perhaps in the reign of Ptolemy IV Philopator (221–204 BC). Furthermore, ps.-Plutarch elsewhere attributes a report to Archelaus, albeit from a work on stones,[39] a report found anonymously also in ps.-Aristotle.[40] Apart from *On Rivers* Archelaus, the paradoxographical author, is not known to have composed works on rivers or stones, and his *Peculiar Natures* appears to have dealt almost entirely with animals.

With Jacoby's famous study of ps.-Plutarch in mind,[41] we must conclude that Archelaus is a bogus author fabricated on the basis of the well-known paradoxographer. Alternatively, Archelaus, the Cappadocian client king, who composed a book on stones and died in AD 17 (*FGrHist/BNJ* 123), served as the inspiration for ps.-Plutarch's fictitious paradoxographer.

The second report common to all three works is concerned with the Nile.

Stob. *Flor.* **4.36.18**: Θρασύλλου ἐν τοῖς Αἰγυπτιακοῖς

Νεῖλος ποταμός ἐστι τῆς Αἰγύπτου, ἐκαλεῖτο δὲ τὸ πρότερον Μέλας. γεννᾶται δ' ἐν αὐτῷ λίθος κυάμῳ παρόμοιος, ὃν ἂν κύνες ἴδωσιν, οὐχ ὑλακτοῦσι· ποιεῖ δ' ἄριστα πρὸς τοὺς δαιμονιζομένους· ἅμα γὰρ <τῷ> αὐτὸν προστεθῆναι ταῖς ῥισὶν ἐξέρχεται τὸ δαιμόνιον.

In Thrasyllus' *Aegyptiaca*:

The Nile is a river in Egypt, and it was previously called Melas. In it is produced a stone similar to a bean. When dogs see it, they do not bark. It has the best effects against those possessed by demons, for when it is placed before the nose, the demon exits.

39 [Plut.] *Fluv.* 9.3 = *FGrHist* 1656 F 18.
40 *Mir.* 167.
41 Jacoby (1940).

***Mir.* 166:**Ἐν τῷ Νείλῳ ποταμῷ γεννᾶσθαι λίθον φασὶ κυάμῳ παρόμοιον, ὃν ἂν κύνες ἴδωσιν, οὐχ ὑλακτοῦσι. συντελεῖ δὲ καὶ τοῖς δαιμονί τινι γενομένοις κατόχοις· ἅμα γὰρ τῷ προστεθῆναι ταῖς ρισὶν ἀπέρχεται τὸ δαιμόνιον.

They say that in the Nile a stone is produced similar to a bean. When dogs see it, they do not bark. It is effective against those who have become possessed by a demon, for when it is placed before the nose, the demon departs.

[Plut.] *Fluv.* 16.1–2: (1) Νεῖλος ποταμός ἐστι τῆς Αἰγύπτου κατὰ πόλιν Ἀλεξάνδρειαν. ἐκαλεῖτο δὲ τὸ πρότερον Μέλας ἀπὸ Μέλανος τοῦ Ποσειδῶνος. ὕστερον δὲ Αἴγυπτος ἐπεκλήθη δι᾽ αἰτίαν τοιαύτην. Αἴγυπτος, Ἡφαίστου καὶ Λευκίππης παῖς, βασιλεὺς ὑπῆρχεν τῶν τόπων· δι᾽ ἐμφύλιον δὲ πόλεμον μὴ ἀναβαίνοντος τοῦ Νείλου καὶ λιμῷ συνεχομένων τῶν ἐγχωρίων, ἔχρησεν ὁ Πύθιος τὴν εὐφορίαν, ἐὰν ὁ βασιλεὺς ἀποτρόπαιον θεοῖς τὴν θυγατέρα θύσῃ. θλιβόμενος δὲ ὑπὸ τῶν κακῶν ὁ τύραννος τοῖς βωμοῖς Ἀγανίππην προσήγαγε. τῆς δὲ διασπασθείσης ὁ Αἴγυπτος δι᾽ ὑπερβολὴν τῆς λύπης ἑαυτὸν ἔρριψεν εἰς ποταμὸν Μέλανα, ὃς ἀπ᾽ αὐτοῦ Αἴγυπτος μετωνομάσθη. προσηγορεύθη δὲ Νεῖλος δι᾽ αἰτίαν τοιαύτην. Γαρμαθώνη, τῶν κατ᾽ Αἴγυπτον βασίλισσα τόπων, ἀποβαλοῦσα τὸν υἱὸν αὐτῆς Χρυσοχόαν, [ἐν] ἀκμῇ ἔφηβον, μετὰ τῶν οἰκετῶν ἐθρήνει συμπαθῶς τὸν προειρημένον. Ἴσιδος δὲ αἰφνιδίως ἐπιφανείσης, τὴν λύπην πρὸς καιρὸν ὑπερέθετο, καὶ προσποιητὴν χαρὰν σκηψαμένη τὴν θεὸν ὑπεδέξατο φιλανθρώπως. ἡ δὲ τὴν διάθεσιν τούτων ἀμείψασθαι βουλομένη τῆς εὐσεβείας Ὀσίριδι παρεκέλευσεν, ὅπως ἀναγάγῃ τὸν υἱὸν αὐτῆς ἐκ τῶν καταχθονίων τόπων. τούτου δὲ ταῖς δεήσεσι τῆς γυναικὸς συμπεριενεχθέντος, Κέρβερος, ὃν ἔνιοι καλοῦσιν Φοβερόν, ὑλάκτησεν· Νεῖλος δ᾽ ὁ τῆς Γαρμαθώνης ἀνὴρ αἰφνιδίως ἔνθεος γενόμενος, ἑαυτὸν ἔρριψεν εἰς ποταμὸν καλούμενον Αἴγυπτον, ὃς ἀπ᾽ αὐτοῦ Νεῖλος μετωνομάσθη. (2) γεννᾶται δ᾽ ἐν αὐτῷ λίθος κυάμῳ παρόμοιος, ὃν ἐὰν κύνες ἴδωσιν, οὐχ ὑλακτοῦσι· ποιεῖ δὲ [ἄριστα] πρὸς τοὺς δαιμονιζομένους· ἅμα γὰρ [αὐτὸν] προστεθῆναι ταῖς ρισὶν, ἀπέρχεται τὸ δαιμόνιον. γεννῶνται δὲ καὶ ἄλλοι λίθοι, κόλλωτες καλούμενοι· τούτους κατὰ τὴν ἀνάβασιν τοῦ Νείλου συλλέγουσαι χελιδόνες κατασκευάζουσι τὸ προσαγορευόμενον Χελιδόνιον τεῖχος, ὅπερ ἐπέχει τοῦ ὕδατος τὸν ροῖζον καὶ οὐκ ἐᾷ κατακλυσμῷ φθείρεσθαι τὴν χώραν, καθὼς ἱστορεῖ Θράσυλλος ἐν τοῖς Αἰγυπτιακοῖς.

(1) The Nile is a river in Egypt by the city Alexandria, and it was previously called Melas from Melas, the son of Poseidon. Later it was called Aegyptus for the following reason: Aegyptus, the son of Hephaestus and Leucippe, was king over the lands. When on account of an internecine war the Nile did not swell and the inhabitants were afflicted by hunger Pythia prophesied fertility, if the king should offer his daughter to the gods in order to avert evil. Weighed down by troubles the ruler led Aganippe to the altars. When she had been torn asunder Aegyptus who was exceedingly grieved threw himself into the river Melas, which was renamed Aegyptus after him. It was called the Nile for the following reason: When Garmathone, queen of the lands of Egypt, had lost her son Chrysochoas at the peak of his youth she and her household began to bewail the aforementioned out of sympathy. When Isis had suddenly

appeared, she stopped morning for a time and put forward an assumed happiness and greeted the goddess friendly. Desiring a change in the state of things in return for piety she implored Osiris that he may bring her son up from the areas below. Having been moved by the entreaties of the woman Cerberus, whom some call Phoberus, barked, and Nilus, the husband of Garmathone, immediately became possessed and threw himself into the river called Aegyptus, which was renamed the Nile after him. (2) There is produced in it a stone similar to a bean. When dogs see it, they do not bark. It is effective against those possessed by a demon, for when it is placed before the nose, the demon departs. Other stones are also produced (in it), called *kollotes*. Collecting these at the time of the swelling of the Nile swallows prepare the so-called Chelidonian wall, which restricts the flow of the water and does not allow the country to be destroyed by a flood, just as Thrasyllus records in the *Aegyptiaca*.

Like the report on the river Lycormas, Stobaeus and ps.-Aristotle retain the name of the Nile, but perhaps due to its fame, the latter decides not to mention Egypt. Exceptionally, as noted already, Stobaeus includes the original name of the river, Melas. Furthermore, only Stobaeus copies out the name of the source, Thrasyllus, but in *On Rivers* it is far from clear whether Thrasyllus is to be taken as the author only of the second report on the stones known as *kollotes* or also the report on the stone similar to a bean. Such details are, as we have demonstrated, often ignored by Stobaeus.

Apart from the aforementioned reasons for excluding mythological passages, the length of the myth on the Nile would itself have been an argument for omitting it. Nonetheless, the report on the stone which makes dogs desist from barking finds its explanation in the myth told by ps.-Plutarch: Garmathone, queen of Egypt, mourned the loss of her son, Chrysochoas, but the goddess Isis took pity on her and implored her husband to bring up Chrysochoas from the underworld. Osiris' attempt was, however, foiled by the barking of Cerberus.

Apparently, *On Rivers* appears to be a treasure trove of mythology elsewhere unattested, but at closer inspection it does not seem representative of the large number of texts that have not been transmitted. Rather, the alternative mythologies are just as much the invention of ps.-Plutarch as the bogus authors in *On Rivers*. ps.-Plutarch's myths are mock myths, pastiches, and at times the unknown author of *On Rivers* even engages with the actual Plutarch – in this chapter and elsewhere with the Chaeronean's treatise on *Isis and Osiris*.[42] Thus, the mythological part of this chapter is a mash-up of Demeter's pursuit of Persephone, Orpheus' attempt to bring back Eurydice, and Isis' search for Osiris, or what was left of him. The paradoxographical report on the stone similar to a bean (κύαμος) is an allusion to the somewhat famous Egyptian bean (Αἰγύπτιος κύαμος), according to Diodorus Siculus a foodstuff fundamental to Egyptian civilization.[43] Larger than the common

---

42 See the commentary on Demostratus of Apameia, *FGrHist* 1688 F 1 = [Plut.] *Fluv.* 13.2.
43 Diod. Sic. 1.10.1.

bean,[44] it is recorded as being particularly excrementitious (περιττωματικωτέρων).[45] Being similar to this bean, the stone will have been sufficiently large to jam into the mouth of a dog or simply to be used as a missile. In addition, its excrementitious effect made it apt for driving out not just demons. Unfortunately, the stone is about as fictitious as everything else in *On Rivers*. In his commentary for *Brill's New Jacoby* on Thrasyllus, that is, the bogus author of the present report, Stanley Burstein, who does not comment on the Egyptian bean, writes: "the humor in the passage is probably unintended."[46]

I am of the exact opposite opinion.

Throughout *On Rivers*, ps.-Plutarch is constantly engaging in an intertextual dialogue with established mythology, rhetoric, style, and literary genres. In the case of paradoxography, he associates all phenomena with myth, a first indication that this is not traditional paradoxography. But he does go to lengths to follow the conventions of this genre: he provides an authority and a toponym and leaves out any explanation or doubt. Since myth, source, and report are almost exclusively the invention of ps.-Plutarch, he deliberately seems to be poking fun at this popular genre.

No one in antiquity appears to have seen through this caricature. Indeed, the quotations from *On Rivers* leave us with the impression that Stobaeus and the compiler of the appendix walked into the trap set by the anonymous author.

3. Conclusion

In this chapter, we have examined the use of a number of reports from *On Rivers* in the 5th-century AD compiler Stobaeus as well as in the appendix of the *Mir.* What the two have in common is the exclusion of the mythological material, that is, they skip this part and go straight to a seemingly paradoxographical report. Furthermore, the etymology of a toponym is similarly left out. The result is a very concise report devoid of any context apart from toponym, fitting for a collection of paradoxographical reports.

In our opinion, Stobaeus and the excerptor of the appendix do, albeit unknowingly, walk into a trap set by ps.-Plutarch. Overlooking the subtle and minute relationship between myth, etiology, etymology, and products of a particular locale, they ignore how *On Rivers* resorts to alternative mythologies, from which spring paradoxography-like reports on marvelous phenomena. At least Stobaeus preserves the names of the authors, from which ps.-Plutarch claims to be quoting, but in the appendix of *Mir.* these are, apart from one instance, excluded. This is not uncommon in paradoxographical literature, where sources may deliberately be suppressed to attach greater authority to the author/compiler. Under normal circumstances, a toponym would ensure the reliability of the report, and an unconvinced reader could in theory visit the location and behold the phenomenon with his own eyes. The appendix of *Mir.* does, however, end up not only transmitting mock reports but also taking the honor for them.

---

44 Diosc. 2.106.
45 Gal. *Alim. Fac.* 532.
46 See the commentary on *BNJ* 622 F 1.

Appendix: Stobaeus 4.36.1–32

(1) Εὐριπίδου Ὀρέστῃ.
ὦ φίλον ὕπνου θέλγητρον, ἐπίκουρον νόσου,
  ὡς ἡδύ μοι προσῆλθες ἐν δέοντί τε.
(2) ἐν ταὐτῷ.
Ἰδού· φίλον τοι τῷ νοσοῦντι δέμνια,
  ἀνιαρὸν <ὂν>τὸ κτῆμ', ἀναγκαῖον δ' ὅμως.
(3) ἐν ταὐτῷ.
αὖθις δ' ἐς ὀρθὸν στῆσον, ἀνακύκλει δέμας.
  δυσάρεστον οἱ νοσοῦντες ἀπορίας ὕπο.
(4) ἐν ταὐτῷ.
ἡ κἀπὶ γαίας ἁρμόσαι πόδας θέλεις;
  – Μάλιστα· δόξαν γὰρ τόδ' ὑγιείας ἔχει.
κρεῖσσον δὲ τὸ δοκεῖν, κἂν ἀληθείας ἀπῇ.
(5) Εὐριπίδου.
πρὸς τὴν νόσον τοι καὶ τὸν ἰατρὸν χρεὼν
  ἰδόντ' ἀκεῖσθαι, μὴ 'πιτὰξ τὰ φάρμακα
  διδόντ', ἐὰν μὴ ταῦτα τῇ νόσῳ πρέπῃ.
(6) Φιλήμονος Μύστιδος.
πολὺ μεῖζόν ἐστι τοῦ κακῶς ἔχειν κακὸν
  τὸ καθ' ἕνα πᾶσι τοῖς ἐπισκοπουμένοις
  δεῖν τὸν κακῶς ἔχοντα πῶς ἔχει λέγειν.
(7) †Θελλέροφος.
πρὸς τὴν νόσον τοι καὶ τὸν ἰατρὸν χρεὼν
  ἰδόντ' ἰᾶσθαι, μὴ ἐπιτὰξ τὰ φάρμακα
  διδόντ', ἐὰν μὴ ταῦτα τῇ νόσῳ πρέπῃ.
  νόσοι δὲ θνητῶν, αἱ μέν εἰσ' αὐθαίρετοι
  αἱ δ' ἐκ θεῶν πάρεισιν, ἀλλὰ τῷ νόμῳ
  ἰώμεθ' αὐτάς. ἀλλά σοι λέξαι θέλω,
  εἰ θεοί τι δρῶσιν αἰσχρόν, οὐκ εἰσὶν θεοί.
(8) ἐκ τῶν Ἡλιοδώρου Ἰταλικῶν θαυμάτων.
Ἰταλίης οὐ πολλὸν ὑπερστείχοντι κολώνην
  Γαυρείην χώρη τις ὁδιτάων ἐπὶ λαιὰ
  κέκλιται ἀργήεσσα χιὼν ὥς· ἐκ δέ οἱ ὕδωρ
  ἀΐσσει μάλα πικρὸν ἀναπνεῦσαι πιέειν τε.
κεῖνο πολυστάφυλοι περιναιέται ἀνέρες ὕδωρ
  ὅσσων ἄλκαρ ἔχουσιν· ὁ μὲν λοετροῖο χατίζων
  αὕτως, ὄφρα κε μοῦνον ἐν ὕδατι γυῖα καθήρῃ,
  ὀφθαλμοὺς βλεφάροισι λίην ἀραρῶσι καλύψας
  δεύεται, ὡς μή οἵ τι παραδράμῃ ἕρκεος εἴσω
  ὑγρὸν ἐπὶ γλήνης· τὸ γὰρ ἄλγεος <ἄν τι τιθ>είη.
ὃς δέ κε λημηρῇ νεφέλῃ εἰλυμένος ὄσσε
  ἀσχάλλῃ ὀδύναις, κεροειδέα δ' ἀμφὶ χιτῶνα
  οἶδος πιαλέοισι περιβρίθῃ πελάνοισι,
κείνῳ καίριόν ἐστι καὶ ἀσφαλὲς ὄμμα διῆναι
  ἀμπετὲς ἀκλήιστον· ἄφαρ δ' ἀπὸ πᾶσα τελέσθη
  θυμοδακὴς ὀδύνη, ῥέα δ' ἄλθεται ὕδατι νοῦσος.

(1) In Euripides' *Orestes* (211–212):
Sweet charm of sleep, savior in sickness, how
  sweetly you came to me, how needed!
(2) In the same (229–230):
There! His couch is welcome to the sick man, a
  painful possession, but a necessary one.
(3) In the same (231–232):
Set me upright once again, turn my body round; it
  is their helplessness that makes the sick so hard
  to please.
(4) In the same (233, 235–236):
Will you set your feet upon the ground and take a
  step at last? – Oh, yes; for that has a semblance of
  health; and the semblance is preferable, though it
  is far from the truth.[47]
(5) From Euripides (F 286b Kannicht):
As to illness, a doctor too must cure it after
  examining it, not by giving remedies by rote, in
  the case these do not suit the illness.[48]
(6) From Philemon's *Mystis* (*PCG* F 47):
A much greater illness than feeling ill is that being
  ill it is necessary to say to each and every one,
  who enquires, how you are feeling.
(7) Thellerophus (Eur. F 286b Kannicht):
As to illness, a doctor too must heal it after examining
  it, not by giving remedies by rote, in the case these
  do not suit the illness. Human illnesses are some
  of them self-inflicted, others come from the gods,
  but we treat them by rule of practice. This is what
  I want to say to you, however: If gods do anything
  shameful, they are not gods.[49]
(8) From Heliodorus' *Italian Wonders* (*SH* 472):
Passing over Mons Gaurus of Italy – only a small
  distance – a country of wanderers lies on rocks
  white like snow. Whence water sprouts, very bitter
  to inhale and to drink. The inhabitants, who are rich
  in vines, use the water as an eye remedy. Thus, he
  who is need of a bath, just as if he should only wash
  the knees in water, wets his eyes having closed
  his lids firmly that not any fluid should escape the
  barrier onto the pupil. For it is the cause of pain.
  He, whose eyes are covered in a foggy mist vexes
  at the pain. From around the cornea a swelling with
  much gum drips. At this point it is safe and the right
  time for him to wet his eyes that are open and not
  closed. Now, every heart-breaking pain has ceased,
  and the disease is easily cured with water.

(Continued)

---

47 Translated by Coleridge (1938).
48 Translated by Collard and Cropp (2008).
49 Translated by Collard and Cropp (2008), modified.

(Continued)

---

(9) Σωκράτους.

Σωκράτης ἐρωτηθεὶς τί νόσος ἔφη 'ταραγμὸς σώματος'.

(9) From Socrates (Giannantoni I C 310):

When asked what a sickness is, Socrates replied "a disturbance of the body."

(10) Διογένους.

Διογένης πρός τινα τῶν συνήθων τὸ σῶμα βεβλαμμένον καὶ ποτνιώμενον, 'εὖ' ἔφη 'φίλος, ὅτι πονεῖς, ἵνα μὴ πονῇς'.

(10) From Diogenes:

Diogenes said to a friend whose body was in pain and who was crying loudly: "My friend, it is good that you are in pain (now), so that you may (later) not be in pain."

(11) Δημοσθένους Ὀλυνθιακῶν β΄.

ὥσπερ γὰρ ἐν τοῖς σώμασιν, ἕως μὲν ἂν ἐρρωμένος ᾖ τις, οὐδὲν ἐπαισθάνεται τῶν καθ' ἕκαστα σαθρῶν, ἐπὰν δὲ ἀρρώστημά τι συμβῇ, πάντα κινεῖται, κἂν ῥῆγμα, κἂν στρέμμα, κἂν ἄλλο τι τῶν ὑπαρχόντων σαθρὸν ᾖ.

(11) From Demosthenes' *Second Olynthiacs* (2.21):

For just as in our bodies, so long as a man is in sound health, he is conscious of no pain, but if some malady assails him, every part is set a-working, be it rupture or sprain or any other local affection.[50]

(12) Ἀγάθωνος Σαμίου ἐν δευτέρῳ Περὶ ποταμῶν.

Ἴναχος ποταμός ἐστι τῆς Ἀργείας χώρας. γεννᾶται δ' ἐν αὐτῷ βοτάνη κύουρα καλουμένη, πηγάνῳ παρόμοιος, ἣν αἱ γυναῖκες, ὅταν ἀκινδύνως ἐκτρῶσαι θέλωσιν, ἐν οἴνῳ βεβρεγμένην τοῖς ὀμφαλοῖς ἐπιτιθέασιν.

(12) In Agathon of Samos' *On Rivers* 2 ([Plut.] *Fluv.* 18.1.2):

Inachus is a river in the territory of Argos. In it grows a plant called *kyura*, similar to a rue, which women, when they wish to have an abortion without trouble, place on their navels after it has been steeped in wine.

(13) Χρυσέρμου Κορινθίου ἐν τῷ ιγ΄ Περὶ ποταμῶν.

Εὐφράτης ποταμός ἐστιν τῆς Παρθίας. γεννᾶται δ' ἐν αὐτῷ λίθος ἀετίτης καλούμενος, ὃν αἱ μαῖαι ταῖς δυστοκούσαις ἐπὶ τὰς γαστέρας ἐπιτιθέασι, καὶ παραχρῆμα τίκτουσιν ἄτερ ἀλγηδόνος. εὑρίσκεται δ' ἐν αὐτῷ καὶ βοτάνη ἄξαλλα, μεθερμηνευομένη θερμόν· ταύτην οἱ τεταρταΐζοντες, ὅταν ἐπὶ τοῦ στήθους τιθῶσιν, ἀπαλλάσσονται παραχρῆμα.

(13) In Chrysermus the Corinthian's *On Rivers* 13 ([Plut.] *Fluv.* 20.2–3):

The Euphrates is a river in Parthia. In it is produced a so-called eagle-stone, which midwives place on the stomachs of women suffering in childbirth, and immediately they give birth without pain. There is found in it also a plant, *axalla*, which translates as *thermos*. Those suffering from quartan fever place this one on their breast and are immediately relieved.

(14) Νικίου ἐν τοῖς Περὶ λίθων.

ὄρος ἐστὶ Δρίμυλον καλούμενον τῆς Παρθίας, ἐν ᾧ γεννᾶται λίθος σαρδόνυχι παρόμοιος· ποιεῖ δ' ἄριστα πρὸς ἀμβλυωπίας εἰς ὕδωρ θερμὸν βαλλόμενος.

(14) In Nicias' *On Stones* ([Plut.] *Fluv.* 20.4):

There is a mountain called Drimylon in Parthia, in which is produced a stone similar to sardonyx. It has the best effects against dim-sightedness when thrown into hot water.

(15) ἐκ τῆς Ἀριστοτέλους Συναγωγῆς ἀκουσμάτων θαυμασίων.

ἐν Τραπεζοῦντι τῇ ἐν τῷ Πόντῳ γίνεται τὸ ἀπὸ τῆς πύξου μέλι βαρύοσμον· καὶ φασὶ τοὺς μὲν ὑγιαίνοντας ἐξιστάναι, τοὺς δ' ἐπιλήπτους καὶ τελέως ἀπαλλάσσειν.

(15) From Aristotle's *Collection of Marvelous Things Heard* (*Mir.* 18):

In Trapezus in Pontus there is a foul-smelling boxwood honey, and they say that it makes healthy persons go mad but cures completely those suffering from epilepsy.

(16) Καλλισθένους Συβαρίτου ἐν ιγ΄ Γαλατικῶν.

Ἄραρ ποταμός ἐστι τῆς Κελτικῆς, τὴν προσηγορίαν εἰληφὼς παρὰ τὸ ἡρμόσθαι τῷ Ῥοδανῷ· καταφέρεται γὰρ εἰς τοῦτον κατὰ τὴν χώραν τῶν Ἀλλοβρόγων. γεννᾶται δ' ἐν αὐτῷ μέγας ἰχθὺς κλουπαῖα προσαγορευόμενος ὑπὸ τῶν ἐγχωρίων· οὗτος αὐξομένης μὲν τῆς σελήνης λευκός ἐστι, μειουμένης δὲ μέλας

(16) In Callisthenes of Sybaris' *Galatica* 13 ([Plut.] *Fluv.* 6.2–3):

Arar is a river in Gaul, which has received its name from the fact that it is joined to the Rhodanus. For it enters this (river) in the territory of the Allobroges.

---

50 Translated by Vince (1930).

γίνεται παντελῶς· ὑπεραυξήσας δὲ ἀναιρεῖται ὑπὸ τῶν ἰδίων ἀκανθῶν. εὑρίσκεται δ' ἐν τῇ κεφαλῇ αὐτοῦ λίθος χόνδρῳ παρόμοιος ἁλός, ὃς κάλλιστα ποιεῖ πρὸς τεταρταίας νόσους τοῖς ἀριστεροῖς μέρεσι τοῦ σώματος προσδεσμευόμενος τῆς σελήνης μειουμένης.

In it is born a large fish called Clupaia by the locals. When the moon waxes it is white, when it wanes it becomes completely black. When it has grown exceedingly it is killed by its own thorns. In its head is found a stone similar to a lump of salt, which has the best effects against quartan fever when tied to the left part of the body during the waxing of the moon.

(17) Ἀρχελάου ἐν α΄Περὶ ποταμῶν.
Λυκόρμας ποταμός ἐστι τῆς Αἰτωλίας, μετωνομάσθη δ' Εὔηνος. γεννᾶται δ' ἐν αὐτῷ βοτάνη σάρισα προσαγορευομένη, λόγχῃ παρόμοιος, ποιοῦσα πρὸς ἀμβλυωπίας ἄριστα.

(17) In Archelaus' *On Rivers* 1 ([Plut.] *Fluv.* 8.1–2): Lycormas is a river in Aetolia, and it was renamed Euenus. In it grows a plant called *sarisa*, similar to a spear, which has the best effects against dim-sightedness.

(18) Θρασύλλου ἐν τοῖς Αἰγυπτιακοῖς.
Νεῖλος ποταμός ἐστι τῆς Αἰγύπτου, ἐκαλεῖτο δὲ τὸ πρότερον Μέλας. γεννᾶται δ' ἐν αὐτῷ λίθος κυάμῳ παρόμοιος, ὃν ἂν κύνες ἴδωσιν, οὐχ ὑλακτοῦσι· ποιεῖ δ' ἄριστα πρὸς τοὺς δαιμονιζομένους· ἅμα γὰρ <τῷ> αὐτὸν προστεθῆναι ταῖς ῥισὶν ἐξέρχεται τὸ δαιμόνιον.

(18) In Thrasyllus' *Aegyptiaca* ([Plut.] *Fluv.*16.1–2): The Nile is a river in Egypt, which was previously called Melas. In it is produced a stone similar to a bean. If dogs see it, they do not bark. It is most useful against those possessed by demons. For when it is placed before the nose, the demon exits.

(19) Τιμαγόρου ἐν α΄ Περὶ ποταμῶν.
Κάικος ποταμός ἐστι τῆς Μυσίας. φύεται δ' ἐν αὐτῷ βοτάνη ἠλιφάρμακος καλουμένη, ἣν οἱ ἰατροὶ τοῖς αἱμορραγοῦσιν ἐπιτιθέασι καὶ τῶν φλεβῶν μεσολαβοῦσι τὴν ἔκρυσιν.

(19) In Timagoras' *On Rivers* 1 ([Plut.] *Fluv.* 21.1.3): Caicus is a river in Mysia. In it grows a plant called *elipharmakos*. Doctors apply it in cases of heavy bleedings, and so hinder the flow of the veins.

(20) Κτησίου Κνιδίου ἐν β΄ Περὶ ὀρῶν.
ὄρος ἐστὶ τῆς Μυσίας Τεύθρας καλούμενον. γεννᾶται δ' ἐν αὐτῷ λίθος ἀντιπαθὴς προσονομαζόμενος, ὃς κάλλιστα ποιεῖ πρὸς ἀλφοὺς καὶ λέπρας δι' οἴνου τριβόμενος καὶ τοῖς πάσχουσιν ἐπιτιθέμενος.

(20) From Ctesias' *On Mountains* 2 ([Plut.] *Fluv.* 21.4–5): There is a mountain in Mysia called Teuthras. In it is produced a stone which goes by the name *antipathes*, which is most effective against white leprosy and leprosy when dissolved with wine and placed on those suffering.

(21) Σωστράτου ἐν α΄ Μυθικῆς <συν>αγωγῆς.
Τίγρις ποταμός ἐστι τῆς Ἀρμενίας. γεννᾶται δ' ἐν αὐτῷ βοτάνη κριθῇ παρόμοιος ἀγρίᾳ. ταύτην οἱ ἐγχώριοι θερμαίνοντες ἐν ἐλαίῳ καὶ ἀλειφόμενοι οὐδέποτε νοσοῦσι μέχρι τῆς ἀνάγκης τοῦ θανάτου.

(21) In Sostratus' *Mythical Collection* 1 ([Plut.] *Fluv.* 24.1.4): The Tigris is a river in Armenia. In it is produced a plant similar to barley. The locals heat it with olive oil, anoint themselves with it and are never sick until (they reach) the pains of death.

(22) Κλειτοφῶντος Ῥοδίου ἐν α΄ Ἰνδικῶν.
Ἰνδὸς ποταμός ἐστι τῆς Ἰνδίας. φύεται δ' ἐν αὐτῷ βοτάνη καρπύλη καλουμένη, βουγλώσσῳ παρεμφερής· ποιεῖ δ' ἄριστα πρὸς ἰκτέρους διὰ ὕδατος χλιαροῦ διδομένη.

(22) In Clitophon the Rhodian's *Indica* 1 ([Plut.] *Fluv.* 25.1.3): The Indus is a river in India. In it grows a plant called *karpyle*, somewhat similar to *oxtongue*. It is most efficient against jaundice when given with lukewarm water.

(23) Πορφυρίου ἐκ τοῦ Περὶ Στυγός.
ἥ τε ἰτέα αὕτη τὸν καρπὸν ἀποβάλλει πρὶν ἐκθρέψαι· διὸ ὠλεσίκαρπον αὐτὴν ὁ ποιητὴς (*Od.* 10,510) ὀνομάζει. καὶ μέντοι ἱστόρηται ὡς μετὰ οἴνου δοθεὶς ὁ ταύτης καρπὸς ἀγόνους ποιεῖ τοὺς πιόντας καὶ κατασβέννυσι τὸ σπέρμα καὶ μαραίνει τὴν γόνιμον ὁρμήν.

(23) From Porphyry's *On Styx* (F 9 Castelletti): The willow sheds its fruit before ripening. For this reason the poet called it *the one that loses its fruit*. Indeed, it is reported that when given with wine its fruit makes those drinking it sterile, dries up the semen and extinguishes the libido.

(24) ἐκ τῆς Τροφίλου . . .

(*Continued*)

(Continued)

---

(25) <ἐκ τῆς Ἀριστοτέλους> Συναγωγῆς ἀκουσμάτων θαυμασίων.

τὸ τῆς ἴκτιδος λέγεται αἰδοῖον εἶναι οὐχ ὅμοιον τῇ φύσει τῶν λοιπῶν ζῴων, ἀλλὰ στερεὸν διὰ παντὸς οἷον ὀστοῦν. φασὶ δὲ καὶ στραγγουρίας αὐτὸ φάρμακον εἶναι ἐν τοῖς ἀρίστοις, καὶ δίδοσθαι ἐπιξυόμενον.

(26) ἐν ταὐτῷ.

φασὶ τὸν γαλεώτην, ὅταν ἐκδύῃ τὸ δέρμα καθάπερ οἱ ὄφεις, ἐπιστραφέντα καταπίνειν· τηρεῖσθαι γὰρ ὑπὸ τῶν ἰατρῶν διὰ τὸ χρήσιμον εἶναι τοῖς ἐπιλήπτοις.

(27) ἐν ταὐτῷ.

φασὶ τὴν φώκην ἐξεμεῖν τὴν πιτύαν, ὅταν ἁλίσκηται· εἶναι δὲ φαρμακῶδες καὶ τοῖς ἐπιληπτικοῖς χρήσιμον.

(28) ἐν ταὐτῷ.

ἐν Ἄργει φασὶ γίνεσθαι ἀκρίδος τι γένος, ἣν καλεῖσθαι σκορπιομάχον· ὅταν γὰρ ἴδῃ τάχιστα σκορπίον, ἀνθίσταται αὐτῷ. ἀγαθὸν δέ φασιν εἶναι καὶ πρὸς τὰς πληγὰς τοῦ σκορπίου τὸ ἐπιφαγεῖν αὐτήν.

(29) Ἀλκμαίωνος.

λέγει δὲ τὰς νόσους συμπίπτειν ὡς μὲν ὑφ’ οὗ δι’ ὑπερβολὴν θερμότητος ἢ ξηρότητος, ὡς δὲ ἐξ οὗ διὰ πλῆθος τροφῆς ἢ ἔνδειαν, ὡς δὲ ἐν οἷς <δι’> αἷμα ἢ μυελὸν ἢ ἐγκέφαλον· γίνεσθαι δέ ποτε καὶ ὑπὸ τῶν ἔξωθεν αἰτιῶν, ὑδάτων ποιῶν ἢ χώρας ἢ κόπων ἢ ἀνάγκης ἢ τῶν τούτοις παραπλησίων.

(30) Διοκλέους.

Διοκλῆς τὰς πλείστας τῶν νόσων δι’ ἀνωμαλίαν γίνεσθαι ἔφη.

(31) Ἐρασιστράτου.

Ἐρασίστρατος ‘πλῆθος καὶ διαφθορὰ τἀνωτάτω αἴτια’.

(32) Πλάτωνος ἐκ τοῦ Συμποσίου.

ἐὰν μέν σοι ἐθέλῃ ἀπνευστὶ ἔχοντι πολὺν χρόνον παύσασθαι ἡ λύγξ· εἰ δὲ μή, ὕδατι ἀνακογχυλίασον. εἰ δ’ ἄρα πάνυ ἰσχυρά ἐστιν, λαβών τι τοιοῦτον οἵῳ κνήσαις ἂν τὴν ῥῖνα, πτάρε· καὶ ἐὰν τοῦτο ποιήσῃς ἅπαξ ἢ δίς, καὶ εἰ πάνυ ἰσχυρά ἐστι, παύσεται.

---

(24) From Trophilus (*FGrHist* 1677) . . .

(25) (From Aristotle’s) *Marvellous Things Heard* (*Mir.* 12):

The penis of the marten is said by nature not to be like that of other animals, but continuously erect like a bone. They say that it is among the best medicine for strangury and is given in grated form.

(26) In the same (*Mir.* 66):

They say that the gecko turns around and swallows its skin when like snakes it changes it. It is watched for by doctors since it is useful to epileptics.

(27) In the same (*Mir.* 77):

They say that the foal vomits forth rennet when it is overtaken. It has medicinal use for epileptics.

(28) In the same (*Mir.* 139-b):

In Argos they say that there is a species of locusts, which is called *scorpion-killer*. As soon as it sees a scorpion, it takes a stand against it. They say that eating it is beneficial against scorpion stings.

(29) From Alcmaeon (F 4 DK):

He says that diseases occur, as to the cause, by an excess of food or a lack thereof, as to the location, in the blood, the marrow or the brain. Occasionally, they also happen for external reasons, e.g. the quality of water, location, work, force etc.

(30) From Diocles (F 30 Wellmann):

Diocles said that most sickness is caused by an anomaly.

(31) From Erasistratus (Diels, *Dox. Graec.* p. 443):

Erasistratus (says): “The body and disorder are the foremost reasons.”

(32) From Plato’s *Symposium* (185 d-e):

But during my speech, if on your holding your breath a good while the hiccough chooses to stop, well and good; otherwise, you must gargle with some water. If, however, it is a very stubborn one, take something that will tickle your nostrils, and sneeze: do this once or twice, and though it be of the stubbornest, it will stop.[51]

---

51 Translated by Fowler (1925).

Works cited

Borghart, P. and De Temmerman, K. 2010. *Biography and Fictionality in the Greek Literary Tradition.* Ghent: Academia Press.

Boulogne, J. 2002. *Plutarque. Œuvres morales, IV. Conduites méritoires de femmes. Étiologies romaines – Étiologies grecques. Parallèles mineurs.* Paris: Belles Lettres.

Bowersock, G.W. 1994. *Fiction as History. Nero to Julian.* Berkeley: University of California Press.

Brodersen, K. 2022. *Plutarch De fluviis. Zweisprachige Ausgabe.* Kartoffeldruck-Verlag: Speyer.

Cameron, A. 2004. *Greek Mythography in the Roman World.* Oxford: Oxford University Press.

Coleridge, E.P. 1938. "Orestes." In Oates, W.J. and O'Neill, E. eds. *The Complete Greek Drama. 2. Tragedies.* New York: Random House. 111–166.

Collard, C. and Cropp, M. 2008. *Euripides Fragments. Aegeus – Meleager.* Cambridge, MA: Harvard University Press.

Delattre, C. 2011. *Pseudo-Plutarque, Nommer le monde. Origine des noms de fleuves, de montagnes et de ce qui s'y trouve, traduit, présenté et annoté.* Villeneuve d'Ascq: Presses Universitaires de Septentrion.

Delattre, C. 2017. "L'alphabet au secours de la géographie. (Dés)organiser le De fluviis du pseudo-Plutarque." *Polymnia* 3: 53–82.

De Lazzer, A. 2000. *Plutarco. Paralleli minori. Introduzione, testo critico, traduzione e commento.* Naples: D'Auria.

Dorda, E.C. *et al.* 2003. *Plutarco. Fiumi e monti.* Naples: M. d'Auria.

Flashar, H. 1972. *Aristoteles. Mirabilia, übersetzt von H. F. De audibilibus, übersetzt von U. Klein.* Berlin: Akademie-Verlag; ²1981 with corrigenda to *Audib.*, but not to *Mir.*

Fowler, H.N. 1925. *Plato.* Cambridge, MA: Harvard University Press.

Giacomelli, C. 2021. *Ps.-Aristotele, De mirabilibus auscultationibus. Indagini sulla storia della tradizione e recenzione del testo.* Berlin: Walter de Gruyter.

Giannini, A. 1964. "Studi sulla paradossografia greca, II. Da Callimaco all'età imperiale. La letteratura paradossografica." *Acme* 17: 99–140.

Henderson, I. 2011. "The Second Sophistic and Non-Elite Speakers." In Schmidt, T. and Fleury, P. eds. *Perceptions of the Second Sophistic and Its Times – Regards sur la Seconde Sophistique et son époque.* Toronto: University of Toronto Press. 23–35.

Hercher, R. 1851. *Plutarchi libellus de fluviis.* Leipzig: Weidmann.

Hose, M. 2008. "Ptolemaios Chennos und das Problem der Schwindelliteratur." In Heilen, S. *et al.* eds. *Pursuit of Wissenschaft. Festschrift für Wilhelm M. Calder III zum 75. Geburtstag.* Hildesheim: Olms. 177–196.

Hughes, D.D. 1991. *Human Sacrifice in Ancient Greece.* London: Routledge.

Ibáñez Chacón, Á. 2008/2009. "La violación como tópico en los Parallela minora." *Ploutarchos* 6: 3–13.

Ibáñez Chacón, Á. 2014. *Los Parallela Minora atribuidos a Plutarco (Mor. 305A-316B). Introduccíon, edición, tradución y comentario.* Málaga: Universidad de Málaga.

Jacoby, F. 1940. "Die Überlieferung von Ps. Plutarchs Parallela Minora und die Schwindelautoren." *Mnemosyne* 8: 73–144.

Malheiro Magalhães, J. 2022. "Human-Animal Sex in Ancient Greece." In Serafim, A. *et al.* eds. *Sex and the Ancient City. Sex and Sexual Practices in Greco-Roman Antiquity.* Berlin: De Gruyter. 307–322.

Pajón-Leyra, I. 2009. *Paradoxografía griega. Estudio de un género literario*. Madrid: Universidad Complutense de Madrid.

Poidomani, C. 2016. "Il De fluviis pseudoplutarcheo nella redazione del codice Paris, Bibliothèque Nationale de France, Supplément grec 443A." *Commentaria Classica* 3: 57–82.

Robson, J.E. 2002. "Bestiality and Bestial Rape in Greek Myth". In Deacy, S. and Pierce, K.F. eds. *Rape in Antiquity. Sexual Violence in the Greek and Roman Worlds*. London: Bloomsbury Academic. 65–96.

Rosenkranz, B. 1966. "Die Struktur der ps.-isokratischen Demonicea." *Emerita* 34: 95–129.

Schmitz, T.A. 2014. "Plutarch and the Second Sophistic." In Beck, M. ed. *A Companion to Plutarch*, Chichester: Blackwell. 32–42.

Syme, R. 1983. *Historia Augusta Papers*. Oxford: Clarendon Press.

Ursin, F. 2019. *Freiheit, Herrschaft, Widerstand. Griechische Erinnerungskultur in der Hohen Kaiserzeit (1. – 3. Jahrhundert n.Chr.)*. Stuttgart: Franz Steiner.

Vanotti, G. 2007. *Aristotele. Racconti meravigliosi*. Milan: Bompiani.

Verhasselt, G. forthcoming. "*De mirabilibus auscultationibus* 139–151: Theophrastus' *On Creatures that Bite and Sting* and Aristotle's *Nomima Barbarica*." In Zucker, A., Hellmann, O. and Mayhew, R. eds. *The Aristotelian* Mirabilia *and Early Peripatetic Natural Science*. New York: Routledge.

Vespa, M. 2020. "Presentifying the Divine in Ancient Greek Tales: Human Voices in Animal Bodies." In Schmalzgruber, H. ed. *Speaking Animals in Ancient Literature*. Heidelberg: Winter. 401–425.

Vince, J.H. 1930. *Demosthenes Orations*, 1. Cambridge, MA: Harvard University Press.

Whitmarsh, T. 2013. "Resistance Is Futile? Greek Literary Tactics in the Face of Rome." In Schubert, P. *et al.* eds. *Les Grecs héritiers des Romains. Huit exposés suivis de discussions. Entretiens sur l'Antiquité classique*. Genève: Vandoeuvres. 59–85.

Winter, B.W. 2002. *Philo and Paul Among the Sophists. Alexandrian and Corinthian Responses to a Julio-Claudian Movement*. Cambridge: Cambridge University Press.

Ziegler, K. 1949. "Paradoxographoi." *Paulys Realencyclopaedie der classischen Altertumswissenschaft* XVIII(3): 1137–1166.

Ziegler, K. 1951. "Ploutarchos." *Paulys Realencyklopaedie der classischen Altertumswissenschaft* XXI(1): 636–962.

Index locorum

146 47
154 69

Inscriptions
CIL XI 3281–3284 23
IG II² 127 105
IG II² 7015 96
IG IX 2 397 99
IG XII (9) 1189.10 96
PHI Chios no. 291 291
SEG XLII 533 99

Isidorus of Sevilla
Etymologiae
16.4.8 93

Isigonus of Nicaea
FGrHist 1659
F 5 98–99
F 6 173
F 10 112

Juba of Mauretania
FGrHist 275
F 5 259

Lactantius
Divinae institutiones
1.6.8 150
1.6.12 150

Leo Choerosphactes
De thermis (ed. Gallavotti)
43–56 240
84–87 130
92–95 97
160–165 92

Leonidas of Tarentum (in *Anth. Pal.*)
16.182 228

Lucian
Gallus
19 181
De saltatione
57 181

Lycophron
592–632 200
633–647 62
704 156
712–737 194
874–875 54
913–926 198–199

930 185
946–950 185
968–969 195
1137–1140 39
1137 195
1278–1280 152
1464 152

Lycus of Rhegium
FGrHist/BNJ 570
T 2 63–64
F 3 5, 40, 63
F 4 106–108
F 6 41

Lydus
De mensibus (ed. Wünsch)
4.137 47

Lysias
7.7 247

Myrsilus of Methymna
FGrHist 477
F 1b 58
F 2 58
F 4 60
F 5 58
F 6 58
F 11 58
FGrHist 1654
F 1 59

Mythographus Vaticanus
2.115 47

Nicander
Theriaca
45–50 92–93
145–165 236

Nicolaus of Damascus
FGrHist 90
F 67–68 150

Oribasius
Collectiones medicae
5.3.29 112

Ovid
Fasti
1.543–582 21
4.417–454 47
4.495–498 47

Index nominum

For Product Safety Concerns and Information please contact our EU
representative GPSR@taylorandfrancis.com
Taylor & Francis Verlag GmbH, Kaufingerstraße 24, 80331 München, Germany

www.ingramcontent.com/pod-product-compliance
Lightning Source LLC
Chambersburg PA
CBHW060447240326
41598CB00088B/3903